VIBRATIONAL MEDICINE FOR THE 21ST CENTURY

ALSO BY RICHARD GERBER, M.D.

Vibrational Medicine:
New Choices for Healing Ourselves

VIBRATIONAL MEDICINE FOR THE 21ST CENTURY

~

The Complete Guide to Energy Healing and Spiritual Transformation

Richard Gerber, M.D.

EAGLE BROOK
An Imprint of HarperCollins*Publishers*

HarperCollins books may be purchased for educational, business, or sales promotional use. For information please write: Special Markets Department, HarperCollins Publishers, Inc., 10 East 53rd Street, New York, NY 10022.

FIRST EDITION

Designed by Richard Oriolo

Library of Congress Cataloging-in-Publication Data

Gerber, Richard, 1954–
Vibrational medicine for the 21st century : the complete guide to energy healing and spiritual transformation / by Richard Gerber.
p. cm.
Includes bibliographical references and index.
ISBN 0-688-16403-X
1. Mental healing. 2. Vital force—Therapeutic use. I. Title.
RZ401.G348 2000
615.8'52—dc21 99–37074
CIP

00 01 02 03 04 10 9 8 7 6 5 4 3 2 1

To all who seek healing of body, mind, and spirit this book is sincerely dedicated.
May they find peace, love, joy, and a deeper respect for the divine spirit within.

Acknowledgments

THERE ARE MANY individuals who have contributed in various ways to the "birthing" of this book. I wish to thank the many unsung pioneers of healing research and vibrational medicine who have paved the way through their tireless efforts to find better ways of healing human suffering. Although a great number of these individuals were misunderstood or cast as heretics by mainstream science during their own lifetimes, it is only now that we have acquired the wisdom to truly understand the value of their work. While many of these scientists, healers, and researchers have been mentioned in these pages, there are still countless others whose names do not appear but whose work and writings are equally important. Without their pioneering efforts, this book might not have been possible.

I wish to thank Joann Davis of Eagle Brook and William Morrow for convincing me to create a self-help guide to vibrational healing approaches that would be more accessible to all people.

While my first book, *Vibrational Medicine,* sought to teach the scientific principles underlying this futuristic energetic healing approach, the level of technical information presented there was sometimes a bit more than some could handle. Through Joann's feedback and constructive comments on my outline, she helped me to create the kind of book that I believe will bring healing and spiritual awakening to many more people who seek "a better way."

I also wish to thank my wife, Lyn, for her patience, love, and understanding and the many hours she spent typing and retyping corrected pages of my manuscript. Without her support, this book would not have been possible.

Also, I wish to acknowledge all of the countless teachers and "guides" who helped to direct me along the path of exploration and discovery that has helped to make this book a reality.

Contents

List of Diagrams

Introduction

WE LIVE IN interesting times. The nature of health care is shifting dramatically as we approach the end of the twentieth century. The rapid growth of technology and science has made a great impact upon health care by changing the way medicine is practiced and even altering our views of what makes human beings healthy or sick. But modern medicine sometimes falls short of solving our health problems. Is this failure the fault of modern medical science, or is it perhaps a reflection of our still not fully understanding the nature of chronic disease? Sometimes it seems today's medical science is so sophisticated that we should already have all the answers to the questions of healing illness and staying healthy. We keep pumping millions and millions of dollars into medical research, yet progress in healing many of the chronic illnesses of our times can be very slow. Are scientists asking the right questions about health and illness? Or are they locked into a medical model that is behind the times?

This book focuses on just such questions, especially with regard to the outdated model of medicine to which many health-care practitioners still subscribe. There is a kind of cultural bias, a scientific ethnocentrism, if you will, tending to follow research only in the mainstream sciences. Perhaps this reflects the fact that scientists are still people, with personal biases, egos, political affiliations, economic needs, and personal belief systems. And scientists, just like people everywhere, are slow to accept change. In our culture, one of our biases is that newer information and techniques are always better than older belief systems and technologies. Doctors often dismiss older medical approaches as outdated and useless. For example, many physicians don't fully appreciate that the roots of modern pharmacology lie in the older approaches of herbal medicine and homeopathy. Most doctors would rather prescribe a synthetic derivative of an herb's active ingredient than give the patient an herbal capsule, even if the two possess similar therapeutic action. Modern scientists assume that older approaches are too outdated to be valuable. However, mounting evidence supports a variety of ancient healing systems based upon views of physiology very different from the current mainstream healing paradigm. Perhaps revisiting ancient wisdom may supply important missing information that could result in a revolution in the field of medicine and healing.

Ancient approaches to understanding disease and body healing often viewed illness from the perspective of the human spirit, or the body's life-force energy. These somewhat mystical viewpoints may now hold the key to understanding why people become ill and how they can regain their health. Yet modern medicine tries to distance itself from ideas of spirit and life energy. Mainstream healers long ago gave up the belief system referred to as "vitalism" or the theory of vital energy. But is vitalism really such an outdated concept when we begin to factor into the human equation some of the newer discoveries in the field of quantum and Einsteinian physics that describe the underlying energetic nature of the physical world? Our understanding of the very structure of the physical body, when viewed from the subatomic level of our molecular makeup, has recently led many scientists and leading-edge thinkers to believe that human beings are more than just biological machines with parts that age and wear down. If we were just sophisticated machines, we might very well need only a better knowledge of replacement parts and techniques for biological engine retuning in order to heal from illness and the ravages of time. However, the model of humans as mere biomachines has not led doctors to the anxiously awaited breakthroughs in healing that the chronic diseases of our modern industrialized societies require.

A number of pioneering researchers in science and medicine are embracing a blended worldview of healing that combines the best of ancient and modern viewpoints of the human body and the varied approaches to achieve wellness. This book is an exploration of some of the ideas gradually working their way into the mainstream of medical thinking. They differ from the current model of medicine. Whereas the existing worldview of medicine sees the body as a great clockwork machine of biological gears and parts, the new worldview

Major Differences Between Conventional Medicine's and Vibrational Medicine's Worldviews

CONVENTIONAL MEDICINE MODEL	VIBRATIONAL-MEDICINE MODEL
Based on Newtonian Physics	**Based on Einsteinian and Quantum Physics**
Views the body as a biomachine	**Views the body as a dynamic energy system**
Sees the brain as a biocomputer with consciousness as a by-product of the brain's electrical activity	**Mind and Spirit are true sources of consciousness (the actual operator who runs the brain/biocomputer)**
Emotions thought to influence illness through neurohormonal connections between brain and body	**Emotions and spirit can influence illness via energetic and neurohormonal connections among body, mind, and spirit**
Treatments with drugs and surgery to "fix" abnormal biomechanisms in the physical body	**Treatment with different forms and frequencies of energy to rebalance body/mind/spirit complex**

of healing sees the body as a complex energy system. The present worldview of medicine sees illness as something that happens to people because of exposure to an infectious bacterium or virus, a trauma to the body, a toxic chemical ingestion, or an inherited abnormal gene. The new paradigm of healing digs deeper into what makes us holders of disease. Our current worldview of medicine does not consider consciousness to be an important causative factor in illness but, instead, merely a by-product of the neurochemical and electrical reactions in a person's brain. In the old school, consciousness was seen not as a cause but as something that was affected by a person's state of illness. True, the existing medical model recognized that being chronically ill could make someone feel depressed, but depression was never seen as playing a contributing role in bringing about physical illness. Newer insights gleaned from research into the effects of emotional stress on the body and its potential to inhibit the body's immune system are gradually affecting medicine, though. As the reader will see, stress, consciousness, and our attitudes toward life play key roles in the new world theories of who becomes sick and who stays well. Just exactly how consciousness and the human spirit affect our susceptibility to illness by enhancing or inhibiting our natural immune defenses is a complex subject, but it is one we will endeavor to tackle in the pages of this book.

The new model of medicine is part of an emerging field of what has been called "vibrational medicine," a science that is part old and part new. It is a synthesis of the best of ancient healing lore combined with the latest discoveries in science to produce an entirely new approach to diagnosing and treating illness. Vibrational medicine is based upon modern scientific insights into the energetic nature of the atoms and molecules making up our bodies, combined with ancient mystical observations of the body's unique life-energy systems that are critical but less well understood aspects of human functioning. Rather than seeing the body as a sophisticated machine, animated only by electrochemical reactions, vibrational medicine views the body as a complex, integrated life-energy system that provides a vehicle for human consciousness as well as a temporary housing for the creative expression of the human soul. In the world of vibrational medicine, illness is thought to be caused not only by germs, chemical toxins, and physical trauma but also by chronic dysfunctional emotional-energy patterns and unhealthy ways of relating to ourselves and to other people. Rather than relying on drugs and scalpels to treat illness, the vibrational-medicine approach to healing employs the use of different forms of energy, both electromagnetic and subtle life energy, to bring about healing changes in the body, mind, and spirit of the sick individual.

Throughout this book, we will endeavor to understand and explain the vibrational view of the human body and how consciousness and even the human soul are critical in understanding why people become ill. In addition, we will examine new ways of maintaining health and wellness as well as energetic methods of rebalancing ourselves when we become ill. An integrated view of medicine and healing is part of an emerging revolution in health care that will take us to the leading edge of discovery by blending ancient healing wisdom with modern-day medical insights. Settle in for this journey of exploration toward a better understanding of the integral relationship between body, mind, and spirit in health and illness. Join us as we take our first steps toward a new medicine for the twenty-first century.

One

WHAT IS VIBRATIONAL MEDICINE?

FOR MANY PEOPLE, the term "vibrational medicine" will be unfamiliar. The word "vibrational" conjures up an image of something vibrating or making sound. Sound waves are the most familiar form of vibration people think about when they use the term "vibrational." But sound waves are actually just one form of vibrating energy. Light is another form of vibration or oscillating energy. They are two examples of the many types of energy that make up the so-called electromagnetic spectrum, which also includes radio waves, television broadcasts, X rays, cosmic rays, ultrasonic waves, and microwaves. Modern physics tells us that the only difference between these forms of energy is that each oscillates at a different frequency or rate of vibration. Hence, vibrational medicine refers to an evolving viewpoint of health and illness that takes into account all the many forms and frequencies of vibrating energy that contribute to the "multidimensional" human energy system.

The most basic such form is the matter that composes the physical body. According to the new perspective of Einsteinian and quantum physics, the biochemical molecules that make up the physical body are actually a form of vibrating energy. During the early part of the twentieth century, Albert Einstein came to the startling conclusion that matter and energy were actually interconvertible and interchangeable. His famous $E = mc^2$ mathematically described how matter and energy were interrelated. Einstein said matter and energy were, in fact, two different forms of the same thing. At the time Einstein came up with this conclusion, few scientists could entirely understand its magnitude. But it was this very realization of matter's interconvertibility into energy that led to the development of the first atomic bomb, in which a few grams of uranium were converted directly into energy, proving Einstein's theory in a most unforgettable demonstration. With the example of the atomic bomb, more scientists came to believe in Einstein's assertion that matter and energy were two expressions of the same thing. In a variety of experiments in particle physics, in which scientists hurl speeding subatomic particles at targets in atom smashers to study the tiny fragments making up the structure of matter, additional evidence has been gathered confirming that all matter is really a form of frozen energy.

The Human Body as Energy Field

If this is so, then all the atoms and subatomic particles making up the human body are also a kind of frozen energy as well. This means people can be considered complex bundles of frozen energy! Since all energy vibrates and oscillates at different rates, then, at least at the atomic level, the human body is really composed of different kinds of vibrating energy. The term "vibrational medicine" comes from this fact. More specificially, vibrational medicine is an approach to the diagnosis and treatment of illness based upon the idea that we are all unique energy systems. By using a vibrational-medicine approach, it is possible to diagnose different types of illnesses based on a knowlege of the different frequencies of energy that can be measured coming from the human body. This idea is certainly not new. For example, many doctors routinely order electrocardiograms (EKGs) on patients as a part of their yearly exams. The electrical energy coming from the heart can give doctors information about whether the heart is functioning properly. The EKG "energy patterns" can reveal to physicians if the heartbeat is regular or erratic and can also indicate whether a patient's heart is receiving enough vital blood flow through the nutrient-carrying coronary arteries. All this information comes from interpreting the patterns of vibrating electrical energy put out by the beating human heart (as picked up by EKG electrodes attached to the patient's skin). The squiggly lines of the electrocardiogram are a diagnostic tool for the physician who looks to detect early heart disease (even in individuals without noticeable symptoms of chest pain or palpitations). Thus, the measurment of the heart's electrical energy, even in a simple EKG, is actually a form of vibrational-medical diagnosis. But even though such medical instruments as EKG machines might provide a very basic form of vibrational diagnosis, the paradigm of healing that guides most doctors in their interpretation of this energetic information is one based upon a viewpoint of the body as a biomechanism, not a vibrational-energy system.

The concept of the body as a complex energetic system is part of a new scientific worldview gradually gaining acceptance in the eyes of modern medicine. The older, yet prevailing, view of the human body is still based upon an antiquated model of human functioning that sees the body as a sophisticated machine. In this old worldview, the heart is merely a mechanical pump, the kidney a filter of blood, and the muscles and skeleton a mechanical framework of pulleys and levers. The old worldview is based upon Newtonian physics, or so-called billiard-ball mechanics. In the days of Sir Isaac Newton, scientists thought they had figured out all the really important laws of the universe. They had discovered laws describing the motion of bodies in space and their momentum, as well as their actions at rest and in motion. The Newtonian scientists viewed the universe itself as a gigantic machine, somewhat like a great clock. It followed, then, that the human body was probably a machine as well. Many scientists in Newton's day actually thought that all the great discoveries of science had already been made and that little work was left to be done in the field of scientific exploration.

The Clash of the Paradigms: The Newtonian Mechanists Versus the Quantum Mechanics

As science grew more sophisticated, so did the nature of the biomachine we were thought to be. That is, as our technologies became more powerful with the discovery of the optical and electron microscopes, the parts and gears of the human machine were studied at smaller and smaller levels. While early European physicians could analyze the human body only in terms of dissection of organs at the time of autopsy, today's medical researchers have the tools to study our physical makeup at the cellular and molecular levels. Modern medicine's current Newtonian biomechanistic viewpoint suggests that if we could only understand how all the different tiny parts fit together in the human body, we could develop better ways of fixing and repairing the body in the event of illness. This mechanical approach to fixing the body is nowhere more evident than in the field of surgery. Surgeons are the ultimate biomechanics. Orthopedic surgeons work with unique surgical "carpentry tools," which include drills, saws, screwdrivers, and screws that allow them to replace arthritic joints with better, synthetic joints of metal, Teflon, and plastic. Vascular surgeons work to cut out clogged arteries and replace them with newer synthetic Dacron grafts to restore adequate blood flow to the oxygen-starved limbs of individuals with vascular disease. Ophthalmologic surgeons remove lenses clouded by cataracts, only to attach artificial intraocular lens implants in their place. Cardiovascular surgeons work on the delicate human heart itself, replacing narrowed or leaky heart valves with synthetic ones. In addition, cardiac surgeons routinely stitch in place synthetic grafts or transplanted blood vessels from the legs to allow blood flow around blockages in the heart's main arteries. And general surgeons routinely cut out tumors from the various organs of the body in order to treat cancer.

While these surgical approaches do indeed provide a very sophisticated "fix," they do not fully explore the reasons behind "why" diseases occur in the first place. That area of study has been relegated to medical researchers known as molecular biologists, who study the body's most minute parts—the structural molecules, the enzymes, and even the genetic structures that compose and direct the function of the body at the cellular level. The thinking is that if we only knew which enzyme was defective or which gene was abnormal, we could invent a molecular solution that would circumvent the disease process and thus "cure all illness." There have actually been many medical breakthroughs resulting from this line of scientific inquiry. Greater knowledge of the structure of human insulin and the genetics of its manufacture ultimately led to the development of genetically engineered human insulin. This insulin is made by simple bacteria with a genetic structure that has been altered by the insertion of the human insulin gene into the genetic code of the bacteria (so-called gene-splicing technologies). As a result of this manipulation, the bacteria become tiny human-insulin-producing factories. Many diabetics owe their quality of life to this new insulin, which tends to produce much less "insulin resistance" (from immune reactions to insulin by the body) than the older types derived from cows and pigs.

The study of molecular biology has certainly advanced our understanding of the physical causes of many types of diseases. Information acquired from studies in molecular biology has also helped medical researchers and drug companies to develop new drugs that work at very specific sites in the body. For example, molecular biologists have discovered that most of the chemicals and hormones in the body act at specialized binding sites known as "cellular receptors." These receptors, embedded in the outer and inner walls of most cells of the body, function as chemically triggered switches that turn on or off specific chemical reactions and metabolic pathways in each cell. Each cellular receptor is like a kind of switch connected to a lock-and-key mechanism, similar to the ignition system for a car. Only the binding of a specific chemical to the receptor will trigger the molecular switch, just as a car will start only if the correct key is inserted into its ignition. By creating drugs that either trigger these receptor switches or else block the cells' receptors from being triggered by the naturally occurring body chemical, pharmaceutical companies have developed many new drugs for the treatment of hypertension, diabetes, asthma, and heart disease. Recent work in the field of immunology has resulted in the creation of antibodies against specific tumors in the body. Cancer researchers, by hooking up a toxic chemotherapy molecule to a tumor-specific antibody molecule, are beginning to develop drugs that will deliver toxic chemicals selectively to cancer cells without causing major side effects in the other tissues of the body.

In theory, the molecular-biology model suggests that physicians could treat all human illness if only they knew the specific molecular causes of diseases. This, of course, assumes that all human illness has a purely physical, molecular basis. While sophisticated in its approach, the molecular-biology model is still based upon the "old-world" Newtonian viewpoint of the body as a biomachine. Even if one assumes that a particular disease is caused by, say, a particular infectious microbe, this mechanistic view of illness does not fully explain why some people exposed to the bug will get violently ill while others may develop only coldlike symptoms or no illness at all. The old-world Newtonian model assumes that if you know all of the pieces of the machine, then you can fix it or build a new one (i.e., the whole must equal the sum of its parts). In reality, however, we are very far from creating living organisms as complex as human beings from scratch. There is more to the human equation than perfectly functioning biomolecules. The old-world, Newtonian model of medicine lacks an appreciation of seemingly intangible things such as emotion, consciousness, and the energy and life force of soul and spirit. Although most modern physicians have begun to appreciate some of the emotional contributions to illness, the majority of today's doctors deal with such issues primarily by prescribing antidepressants and antianxiety drugs or by referring patients to psychiatrists for additional drug treatment and psychotherapy. Although many pioneering medical researchers in the field of psychomatic medicine have sought to study the physiological links between the brain and the body during various emotional and stressful states, few fully appreciate the bigger picture of human beings as dynamic energy systems of body, mind, and spirit.

As the reader will soon see, the newly emerging vibrational model of human functioning provides physicians with the bridge they need in order to go beyond Newtonian medicine to grasp the contributions of the human mind and spirit in various states of health and illness. What is already known about the body in terms of mechanistic function can be put into the framework of the larger dynamic energy system that more fully describes the "multidimensional" human being. The vibrational or energetic model of healing does not deny the validity of discoveries in molecular biology or the biomechanical functions of the body's organs. It merely puts them into the perspective of the bigger picture. To illustrate, we know that the cells of our bodies are fed by various nutrients derived from the food we eat as well as oxygen from the air we breathe. But our cells are fed also by a continuous stream of life-force energy. We possess a variety of specialized energy-distributing systems that also support the cells and organs of our bodies. These energy systems are affected by different factors that can enhance or inhibit the flow of life-force energy to the cells and organs of our bodies. Among those critical factors are such things as our emotions, our relationships to others, our ability to give and to receive love, and even our relationship with God. While seemingly nebulous and difficult to define in terms of specific physiological effects, these emotional and spiritual factors are of great importance to the sustenance and support of the tissues and organs of our bodies.

The new vibrational model of human beings sees consciousness as playing an integral role in health and illness. Consciousness is not merely a byproduct of electrical and chemical signal-processing in the human brain. Consciousness is a kind of energy itself. In some ways, consciousness is a bit like the "ghost in the machine." We are not simply the sophisticated biological mainframe biocomputer of the brain and nervous system. Human consciousness is more akin to the programmer who sits at the computer workstation, with that fabulous workstation being the human nervous system itself. Consciousness has ghostlike qualities that allow it to reside not only in the brain but beyond the body itself. From the perspective of vibrational medicine, our consciousness is not limited to the brain and central nervous system but is also seen as an integral aspect of the human heart. The old adage of acting from the heart as well as from the brain actually has a basis in science. One might say there exists a form of "heart-based" consciousness that acts from a center of love, compassion, and empathy toward others. In the vibrational-medicine view of human functioning, our emotions are not just the result of neurochemical reactions in the limbic system or the emotional centers of our brain. Our emotions are also influenced by a greater, spiritual energy field that encompasses and influences the entire physical body and nervous system. Our reactions to life are recorded not only in the biochemical patterns of memory storage in the brain but also in the seven major life-energy centers of the body that help to nourish our cells and organs. In the energetic view of human health, we are more than mere biological engines. We possess bodies that are energized and motivated by the forces of our spirit and soul. We are energetic beings whose ills may be healed not only by surgical procedures and drugs but also by different forms and frequencies of energy. It is this energetic viewpoint of human beings, as more than just flesh-and-blood mechanisms, that is embraced and described by the vibrational-medicine model of human functioning.

The Vibrational-Medicine Model

In order to talk about this new model of human health, we need to look at the vibrational-medicine model in greater detail. As mentioned, vibrational medicine recognizes that the human body has certain biomechanical functions. The human heart does act as a great pump to deliver life-giving oxygen and nutrients to the various organs of the body. The kidneys function as filtration systems for the blood, helping to excrete toxic waste material from the body. The muscles and skeletal system form a mechanical framework that allows us to move through space and get from one location to another. However, in addition to the biomechanical functions of the body's organs, the vibrational-medicine model views the organs of the body in terms of their innate intelligence and their ability to process different types of information. Most information that the organs receive is in the form of chemical messages that help to regulate the function of each organ within the context of the body's daily needs. But the organs and the cells making up each organ also communicate with each other using nonchemical forms of information-carrying messengers.

For instance, we now know that the cells of the body actually emit weak pulses of light. Those weak cellular light pulses seem to be part of a light-based communication system that helps to coordinate the actions of the cells within each organ. The pulses of light emitted by cells are just one of the many different informational codes the human body and its individual cells use to regulate the function of organs on a day-to-day and moment-to-moment basis. Our cells communicate through the coded messages carried by hormones and biochemicals, as well as through electrical signals (such as those carried by the nerves of the body), and also through weak light signals. The cells of the body appear to have their own inherent intelligence that allows them to understand and use this coded information in its many forms in order to maintain the body in a state of health.

Another example of a nonchemical form of information transmission happens during digestion. The cells of the digestive tract are coordinated by electrical signals coming from actual pacemaker-type cells. These gut-based pacemaker cells send electrical signals to the muscles of the intestinal tract to produce the coordinated contractions that propel food through the digestive system. Without the coordinating influence of these coded electrical signals, the food we eat would just lie in our stomachs and ferment. The brain also sends coded electrical nerve signals to the digestive system through a long nerve known as the vagus nerve. If the the electrical signals to the gut become too excessive, spasms, abdominal pain, and diarrhea can be the result. If the electrical signals to the gut are insufficient, constipation may occur. Thus, the proper stimulation of the digestive system through coded electrical signals is important to the health of our bodies. Similar types of electrical pacemaker cells exist in the urinary tract as well. The electrical messages of these cells stimulate the movement of urine from the kidneys to the bladder. We also depend on the pacemaker cells of the heart to regulate the rate at which our heart beats. When the electrical stimulation from the pacemaker cells in the heart goes awry, irregular heart rhythms can occur, leading to palpitations, breathlessness, and even fainting or death. For very severe cases of irregular heart rhythms, special man-made pacemaker devices are surgically implanted to deliver the appropriate electrical signal stimulation, which will properly regulate the heartbeat.

These examples of various types of nonchemical energy messages illustrate how the body needs coded signals to maintain normal functioning and optimal health. In some ways, these energy signals are like a timing mechanism that synchronizes the activities of the body's organs. One can see an obvious analogy between the firing of spark plugs in cars and coded body signals. However, the human body is much more sophisticated than a car. The heart is more than just a living engine that pumps blood throughout the body. In fact, some thinkers and philosophers suggest that the heart is the seat of the human soul. Perhaps the soul doesn't exist within our hearts, but our hearts may contain the "path" leading to the rediscovery of our souls. More than just an engine of flesh that beats unceasingly from the electrical stimulation of its pacemaker cells, the heart is also driven by a spark of spiritual energy or life force that would seem to leave the body at the time of death.

The issue of death and spiritual energy also helps highlight the way the vibrational-medicine model differs from the existing mechanistic paradigm of healing. Seen from the mechanistic viewpoint, the human brain and body are merely a collection of chemicals, energized and animated by electrical and electrochemical impulses. To the purely mechanistic thinkers, death means the end of the body as well as the destruction of the personality. But in the vibrational or energetic model, spirit is seen as the motivating force that animates the physical form. That is, our spirit, like a vaporous ghost, inhabits the mechanical vehicle we call the physical body. At the time of death, our spirit moves on, leaving behind only a lifeless shell. In the vibrational-medicine model, it is the beingness of our spirit and its experiential journey through the physical world that creates the real adventure and mystery of a person's life. The vibrational-medicine perspective may also do a better job of acknowledging the contribution of spirit in human health and disease than the current view of medicine does.

Our physical bodies and brains allow us to interact with other individuals, to have meaningful relationships, to gain knowledge, and to manifest our thoughts and ideas into actual physical creations. While our spirit may inhabit our physical bodies during our lives upon the earth, the physical dimension is not the primary realm in which our spirit dwells. In the vibrational universe, it is believed other dimensions exist, beyond the physical plane, that our spirit normally inhabits during its natural state of existence. In fact, we all explore this higher realm when we sleep at night. While our body sleeps and our mind enters the dream state, a higher aspect of our spirit leaves our body and has direct experience of this higher-dimensional plane of existence. Some spiritual philosophies refer to this other dimension as heaven, while others call it the astral plane. The subtle difference between the two terms may be only a matter of semantics, but to the spiritually minded seeker, there is indeed a higher plane of existence beyond the physical, which is the true domain of spirit. In reality, we are spiritual beings, with our roots in these ephemeral higher planes of existence. The vast majority of people have forgotten their spiritual origins. Their conscious minds think only about those matters they can touch or taste or feel with their physical five senses. And yet it is this realm of spirit that is the hidden reality behind the true nature of health and illness when viewed from the perspective of vibrational medicine and its understanding of the multidimensional human being.

Long ago, most orthodox scientists relegated the realm of spirit to the theologians and clergy, having found matters of spirit to be too ephemeral and undefined in scientific terms to have relevance to their nuts-and-bolts explanations of the workings of the physical world. If scientists could only see through the eyes of spirit, they would be able to perceive spiritual energies and the workings of the spiritual domain as the real motivating forces behind the facade of everyday physical reality. Many of the systems of the body that help to integrate spiritual and life-energy forces into our cells and organs are intimately involved in connecting human beings to the realm of subtle energies that are often difficult to appreciate with our gross senses. With time, patience, and a little practice, most people can be trained to perceive subtle energies for themselves and to ground these seemingly far-out beliefs into practical experiences.

The Human Energy System

The vibrational-medicine view of human beings as multidimensional energy systems is more easily understood if we examine the many ways our bodies use different forms of energy in nutrition, information processing, and maintaining our general health. The most commonly recognized form of energy our bodies use is that of "metabolic energy" extracted from the foods we eat. We ingest various combinations of sugars, proteins, and fats through our daily meals that are broken down, absorbed through our stomach and intestines, and converted into chemical forms of metabolic cellular energy. While some foods give us the raw building blocks for repairing and regenerating the aging cells of our bodies, much of the food we eat is used to provide chemical energy that allows us to be active, creative beings.

In addition to the chemical energy our cells use as a primary form of fuel, our bodies also use electrical energy. To better understand the nature of human electrical energy, we must first examine how our bodies use electricity to communicate messages from one location to another, primarily through our nervous system. When we want to move our hand to reach for something, the initial thought of reaching for something triggers a message that travels from our brain to the muscles in our arm and hand via electrical impulses carried by nerves. The nerves of the body are a kind of electrical wiring system that allows our brain to telegraph messages to the organs and muscles of our bodies along an electrical message system. The brain uses a kind of electrical code in order to send and receive different forms of information from various parts of the body. The more rapid the rate at which a nerve fires, the higher the intensity of the signal interpreted by the brain. The slower the rate of nerve firing, the lower the signal intensity perceived by the brain. For example, when someone goes to move a hot metal cooking pan sitting on a warm stove, the temperature sensors in the person's finger relay temperature information about the pan directly to the brain. A slow rate of nerve firing from the finger's temperature sensor tells the brain that an individual's hand is touching something only mildly warm. If that same temperature sensor were to send a message to the brain consisting of very rapid nerve firing, the brain would interpret this information as a pan too hot to touch, triggering a reflex withdrawal reaction that moves the hand away from the hot pan. Our nervous system uses a Morse code of electrical dots and dashes that functions as a specialized language to carry different types of biological information to different parts of the body, while also relaying information from the body back to the brain.

More recent studies of the body's electrical functioning have shown that human beings have even more complex electrical control systems than the electrical signals that move information throughout the nervous system. Leading-edge scientists have discovered that living cells seem to have internal elements and membrane structures that give them the capacity to function as tiny integrated circuits much like those found in a computer or other electronic device. These same scientists have proposed that there are certain levels of cellular communication that behave very similarly to electronic control systems. In other words, the cells of our body may communicate with each other not only using chemical and light-based signals but electronically as well. In fact, the cell may use these electronic communication systems as an additional means of controlling and regulating internal cellular processes. This particular field of electronic cellular communication has been nicknamed "bioelectronics," short for biologically based electronic systems. These bioelectronic information-transmission systems may be part of a little-known pathway of communication within and between the cells of our bodies.

In fact, it now appears possible that the type of message that ultimately controls whether or not cells divide and reproduce might be carried not by a genetic or chemical signal but by a biological-electronic or bioelectronic signal. If this is true, then the theory that each human cell can divide and reproduce only so many times, and then no more, could eventually be displaced from its position of eminence in cell biology. And with this displacement go some beliefs about cancer. Scientists working to study bioelectronics have suggested that, in addition to genetic and carcinogenic causation factors, cancer might actually be a disease in which the normal electronic signal regulating cell division has somehow gone wild. The bioelectronic message that should be limiting cell growth might somehow be replaced by an overgrowth signal that tells the cancerous cell to keep dividing and growing. If this type of reasoning turns out to be correct, then perhaps an alternate method of treating cancer might be based on learning the bioelectronic communication signals that control cellular division. If the correct electronic signal could be sent to the cancer cells, it might cause them to stop dividing, thus arresting the growth of the cancer. Such an approach would be primarily an energetic treatment, as opposed to a surgical procedure or administration of a toxic chemotherapy drug to the body. In other words, a greater understanding of bioelectronic cellular-control systems might eventually lead to a purely electromagnetic or vibrational treatment for diseases like cancer.

Besides these chemical and electrical communication and control systems, our cells have light-communication systems as well. Light-communication systems fit into the vibrational-energy model of health. As alluded to earlier, the cells of our body actually emit extremely weak bursts of ultraviolet light. This ultraweak light emitted by our cells is part of a light-based control system that living cells appear to use to communicate messages to one another. Nearly fifty years ago, Russian researchers discovered that when cell cultures in quartz dishes were placed side by side and a poison was added to one cell culture, the cells in both culture dishes would die a "mirror image" death. This meant that when the quartz dishes containing cells were close to each other, yet physically separated, some type of cellular information signal jumped from one quartz container to the next. The cellular information signal was transmitted from one cell culture to another without any "physical" transfer of toxins or chemical messages or hormones between the two cell cultures. Oddly enough, this "death message" would be transmitted only if the dishes containing the cell cultures were made of quartz. If glass culture dishes were substituted, no death message would travel between the culture dishes. It is well known that glass blocks the transmission of ultraviolet light, while quartz allows ultraviolet light to pass through it easily, suggesting the possibility that the death message is carried by UV light. More recently, researchers in Germany used highly sensitive light-measuring equipment to monitor the light emitted by living cells. They were able to confirm finally that living cells emit small bursts of light in the ultraviolet range, especially when a toxin or poison was added to the cell culture. The fact that bursts of ultraviolet light are emitted by dying cells explains why the death message was blocked by the glass cell-culture dishes but consistently produced cell death within adjacent quartz-held cell cultures. The message bringing death to the nearby cells was transmitted by weak bursts of ultraviolet light, similar to the Morse code of light flashes formerly used by sailors to transmit messages from ship to shore.

So far, the energy systems of the body we have been describing are within the realm of conventional electromagnetic, electrical, and chemical energy systems. The neuroelectrical and biochemical energy systems of the body are fairly well established in scientific circles, and in time, the bioelectronic and biophotonic, or light-based, energy systems of the body will be validated by further scientific research. Most conventional medical practitioners acknowledge at least some of these energy systems. Perhaps certain doctors will be comfortable with accepting these specific energy systems as components of a new vibrational-medicine overview of health and wellness. However, our bodies also use other energy systems involving the flow of specific types of energy that have not yet been accepted by conventional or "scientific" medicine. These other types of energy systems are perhaps even more critical to life than the aforementioned energies, yet they still lack "official" recognition by most Western physicians. These other energy systems we have been referring to are the important life-energy and spiritual-energy systems of the multidimensional human being. Much of our knowledge of such "unconventional" energy systems, sometimes referred to as "subtle-energy" systems, comes from the sacred and spiritual knowledge of the Far East and India.

The Acupuncture-Meridian System: The Body's Biocircuitry

"Life-energy systems" are specialized systems of the body that absorb and distribute life energy to our cells, tissues, and organs. One life-energy system that Western medicine dimly recognizes is the acupuncture-meridian system, which consists of a series of conduits or channels for life energy that acupuncturists attempt to manipulate through the insertion of extremely fine needles into special points upon the skin. The art and science of healing with acupuncture is hundreds (if not thousands) of years old, although no one seems to have precise information on its exact origins. These critical body points are known as acupuncture points or simply acupoints. Acupuncture points are unique zones on the skin located along the specialized life-energy pathways or "circuits" known as "meridians." According to the Chinese acupuncturists, meridians carry a type of environmental life energy called *ch'i*. This special type of life energy is said to come from three primary sources. Part of our ch'i energy comes from the vital-energy reserve we inherit from our parents. This type of "inherited" life energy is referred to as ancestral ch'i. The second source of ch'i is absorbed (and produced) from the foods we eat. The third (and possibly the most important) source of ch'i comes directly from the environment. A certain amount of ch'i is absorbed from our surroundings and taken into the body and the meridian system via the acupuncture points themselves. Acupoints appear to function like tiny energy pores in the skin that absorb this unique environmental subtle energy directly into the meridians, where it is then distributed to the organs of the body.

According to theories of traditional Chinese medicine (TCM), illness is mainly the result of an imbalance in the flow of ch'i energy to the organs of the body. Acupuncture treatments are one way of rebalancing the flow of vital life energy to the organs, which may either be deficient in ch'i or overloaded with excessive ch'i energy. The acupuncture points located on the skin also seem to function like miniature electrical relay stations along a vast power line, helping to maintain the flow of energy along each meridian. Thus, in the world of life-energy systems, life and health are dependent not only upon the proper balance of metabolic, electrical, and light-energy flow throughout the cells of our bodies but also upon the proper flow of life energy in the form of ch'i to our bodies' cells and organs. This vibrational viewpoint of health, a perspective that acknowledges the need for appropriate levels of life-energy flow throughout the body, requires a great leap in thinking beyond the limited biomechanistic paradigm of traditional medicine.

The Seven Chakras and the Spiritual Connection to Human Health

There is another form of energy that is also important to human health. This type of energy might be referred to as "spiritual energy." Through a variety of different forms and pathways, spiritual energy flows into the cells and organs of the physical body. One pathway of spiritual energy flow of critical significance to human health is a unique system of seven major energy centers known as the body's *chakras.* But what exactly is a chakra? The word comes from the ancient Sanskrit word for "wheel." To individuals with clairvoyant abilities who can actually see chakras, they are said to look like spinning wheels of color and light. They possess certain (superficial) similarities to acupuncture points. Like acupoints, chakras are specialized energy centers throughout our bodies where a unique form of subtle environmental (life) energy is absorbed and distributed to our cells, organs, and body tissues. However, the flow of this subtle energy (called *prana* by the Hindus) through our chakras is strongly affected by our personality structure and by our emotions, as well as by our state of spiritual development. The yogic tradition of ancient India was the first to describe, many hundreds of years ago, the presence of specialized spiritual-energy centers in the body. Although Hindu and yogic teachers in India have recognized their existence for centuries, knowledge of the chakras has only recently made its way to the West. According to ancient wisdom and current research, each of the major chakras is directly linked to a different region of the body, as well as to a particular hormone-producing endocrine gland and system of bodily nerves.

Seven main chakras exist in the body, along with many other minor chakra centers. For the purpose of our discussion, we will focus primarily on the seven main chakras of the body. The first or root chakra is located at the base of the spine and is connected with the skeleton and the bone marrow at the core of the bones. The bone marrow is actually a source of our body's stem cells. Stem cells are the precursor cells that develop into both red blood cells and white blood cells in the body. As such, the bone marrow also produces the precursors to lymphocytes, the type of white blood cells that seed the thymus gland and immune tissues of the body during early childhood. Because of the root chakra's connection to the bone marrow, it is thought to have some effect upon the body's immunity against different types of illnesses. The second chakra can be found over the lower pelvic region and is often referred to as the sacral center. The second chakra is located directly over the ovaries in women and the testes in men. Since both the testes and the ovaries are endocrine glands and are related to human reproductive function, the sacral chakra is said to be associated with various aspects of sexuality and procreation.

The third or solar plexus chakra, located over the pit of the stomach, is linked to the pancreas and adrenal glands, which contribute to digestive functioning and the body's acute response to stress. The fourth or heart chakra, found in the middle of the breastbone, is associated with the thymus gland, the master gland that plays an important role in regulating the level of immune functioning in the body. Located directly over the thyroid gland is the fifth or throat chakra. Our thyroid gland regulates our body's metabolic rate through the production and release of thyroid hormone into the circulation. The sixth or brow chakra is centered in the forehead directly above the eyes. The brow chakra appears to be strongly linked to the pituitary, the master gland of the brain. The pituitary gland produces special stimulating hormones that regulate and orchestrate the activity and level of function in the various endocrine glands throughout the body. Lastly, the seventh or crown chakra is linked with the activity of the pineal gland, an endocrine gland that produces certain pituitary-regulating hormones (such as melatonin).

From the vibrational-medicine perspective, the seven chakras are more than mere regulators of physiological functioning in the body's endocrine glands. They are actually considered emotional- and spiritual-energy processors. Each of the seven chakras is individually linked to small nerve bundles known as ganglia. Each ganglion is like a little brain center. It appears that each of the seven chakras (and their associated nerve centers) processes and "remembers" different emotional events and traumas that affect us throughout our lifetime. We seem to store specific types of emotional memories in these centers. Perhaps this is one of the reasons we remember things not only with our brains but with our bodies as well. This might also explain why different types of emotional distress seem to affect one part of the body preferentially over another.

The flow of nutritive life energy from the environment to each of the seven chakra-linked body regions is strongly affected by the way we process different emotional and spiritual issues of human development. When a person has a chronic problem in dealing with the emotional and spiritual issues associated with a particular chakra, the resulting constriction of life-energy flow to the body zone linked with that chakra can sometimes manifest as a "dis-ease" or "health challenge." Each of the seven major chakras is associated with a variety of emotional and spiritual issues that are influenced by how we relate to others, as well as to ourselves. The first or root chakra is often likened to a root system that connects us to the earth. Our root chakra is influenced by our sense of connectedness to the soil, the flowers, the trees, and the environment. To illustrate, someone who feels very centered and peaceful while gardening will often have a good flow of energy coming through the root chakra. The first chakra is also said to be the energy center where concerns over our survival and our personal safety are processed. Someone who always feels threatened and unsafe in the world may eventually develop an associated imbalance in the root center.

The second chakra is linked with the gonads and reproductive function. It is the emotional energy center that relates to how we deal with others in regard to our sexual nature. The second chakra is also thought to be the center connected with our personal and working relationships with others. For instance, an individual who has been repeatedly hurt in a relationship, or has been the victim of rape or incest, might carry the scars of such emotional memories in the second chakra center. Since an imbalanced second chakra may result in constricted life-energy flow to the pelvic organs, physical symptoms associated with sacral-chakra dysfunction may manifest as chronic pelvic infections or recurrent bladder problems. The third emotional-energy center, also known as the solar plexus chakra, is affected by the way we deal with the issues surrounding our personal power and our self-esteem. Third-chakra issues usually relate to whether we feel "empowered" (versus "powerless") in our jobs and our home life. Do we have a sense of control over our lives, or do we feel that others have control over us? Do we have a low sense of self-esteem? Such emotional conflicts affect the functioning of the third chakra, which feeds energy to the stomach, pancreas, and digestive organs. Someone with chronically low self-esteem and feelings of lack of personal power might suffer from recurrent ulcers due to an energy imbalance in the third chakra.

The fourth emotional-energy center, also known as the heart chakra, is one of the most critical energy centers because this is where the issues surrounding love and nurturance most strongly affect us. The amount of life energy flowing through the heart centers is influenced by the way we express love to other individuals as well as by our ability to feel love toward ourselves. Whether or not we consider ourselves to be "lovable" actually affects the flow of nurturing life energy to our heart and lungs. The thymus gland, the master immunity-regulating gland, is also said to be connected to the fourth major chakra. Chronic difficulties in expressing love toward others may eventually contribute to physical problems affecting the heart and lungs, as well as to immunity-related disorders that can make us more susceptible to infections and other serious health problems. Even breast cancer may be influenced by the heart-chakra connection. As an illustration, an inability to nurture others, or a lack of nurturance during childhood, might eventually result in developing diseases of the breast (symbolically considered the "organ of nurturance") because of resultant imbalances in energy flow through the fourth emotional-energy center.

The fifth emotional-energy center, also known as the throat chakra, is associated with communication of our thoughts, ideas, and opinions. The throat chakra is also the center linked with issues surrounding the expression of our will, either by voicing our own concerns and opinions or by expressing creative ideas. People who are always afraid to speak their minds or to voice their ideas and concerns can sometimes develop imbalances in their fifth emotional-energy center, resulting in chronic throat problems, recurrent laryngitis, or even thyroid disease. The sixth or brow chakra is associated with various aspects of mental functioning, thought processes, as well as our ability to see. Whether or not we have a "clear vision" of where we are headed and what we are doing with our lives directly influences energy flow through our brow chakras. Since this emotional-energy center is situated in the region of the eyes, our ability to clearly observe the patterns affecting our lives may indirectly affect how clearly our physical eyes are can actually see the world around us. This means that cataracts and eye diseases may be influenced not only by aging, diet, and exposure to ultraviolet light but also by how well our sixth emotional-energy center is functioning. The sixth chakra is also concerned with mental activities and the intellect. A person's clarity of thought may be indirectly influenced by the degree of dysfunction that exists in the sixth emotional-energy center. Lastly, God is linked with the seventh or crown chakra. Here resides our sense of purpose in life as well as our sense of spiritual connection with God. The seventh chakra is intimately related to our search for purpose and meaning in life, especially with regard to our link with the divine. Too often we come to explore seventh-chakra issues only in the latter portion of our lives, especially when facing life-threatening illnesses.

The amount and quality of life energy moving through our seven major emotional-energy centers is a powerful factor influencing the health of the tissues of various parts of our bodies. When we experience difficulties in dealing with one or more of these seven chakra-linked psychospiritual issues, such difficulties may lead to an energy weakness or imbalance in the chakra-associated body region. The chakra-energy factor is an important (but often unrecognized) contributing factor that may partly determine why people become ill. This does not mean that our inherited genetics, our diet, and our exposure to environmental toxins, pathogenic bacteria, and viruses are not important. Each of these factors does contribute to illness in a variety of ways. However, the energy flow through our chakras may be the underlying factor that determines the "weakest link" in the chain that is our body. When a chain is stressed and pulled, it always breaks at the point of the weakest link. Because the chakras are connected to our emotions and our desires, they play an important role in the translation of emotional problems into physical problems.

The evolving field of mind-body medicine known as "psychoneuroimmunology" has provided many new insights into the various hormonal and chemical connections between emotional stress and illness causation. The term "psychoneuroimmunology" (or PNI) refers to the link between the psyche (the emotions), the brain (neurology), and the function of our body's immune system (immunology). Scientists in this developing field have discovered a wide variety of hormonal messages being passed throughout the body that influence how well our immune systems protect us from illness. Until recently, most PNI researchers had focused primarily on the chemical codes and languages the brain and body use to regulate (or inhibit) the function of our immune defenses. By adding the contribution of the chakras to the total energy equation of the body's physiological systems, scientists may eventually gain a clearer understanding of the diverse energy factors that regulate our organs and our immune system and that may lead to stress-related illnesses. The chakras seem to have a powerful function as emotional- and spiritual-energy transformers. The energy inputs of the chakras are merely a different type of coded message our bodies use to pass along biological information that contributes to either illness or wellness.

The chakras are also a kind of unique emotional- and spiritual-feedback system. If we can become educated in the language of the chakras and their emotional and spiritual links, the area that illness settles into in our body can sometimes provide useful feedback to us by revealing whether or not we may be consciously (or unconsciously) blocking some aspect of our emotional and spiritual maturation as human beings. The key to understanding how chakra dysfunction affects us relates to the symbolic nature of the information that chakras process and pass on to the body. We will examine this symbolic communication system in greater detail in a later chapter, after we have built up a fuller understanding of the total multidimensional human-energy system.

The Various Forms of Bioenergy in the Multidimensional Human Being

TYPES OF BIOENERGY	SOURCE OF BIOENERGY	LOCATION OF BIOENERGY ACTIVITY	FUNCTION OF BIOENERGY
Metabolic energy	**Fats, sugars, proteins essential nutrients (converted by body into ATP—cellular-energy currency)**	**Throughout the cells of the physical body**	**Metabolic fuel to power basic cellular processes**
Bioelectrical Energy	**Nerves, muscles, heart, GI tract, brain Triggered by flow of electrically charged ions across cell membranes**	**Transmitted through cells of nervous system and muscular system or organs and body muscle groups**	**Movement through space (body) Movement of blood, food, urine through body Communication between nervous system and the body's organs (autonomic regulation) Information processing and perception (brain)**
	Bone matrix affected by gravitational forces (piezoelectric effect)	**Throughout the bones of the skeleton (especially weight-bearing bones)**	**Assists in bone remodeling for optimal bone strength**
	Current of injury (battery-like effect created by injury, tumors, infections, trauma)	**Throughout the physical body (Number one component of a larger body-wide electrical circulatory system)**	**Local repair of tissue damage, local healing response to tumors, infections, etc.**
	Bioelectronic energy (flow of intracellular electrons, probably from mitochondria in all cells of the body)	**Bioelectronic currents may exist both within the cells of the body and possibly between cells (through intercellular matrix/ connective tissue)**	**May regulate internal cellular processes (such as cell division)**
Biophotonic Energy	**UV biophotons, possibly emitted by DNA in cell nucleus**	**Within the nuclei of all cells of the body**	**Cell-to-cell communication**
Subtle Bioenergies (Subtle Magnetic Life Energies)	**Ch'i—an environmental nutritive subtle energy absorbed directly from the environment, created from processing food, and inherited from parents—a form of life energy**	**Flows through the acu-puncture meridians of the body (Certain types of ch'i also flow through muscles, tendons, and skin)**	**Nutritive subtle-energetic support, coordination of organ activities Defense against illness**

TYPES OF BIOENERGY	SOURCE OF BIOENERGY	LOCATION OF BIOENERGY ACTIVITY	FUNCTION OF BIOENERGY
Subtle Bioenergies (Subtle Magnetic Life Energies)	**Prana—another environmental nutritive subtle energy (carried by oxygen and sunlight)**	**Flows thru chakras and threadlike nadis**	**Nutritive subtle-energetic support for the organs and tissues of the body**
	Etheric energy (possibly equivalent to basic life energy)	**Flows through chakras of etheric body and through ethereal fluidium (Ground substance/ connective tissue of the etheric body) flows throughout entire etheric body**	**Provides subtle-energetic growth template for physical body Aids in growth, development, and repair of trauma to physical body Distortion in etheric body leads to cellular distortion in the physical body**
	Astral energy	**Flows through astral body and astral chakras Astral thought forms in auric field**	**Emotional-energy processing**
	Mental energy	**Flows through mental body and mental chakras Mental thoughtforms in auric field**	**Intellectual functioning, creativity, abstract thought**
	Higher spiritual energies	**Flows through higher spiritual bodies (casual body and higher bodies of light)**	**The flow of soul energy into physical form Repository of soul's memories (from lifetime to lifetime)**

The Human Subtle Bodies: The Invisible Energy Templates Behind the Physical Form

In addition to the spiritual- and life-energy inputs to the body from the chakras and the acupuncture-meridian system, other energy and information systems influence our health and well-being. Surrounding and interpenetrating our physical body is an energy field or structure known as the "etheric body." The etheric body is a kind of invisible duplicate of the physical body that actually occupies the same space as the physical body (but at a higher vibratory rate or energy frequency than the physical). It is the first in a series of what are called the "higher spiritual bodies." In a very real sense, our soul, our "true self," expresses itself through a physical body that is subtly influenced and molded by these various higher spiritual bodies. Each of our spiritual bodies is formed from vibrating life-energy fields of progressively higher and finer levels of energy and matter.

The first of these higher spiritual bodies, known as the etheric body, is actually a highly structured energy field that is invisible to the naked eye (but easily visible to clairvoyant sight). It provides a unique form of energetic information to the cells of the body that helps to guide human growth and development. The etheric body functions as the invisible scaffolding upon which grow the primitive cells of the developing human fetus. Clairvoyants who have observed fetal development during normal human pregnancies have noted that the higher spiritual bodies actually appear first, like an invisible "mold," that subtly guides the process of unfoldment of the fetus's physical form. While working in concert with unfolding patterns of genetic information, the etheric template helps to spatially guide the developing fetal cells to their proper location in the physical body, almost like a road map that the cells follow to find their final destination in the finished human form. Even though our knowledge of DNA has grown by leaps and bounds over the last few decades, biologists are still unable to explain exactly how the cells of the embryo actually do find their final resting places in the appropriate parts of the "finished" body. Although molecular biologists have been able to figure out how our genetic code instructs primitive fetal cells to develop into nerve or muscle or bone cells, they have never fully explained how each cell gets to its "correct" location in the body. The etheric body appears to be that missing link of guiding information that helps to solve the puzzle of "how" a fertilized egg is magically transformed into a complete human being.

While the etheric body of human beings is invisible to the naked eye and as such has never been "directly" photographed, an electrical photographic technique known as Kirlian photography has provided "indirect" electrographic images of the etheric body of plants. Kirlian photography takes pictures of an object (living or nonliving) in the presence of a high-frequency electrical field. This form of photography uses a special electrical power source to generate a spark discharge around and through the object of interest. The weak spark discharge around the object generated by the Kirlian power source is sometimes referred to as the "Kirlian aura." Unlike a traditional camera that captures light through a glass lens, the Kirlian technique is closer to a contact sun print made by placing a leaf directly against sun-sensitive photographic paper while exposing the leaf to sunlight. In a sun print, the light from the sun directly exposes the film, so that the leaf lying on the photographic film produces an image that captures the outline of the leaf. Kirlian photography creates a similar contact print when a plant is photographed after being placed directly upon film. But instead of using sunlight as a light source, Kirlian photography uses the electrical spark discharge around the leaf (produced in total darkness) to create an image upon photographic film. The Kirlian leaf images are actually quite beautiful, with many colors and lights, like a galaxy of stars, painted around and through the outline of the photographed leaf. But in certain circumstances, Kirlian photography is able to register a unique, ephemeral phenomenon that has relevance to our discussion of the etheric body. Under special conditions, Kirlian researchers slice off the upper portion of a leaf, destroy that amputated portion, then electrically photograph the leaf. The final Kirlian image of the amputed leaf reveals an image of the intact leaf as it looked *before* it was cut (including the "missing" leaf fragment). This phenomenon is known as the "phantom-leaf effect" because the electrophotograph reveals the invisible "phantom" of the destroyed, amputated portion of the leaf. The "phantom" in the phantom-leaf photo looks identical to the missing part of the leaf that had been destroyed prior to the leaf's being photographed, down to the inner vein structure and outer leaf geometry. The Kirlian phantom-leaf effect provides some of the strongest visual evidence yet for the existence of an invisible, doppelgängerlike etheric body possessed by all living organisms.

As previously mentioned, the etheric body of human beings appears to function like an invisible scaffolding to the cells of the physical body during its earliest stages of development. The etheric body acts like a kind of "Jell-O mold" or template that helps to guide the movement and migration of the cells of the body (in concert with the normal genetic mechanisms of embryological development) to their final "resting places" in the completed human infant. In addition to providing structural information to our cells during their early stages of development, our etheric body also serves as an invisible scaffolding and growth guide for our cells during adolescence and adult life. In the event of physical trauma to our physical body, our etheric body carries the energy template that invisibly guides and assists the process of bodily repair, helping to direct the normal biochemical and electrical cellular information systems to heal and "rebuild" the body.

As the etheric body is such a crucial component in the health and growth of living systems, "disturbances" or "distorted patterns of energetic information" in the etheric body could contribute to the development of illness as well. It is likely that many diseases begin not in our physical tissues but as patterns of distortion in our etheric or higher spiritual bodies. The distorted etheric-body patterns create a dysfunctional growth template that might eventually lead to abnormal structures in the cells and tissues of the physical body. In a sense, this concept of an energy imbalance leading to physical illness is mirrored by the traditional Chinese medicine viewpoint which suggests that all illnesses are related to imbalances in the body's ch'i energy. According to classical Chinese acupuncture theory, an imbalance in the flow or character of ch'i (etheric life energy?) within the meridian system will ultimately manifest as disease in the physical body (unless steps are taken to "correct" the imbalance). Since life-energy imbalances or etheric-body disturbances seem to precede the appearance of physical disease, the ideal preventive medicine would be able to detect (and correct) these etheric/life energy disturbances before people actually became sick. (This was actually one of the "ideal" goals of the ancient acupuncturists.)

The Kirlian "Phantom-Leaf Effect"

The Best Scientific Evidence for the Existence of the Etheric Body

KIRLIAN PHOTOGRAPHIC SETUP

leaf is electrically grounded via wire to stem
↓

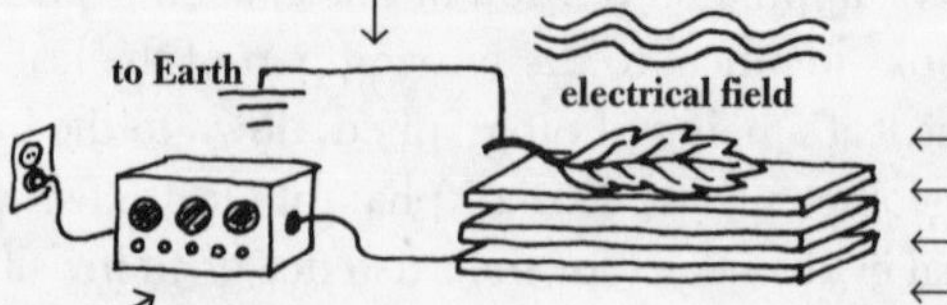

← leaf specimen
← photographic film
← insulator
← Kirlian electrode

Kirlian photography takes place in total darkness

Kirlian power generator (produces high-voltage, high-frequency electrical field that "bathes" the object being electrographically photographed)
↓
object/leaf becomes highly electrically charged (polarized at high electrical potential)
↓
high-voltage electrons go from high potential (leaf) to low potential (ground), creating a "spark discharge" around the leaf (also known as "corona discharge")
↓
spark discharge produces light that exposes photographic film and "Kirlian image" of leaf is registered on film

How can electrons be used to capture the Kirlian "phantom-leaf effect"?

"Capture" of the "phantom" may occur because the frequency of the electrons in the spark discharge (the same frequency as the electrical field produced by the Kirlian power generator) matches a "subharmonic" of the energy frequency of the leaf's etheric body.

This creates "harmonic resonance" between corona-discharge electrons around the leaf and the leaf's etheric body, producing energy interactions with the electrons that "paint" a picture (on film) of the leaf's etheric body (the "phantom-leaf" effect).

CAPTURING THE "PHANTOM" ON FILM

leaf

→ **Kirlian image of uncut, whole leaf**

→ **tip of leaf is severed and destroyed** — **leaf tip is amputated**

→ **remaining leaf is placed on new piece of film, and Kirlian power generator is switched on (in darkness)**

→ **missing tip of leaf still present in Kirlian photo**

Kirlian image shows missing "phantom" (the leaf's etheric body)

Many individuals who are clairvoyant claim to be able to "see" the etheric body. Some clairvoyant investigators who have "psychically" observed human fetal development claim that the meridian system actually forms at a very early embryological stage of human development, possibly because the meridians appear to play a role in connecting the etheric body with the physical body, forming a kind of "physical-etheric interface." Assuming that meridians are truly linked with the etheric body and that most if not all illness does begin at the etheric level, it would make sense that a "pre-illness" disturbance in the etheric body would also be associated with a ch'i-energy disturbance in the body's meridian system, supporting the traditional Chinese medicine notion that all illness may indeed be related to ch'i-energy imbalances.

In support of the idea that etheric imbalances precede physical disorders, a number of gifted medical intuitives describe clairvoyantly observing the "beginnings" of illness as energetic disturbances in a person's etheric body, days or weeks prior to the appearance of actual physical symptoms. Such an ability to detect illness reliably and accurately at a "pre-physical" stage would be truly miraculous. For instance, imagine picking up diseases like cancer before they had ever reached the single cancer-cell stage. Currently, a small number of alternative health practitioners do use clairvoyant diagnosis to help get to the "root" of their patients' medical problems. But to successfully integrate an "etheric-based" disease-detection approach into widespread future medical practice would require the development of some form of technological imaging of the etheric body that could be mass-produced and installed in hospitals and doctors' offices. Theoretically, it might be possible to obtain an image of a person's entire etheric body by creating a kind of futuristic, whole-body Kirlian-imaging system (perhaps a hybrid system using elements of Kirlian technology with MRI or magnetic resonance imaging). Remember, Kirlian photography is one of the few technological innovations to have captured an image of the etheric body (as demonstrated in the case of the phantom-leaf effect).

Both scientific and clairvoyant research seems to indicate that the etheric body is composed of a type of subtle magnetic energy very difficult to measure directly or to observe directly with the naked eye. However, it may be possible to use electrical methods to "indirectly" visualize and record the energy patterns of the subtle magnetic etheric body. Under special conditions of "harmonic resonance" that exist when lower octaves of Kirlian electrical energy resonate with the higher octaves of etheric-body energies, these elusive magnetic etheric energies excite and deflect electrons of the Kirlian spark discharge, which in turn "paints" the phantom leaf's Kirlian aura upon the adjacent photographic film. In other words, we might not be able to directly measure actual etheric energy, but using Kirlian photographic techniques, we can capture "secondary" electrical effects (the spark-discharge image of the phantom), which "indirectly" display the etheric life-energy patterns of the plant's etheric body.

It may be easier to think about the phantom-leaf effect in terms of the analogy of spray-painting H. G. Wells's so-called invisible man. While the paint atoms adhering to the invisible man's spray-painted body are not the actual invisible man, the coating of paint reveals to us his general outline and facial form. The electrons in the Kirlian spark discharge around the energetically stimulated leaf act like the molecules of spray paint upon the body of the invisible man because they show us the outline of the "invisible" etheric phantom leaf. Perhaps Kirlian technology will one day be modified to capture the entire etheric body of human beings as well as plants. The resulting innovations would be able to detect "pre-illness changes" in people while the illness-producing disturbances are still at the etheric stage. It would be wonderful if we could detect (and "vibrationally" rebalance) etheric-energy disturbances that precede such illnesses as heart disease, arthritis, diabetes, or cancer before they actually become "anchored" in the physical body.

There is another available technological alternative to capturing an image of the etheric body on film in order to detect etheric imbalances. It may be possible to use existing electrical technologies to diagnose energy disturbances in the etheric body through tapping into the link between the etheric body and the acupuncture-meridian system. The subtle magnetic energy of the etheric body can influence the flow of electrons and thus affect the electrical characteristics of the body's meridians and acupoints that transmit etheric life energy to the organs of the body. This may be one of the reasons acupuncture meridians have unique electrical characteristics that help distinguish the skin overlying acupuncture points from ordinary skin. Skin over acupoints has been shown to have high electrical "conductivity," conducting weak electrical currents almost ten to twenty times more easily than the surrounding skin. Special diagnostic instruments have been developed that can be used to obtain health information about a person's organs merely by taking electrical measurements from key acupuncture points on the body. Such systems are comparable to the computerized hookup an auto-repair shop would use to electronically diagnose malfunctions in automobiles. The theory behind the usage of such electroacupuncture diagnostic instruments hinges on the detection of an abnormal electrical reading in an organ-linked acupoint. The abnormal reading suggests a problem in the organ at either a physical or an energetic level. For example, a low electrical reading in a person's lung-meridian acupoint might reflect an energy imbalance and potential health problem in the individual's lungs. While the lung-meridian imbalance may reflect ongoing disease in the lungs, a traditional Chinese-medicine approach might interpret the abnormal electrical-acupoint reading as a lung "energy disturbance" (at the etheric level?), which has yet to manifest as physical lung changes.

Following this same line of reasoning, let us assume that most diseases not caused by obvious exposure to toxic chemicals, germs, blunt physical trauma, or "bad" genes could actually be due to an imbalance or distortion in an individual's etheric body. Etheric disturbances are sort of like a hidden factor that predisposes people toward developing certain illnesses. In an etheric-body imbalance, the normal pattern of information flowing to a particular organ in the body has become distorted. The distorted energy information is eventually translated by the body into distorted cellular structure and function and, ultimately, abnormal tissues. Ideally, if we could detect the precursors to disease that exist as distorted patterns of etheric energy, we could attempt to heal the person by correcting imbalances at the etheric level of causation before serious physical illness actually occurs. This is one of the lofty goals of vibrational medicine: to detect illness at the energetic stage of "information distortion."

Besides possessing an etheric body that complements the physical body, humans possess other higher spiritual bodies as well. One of these spiritual bodies, known as the astral body, participates in how we feel, how we express ourselves, and in how we are influenced by our emotions. Some clairvoyants actually refer to the astral body as the emotional body. Like the etheric body, the astral body is a kind of structured energy field that contributes certain types of energy information to the physical body. While the etheric body is strongly attached to the physical body, the astral body appears to be more mobile and can move about independent from the physical body. Many clairvoyant and psychic researchers claim that part of human consciousness is able to move from the physical body into the astral body at different times. It has been said that human beings travel in their astral bodies at night while their physical bodies lie in bed sleeping. However, people can also be traumatically ejected from their physical body into their astral body by major accidents and severe physical illnesses. The extreme example of a traumatic ejection of human consciousness into the astral body is what happens during a so-called near-death experience (NDE). The NDE is a unique phenomenon. During near-death experiences, people who are (temporarily) clinically dead report floating near the ceiling during efforts to resuscitate their dying bodies. The disembodied consciousness of the clinically dead person is able to look down at the physical body, lying in the midst of intensive CPR efforts. Often an individual who comes back from the NDE will accurately provide information about resuscitation procedures that took place, as well as detailed descriptions of the people performing CPR. What seems to occur in the case of the NDE is that the individual's consciousness is traumatically ejected into the astral body, where the person is able to actively observe medical procedures being performed on his or her lifeless physical form. Interestingly, many individuals who undergo a near-death experience frequently report such peaceful, pleasant experiences that they come back having lost all fear of death they might have experienced prior to their NDE.

The astral body is strongly affected by our emotions, primarily lower emotions such as sexual desires, anger, and sadness. Some suggest that the astral body may be linked through special energy connections to the brain and nervous system (the centers normally associated with emotional processing). Science has established that there are certain areas of the brain, such as the limbic system, that appear to be involved with the expression of human emotion. However, we express and experience emotions not only through the centers in our brains that deal with emotions but also through our energy links with the astral/emotional body. If one accepts this astral aspect of human emotion, one must consider the possibility that certain emotional disturbances might not be totally due to neurochemical imbalances, as most psychiatrists assume. Instead, energy disturbances in the astral or emotional body could be a major contributing factor to our emotional ups and downs. This astral-emotional link may be one of the reasons certain psychiatric illnesses and emotional disturbances are sometimes resistant to drug therapies, which deal only with the physico-chemical aspects of emotion. Certain vibrational-medicine modalities are designed to help rebalance the energy patterns of the astral/emotional body, thus providing relief from various resistant emotional problems (which will be discussed in greater detail in later chapters). Different substances from nature, such as flower essences and essential oils, can help people to achieve emotional equilibrium by rebalancing the subtle life-energy patterns of the chakras, the meridians, and even the astral body.

Some clairvoyants and healers who can see the astral form claim that the astral body is made up of a kind of subtle energy with unique magnetic qualities. In addition to being magnetic in nature, the energy making up the astral body is said also to be strongly influenced by our thoughts and our emotions. The magnetic qualities of the astral body create patterns of magnetic attraction in our thoughts and in the emotional-energy patterns of our energy fields, such that we tend to "magnetically attract" more of the same emotional energy to ourselves. Thus it appears our thoughts are not merely neurochemical reactions or electrical discharges in the brain. Strong thoughts and emotions may actually create a kind of energy structure known as a "thoughtform." Again certain clairvoyants claim to see thoughtforms in the auric fields of different individuals. Thoughtforms, being composed of this subtle magnetic astral energy, act like a magnet that attracts more of the same emotional energy quality to itself. This may be one of the hidden reasons behind the old adage "Misery loves company." Frequently, those who focus on the negative things in their lives seem to gravitate to others of a "like mind" who also complain, and they mutually reinforce each other's negative viewpoints on life. But what if the astral body's magnetic attracting ability is considered? The reasons for such clusterings of like-minded negative thinkers could be energetic as well as psychological in origin. In the future, more and more psychologists and physicians may begin to use flower essences and other vibrational remedies to rebalance the disturbed magnetic emotional-energy patterns that contribute to physical and psychological illnesses, allowing them to directly affect illness even up to the astral levels of human functioning.

Beyond the human etheric and astral bodies resides another spiritual body known as the mental body. The mental body, also composed of a subtle magnetic energy, vibrates faster than astral energy. As its name implies, the mental body is intimately involved in the energy of thought, creativity, invention, and inspiration. Just as with the astral body, strong ideas and mental energy patterns create mental thoughtforms that can sometimes be observed by clairvoyants who are able to perceive the higher aspects of the human energy field. Individuals who can see the auric field can sometimes observe pictures and symbols of items in a person's aura related to that person's occupation or creative hobbies. These symbols may represent the mental thoughtforms of projects and ideas that the individual is consciously or unconsciously exploring. Disturbances in a person's mental body can create dysfunctional energy patterns that filter down through the various spiritual bodies and eventually lead to difficulties with thought processes at the level of the physical nervous system. This invisible link between the mental body and the physical body suggests that certain mental problems could actually have their origin at higher-spiritual-energy levels. Vibrational medicine looks to various flower essences and specific high-potency homeopathic remedies to correct higher-spiritual-energy disturbances in the multidimensional human energy system, rebalancing both the physical and the spiritual bodies with certain healing frequencies of subtle energy. Later chapters will further address these unique treatments.

The Higher Spiritual Bodies and Their Link to the Human Soul

Finally, the human energy field extends to an even higher spiritual level known as the causal body or the causal field. The causal body might be considered the closest thing to the soul. The record of all that a soul has experienced upon the physical earth plane, in its current life as well as in past lives, is said to be contained in an individual's causal field. According to a number of spiritual philosophies, the soul lives many different lives though a variety of different physical bodies over the course of earth's history. The causal body would be where an energy recording of the soul's sojourn through physical life in its many varied expressions is actually "encoded" and stored. To truly understand the causal body, one must believe not only that the human soul is immortal but that it moves through a progressive spiritual learning curve by returning to the earth numerous times in different physical bodies. This is the basic tenet of reincarnation, a belief system shared by millions of people throughout many different cultures and spiritual traditions.

The Multidimensional Human Being: A Look at the Many Energy Layers

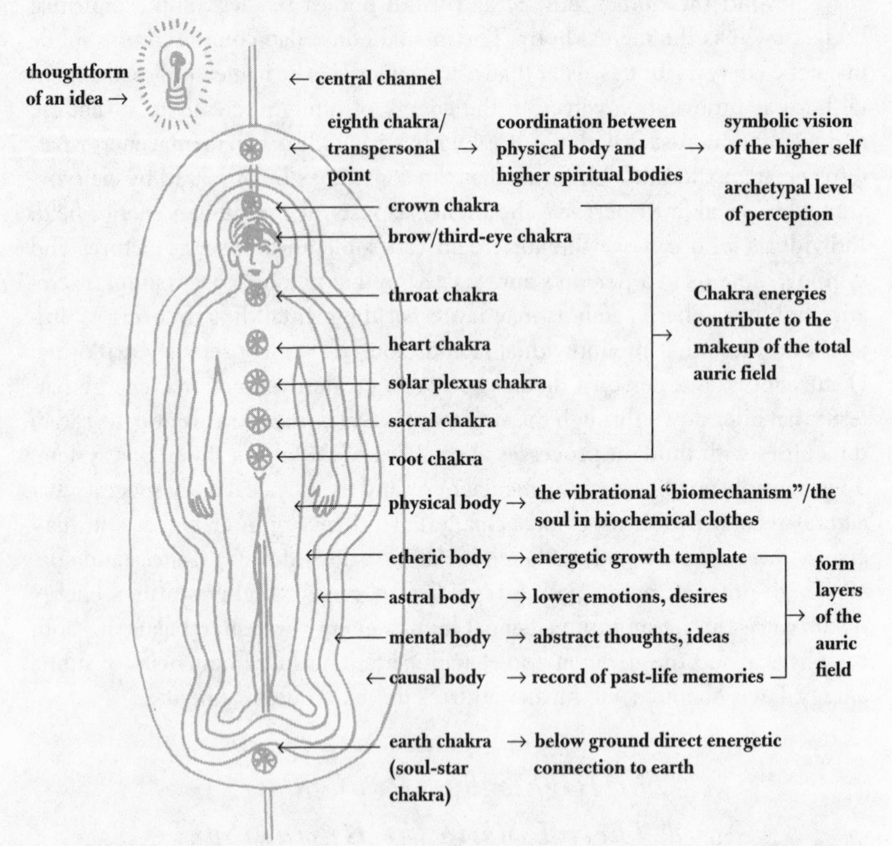

Because the causal body retains a record of our past lives, it carries the memories and energetic recordings of traumatic events experienced during the learning path of a particular soul. While the physical bodies of expression may change over time, the soul maintains the same causal body from lifetime to lifetime. Unresolved traumas and conflicts may be carried over from one lifetime to affect the body and life patterns of another lifetime because of the memory patterns retained by the causal body. Abnormal energy patterns, caused by certain unresolved emotional conflicts from a past lifetime, could ultimately influence the health and development of the physical body in a current life. The unique types of health problems, related directly to the causal body, are known as "karmic illnesses." While karmic illness is difficult to document, it may be more common than previously suspected. The concept of karma is actually the "law of action and reaction" carried to the highest spiritual level, the causal level. The "law of karma," stated in a slightly different way, is "Whatever goes around comes around." According to the law of karma, the good deeds or bad acts perpetrated in one lifetime constitute an action to which there must eventually come some type of appropriate reaction. The reaction could occur during the lifetime of a past karmic event, or it might "carry over" to a future lifetime to manifest as a problem in the health of the current physical body. All of this relates to how life events occurring at the physical level are energetically recorded in our higher spiritual bodies at the causal level.

The causal body, which remembers all that has happened to the soul in previous lives, can produce structural changes in the physical body that may range from insignificant birthmarks to serious physical illnesses. The causal plane or plane of causation is so called because when one observes ordinary reality from that higher spiritual perspective, it becomes possible to see the "true" causes behind seemingly random physical events and ailments. The causal plane is also the level at which our "higher self" operates. Each of us has a wise inner self known in spiritual circles as the higher self. The higher self is the part of us that always knows what's going on in our lives and why things are happening to us even though we may not be consciously aware of these unseen patterns of meaning. The higher self sees the total pattern of our lives from the lofty perspective of the causal-plane level and attempts to share information and insights to the conscious personality, often through the symbolic language of dreams.

Neurologically and energetically speaking, our higher selves seem to be more intimately connected to our right brains, our right cerebral hemispheres, than to our left brains. The right brain speaks in a language that is more symbolic and metaphorical than literal. Unfortunately, we live in a culture that is too left-brained, logical, and literal for many individuals. Therefore, the right-brain-oriented dream messages from people's higher selves are often ignored or misunderstood because they are taken literally and not symbolically. Learning to pay attention to dreams and interpret their unique symbolic language can be one of the best ways for getting in touch with the higher self. One can carry this process a step further by using the same symbolic language of dreams as an interpretive tool for analyzing the hidden meaning behind the events of everyday life. By considering our daytime activites from the right-brain perspective of dream-symbol interpretation, we can begin to see life through the eyes of our higher selves.

The great Chinese philosopher Lao Tzu once remarked upon waking from a dream about a butterfly, "Am I a man who has just dreamt of being a butterfly, or am I a butterfly dreaming I am a man?" In other words, which is the true reality? Is it possible that our dreams may be a reflection of the true nature of reality? It is only by applying the symbolic language of dream interpretation to everyday events that one can begin to penetrate the mystery behind seemingly random events. It is from the perspective of the causal body and higher spiritual levels of observation by which we can see how events that seemed accidental or meaningful coincidences are sometimes hidden messages orchestrated by our higher spiritual selves. The term for this type of meaningful coincidence, coined by psychiatrist Carl Jung, is "synchronicity." Synchronicities are unusual coincidences that are often meaningful to just a single person. Synchronicities can be big or small. A small synchronicity is talking about someone you haven't heard from in years and having that person call you on the phone an hour later. Synchronistic events are often arranged by the higher self (which operates at the higher spiritual level of the causal plane, where space and time do not operate as they do on the physical plane). At the level of the causal plane, our causal body is able to operate from a perspective that transcends space and time. It is the level at which we move closer to the infinite, spiritual energies of the divine in all of us. As our awareness begins to shift to the level of causal-body consciousness, synchronistic events tend to come into our lives with increasing frequency.

Often, the meanings of events interpreted from the causal-level perspective may be totally reversed from how they appear to the logical, literal way of perceiving ordinary reality. When viewed from the higher spiritual level of interpretation, negative events can have potentially positive meanings, because every adversity and obstacle we encounter is really a chance for a new learning experience for the soul. Seeing the world from a causal-plane, symbolic perspective is a bit like learning to find the "silver lining behind every dark cloud" or attempting to "make lemonade when life keeps sending you lemons." When viewed from the symbolic level of the causal body, all life experiences, even physical illnesses, are potential learning experiences for the soul, each carrying different spiritual messages that transcend the literal interpretation of ordinary reality.

If we could allow ourselves to see the world with the symbolic sight of spiritual awareness, we would gradually acquire greater clarity in understanding the deeper mysteries behind the events of our day-to-day lives, including our illnesses. When we operate at the symbolic level of the causal plane, it brings us a step closer to awakening to our divine nature. At the primary, spiritual level of existence, we are really beings of light who have entered into physical bodies, yet we have forgotten our divine roots. We remain connected to these divine roots through energy threads that link our physical bodies to our causal bodies, the level at which the higher self, the true soul, operates. Our higher selves help to remind us of our divinity and of the hidden meaning behind seemingly random life events. Little synchronicities and meaningful coinicidences tend to push us along in the right direction, if, of course, we have learned how to find their hidden meanings.

The trick is learning to pay attention and to correctly interpret the signs and symbols sent by our higher selves to guide us along the way. Each of the positive and negative events of our lives, especially traumatic ones, help to shape us just as a raw diamond is made beautiful by the process of carving and polishing the rough stone until it awakens as a brilliant gemstone. It is only through the testing of our capabilities through trials and challenges that we can learn the deeper meanings of our lives.

From the eternal perspective of our soul and our higher self, it is not how much wealth we physically accumulate that matters in each lifetime. It is the quality of lives and the depth of the relationships and friendships we have developed with others that ultimately hold the greatest value to our souls. Souls often pursue relationships from lifetime to lifetime within a grouping of souls, sometimes reversing roles between parent and child, teacher and student, the tormentor and the tormented, in order to experience both sides of relationships with greater clarity and compassion. It is quite likely that before entering into a physical body, our soul makes "sacred" contracts between itself and other souls to accomplish certain pre-agreed-upon life lessons. The nature of these contracts depends upon the "core curriculum of learning" that a soul chooses for a particular lifetime. If we assume that this is true, it brings up an interesting point. If, for example, each soul has a choice in deciding who its parents will be, wouldn't we all want to incarnate into families with wealth, privilege, and status, where we wouldn't have to struggle to make a living? The idea of someone's consciously choosing to incarnate into a family where there are conflicts or abusive parents or a lack of financial wealth suggests that the soul does not always choose the "path of least resistance."

If there is indeed a higher level to choice and free will, why would someone voluntarily choose to incarnate into a body destined to develop a serious illness, as in the case of a genetically inherited disorder? The torment and tragedy of illness may make sense to us only when we view it from the symbolic level of our higher self. While disease is generally seen as a negative experience, the potential for transformation exists in all illnesses, not only for sick individuals, but also for those around them as well. The outcome all depends upon how those individuals choose to deal with that illness and how they use their experience to relate to others. One sick individual may gain strength and nurturance from the loving support of her family in times of dire need. Conversely, another sick person may choose to exploit his illness for secondary gain and constant attention from others. In certain circumstances, an illness can be transformational for a patient's entire family. Specifically, some parents who have lost children to childhood cancer have gone on to establish support groups for families of children with similar illnesses. While some families might never recover from the loss of a child, there are always exceptional families who become emotionally and spiritually transformed through their experience of dealing with the loss. At the level of causal or soul awareness, "how" you choose to deal with a traumatic event has more to do with the outcome than the actual event itself. Ultimately it is the emotions and attitudes of each individual that determine how life events will lead toward a positive or negative learning experience.

By learning to appreciate the higher-dimensional contribution of our causal bodies, our mental, astral, and etheric bodies, as well as our chakras and our meridians, we can begin to see the bigger picture of why people may become ill. An awareness of our spiritual bodies, as well as of our chakras and meridians, gives us new ways of looking at human beings as well as new ways of approaching the concept of health and wellness from a multidimensional perspective. This larger view of humanity goes beyond the concepts of illness as something that stems from a broken part, a bad gene, or a chemical or germ exposure. Although physical factors do play an important role in health and illness, they are often mediated by other energetic and spiritual factors that modern medicine is only now beginning to appreciate. In the following chapters, we will explore in greater depth how these different spiritual and energetic factors play a role in the causation of illness and the creation of health. We will continue to build a foundation for understanding how we can use vibrational medicine to heal ourselves from the level of our physical bodies all the way up to the higher spiritual levels of our being.

Two

~

VIBRATIONAL MEDICINE AND THE PATHS TO ILLNESS VERSUS WELLNESS

IN THE FIRST chapter, we examined the nature of human beings from the unique perspective of their physical and spiritual anatomy. We will now look more closely at the dynamics of human physical and spiritual anatomy to see how imbalances in these delicate systems may lead to illness in the physical body. As the reader will see, human beings possess many specialized systems that feed energy and information to the cells and tissues of the body at a variety of different levels. These systems include the meridian system, the chakras, the etheric body, and the higher spiritual bodies. Each energy system has special properties and dynamics that affect their proper functioning.

Nutritional-Energy Inputs and Their Influence on Human Health

At both the physical and spiritual levels of our being, we take in various forms of energy in order for our physical body to function properly. When this life-sustaining flow of energy becomes blocked at some level, illness can eventually result. At the physical level, we take in sustenance in the form of metabolic energy from the food we eat. Our daily intake of carbohydrates, proteins, fats, and vitamins provides metabolic energy to the cells of our body as well as raw ingredients for use in repairing and rebuilding damaged and aging cellular components. Over a period of seven years, the new molecules we absorb from the foods we eat will replace all the older molecules in our bodies. Although we look the same on the outside, the atoms making up our bodies today have been largely replaced by billions of new atoms and molecules. We literally create a whole new physical body every seven-year cycle. Therefore, the types and quality of foods we eat can be extremely important in determining whether or not we maintain optimal health or eventually become ill.

At the most basic level, we need some type of metabolic fuel to power the cells of our body. Although someone could survive on a simple form of food such as refined sugar for a limited time, it would not be healthy for the body to maintain such a diet. We need a mixed proportion of different fats, sugars, and proteins as well as the necessary vitamins to maintain our daily nutritional-energy balance. Among the critical vitamin and nutritional factors deemed necessary for optimal health are trace amounts of metals such as iron, zinc, cobalt, molybdenum, copper, selenium, and the rarer elements of boron and even gold. Trace metals often are components in enzyme systems of the body that are vital for extracting metabolic energy from foods. Typical trace-metal atoms also conduct bioelectrical currents at the cellular level and contribute to the proper functioning of the body's cellular-energy systems at both the biochemical and bioelectrical levels. Metallic-vitamin cofactors assist the flow of electrons through different parts of the cells, especially in the tiny mitochondria contained in each cell of the body. Mitochondria are tiny sausage-shaped power plants with batterylike qualities. Our mitochondria play a critical role in extracting biochemical energy from the food we eat. Without oxygen, trace elements, and vitamins to enable the flow of electrons through our cells' mitochondria, life would not be possible. Thus, deficiencies of vitamins and trace elements can cause metabolic and bioelectrical energy blocks, which in turn can contribute to the development of physical illness.

As mentioned, vitamins are important cofactors to our body's enzyme systems. They allow our bodies to optimally use all the fuel we ingest. Individuals with a deficiency in certain key vitamins may not effectively use all the metabolic fuel they've taken in through their diet. But how does one determine the daily dosage for each vitamin that will ensure optimal health? For example, vitamin C, probably the most familiar vitamin, is of essential importance for good health. Yet Linus Pauling, one of the first scientists to advocate taking large amounts of vitamin C for optimal health, was ridiculed by most of his medical peers during his lifetime because of his attitude toward high-dose vitamin-C therapy. But within the last few decades, research has validated the benefits of consuming a higher daily intake of vitamin C in order to maintain health in the face of stress and illness. More recently, doctors were surprised to find that vitamin C helps blood vessels rapidly recover from the effects of fatty meals! It seems that meals with high fat content can cause arteries to become temporarily stiffer and less flexible. However, high-dose ingestion of vitamin C prevents this effect upon arteries. Although the jury is still out on whether vitamin C can actually cure the common cold, it has been shown to enhance different aspects of immune functioning (when taken in the appropriate dosages). Specifically, vitamin C is known to stimulate white blood cells to become more agressive in ridding the body of harmful bacteria and viruses.

Lately, folic acid and vitamin B6 have been getting more attention in medical circles because they appear to offer greater protection from diseases affecting the cardiovascular system than previously thought. While most of the dietary emphasis on the prevention of heart disease in recent years has focused on lowering cholesterol and fat intake in the diet, studies over the last three decades have indicated that even relative deficiencies of vitamins B6 and B12 and folic acid can lead to a buildup of a chemical called homocysteine. In high levels, homocysteine can be just as bad or worse than cholesterol, because this oddly named chemical can cause damage to the coronary arteries that supply the heart muscle with vital blood flow.

Even after a relatively brief perusal of existing vitamin information, the reader would conclude that the type and quantity of vitamins and trace elements we take in through our diets is important with regard to preventing physical illness. A deficiency of vitamins, the vital cofactors to cellular energy metabolism, can greatly impair our body's ability to use chemical and electrical energy at a basic tissue level. Any kind of nutritional stress on the body due to protein and calorie malnutrition or to relative deficiencies of essential nutrients can affect the level of vital energy within the physical body. Vitality is a reflection of the body's functional level of metabolic- and subtle-energy balance. As a person's vitality decreases, from either nutritional or other factors, the ability to fight off infection or cancer cells or to heal from injuries becomes markedly diminished. Perhaps the most basic factor that affects our vital-energy level is our diet. Nutritional stress, caused by either nutritional deficiencies (such as inadequate dietary intake of certain vitamins and trace elements) or by an excess of saturated fats and refined sugar in the diet, can significantly impair our ability to fight off and even to prevent acute and chronic illnesses.

Another basic factor that influences health and vitality is a person's level of physical conditioning. Vigorous aerobic exercise on a daily basis has a variety of health benefits. Exercise improves cardiovascular fitness and is a great stress reducer. Regular aerobic exercise raises the "good" or HDL cholesterol in the blood and provides additional benefits for preventing heart disease. Exercise also works as a natural antidepressant by raising blood levels of neurochemicals that improve mood and fight depression. For postmenopausal women who are subject to osteoporosis, exercise and physical conditioning increase the strength of bones that may be weakened by calcium loss.

Both nutrition and exercise play an important role in maintaining strong immune defenses against exposure to a variety of toxic agents. As a person's energy level and vitality improve through good nutrition and regular exercise, the general level of immunocompetence—the ability to fight off germs and cancer cells—improves as well. The question of who gets sick is partly determined by a person's "host resistance." It is extremely important today to have a good host resistance, because we are increasingly exposed to a wide variety of microbes, ranging from cold viruses, pathogenic E. coli, and salmonella food poisoning to life-threatening microbes such as HIV, flesh-eating bacteria, Hanta-virus, and the Ebola virus. A person who takes a daily multivitamin and extra vitamin C while maintaining a regular exercise regimen is less likely to catch a cold from being around an individual with a simple viral upper-respiratory infection than another individual who never exercises or takes vitamins and eats only fast food. But sickness is also partly determined by the virulence of the particular virus or bacterium to which a person is exposed. Therefore, that same healthy, exercising individual who fended off a cold could die within forty-eight hours of being exposed to a minuscule quantity of Ebola virus. Not that anyone has to worry seriously about exposure to Ebola in the United States, since the incidence is exceedingly rare. This example merely demonstrates how an extremely virulent germ can induce illness even in people who normally maintain a good host resistance. While Ebola and Hantavirus outbreaks are rare, exposure to other toxic microbes through our food supply appears to be increasing at an alarming rate. More and more cases of mass food poisonings through fast-food chains or hepatitis outbreaks from contaminated fruit are being reported. Because of the potential dangers from what we eat, it becomes vitally important to maintain a strong, healthy immune system.

As if contamination weren't enough to worry about, the vigilant individual must also consider the variety of chemical pollutants gradually finding their way into the foods we eat. Dumping of toxic materials into landfills sometimes leads to contamination of local groundwater. This can allow chemical pollutants to seep into the water supply of local communities. One recent epidemiologic study found that pregnant women in one community who drank large quantities of tap water had a higher incidence of spontaneous abortions than women who consumed other sources of water and fluids. Scientists are now theorizing that microquantities of chemical contaminants in common drinking water might be the cause of such alarming findings. Many more people are choosing to drink distilled, bottled, and filtered water, avoiding tap water entirely except for bathing and washing clothes. Water is of major importance to normal human physiology both at the physical level and at the subtle-energy level. Water literally holds the essence of life. Unfortunately, there is evidence to suggest that not all people absorb water equally from their intestinal tracts. Australian health researchers have discovered that there can be such a thing as "relative dehydration." Some people can drink large quantities of water while absorbing little of it into the critical cells of the body. But research also suggests that the consumption of fruit juices, alone or mixed with water, can have an enhancing effect upon water absorption throughout the gastrointestinal tract. Water also helps to flush toxins from the body through the kidneys. As you can see, adequate consumption of uncontaminated water is of great importance to good health.

The increasing presence of petrochemicals and heavy metals like lead and mercury in our environment also poses an insidious, invisible threat to human health and vitality. Exposure to heavy metals from mercury-laden seafood or lead-containing drinking cups can significantly impair development of the nervous system and lead to long-term intellectual deficits. While exposure to toxic substances in our environment is probably more common than previously thought, exposure to environmental chemicals often goes unsuspected as a cause of illness. Today we live in a fast-paced society where pesticide-sprayed foods, hormone-injected livestock, recycled "dirty" water, and reclaimed land have become the norm. The safety standards for a particular chemical, hormone, or pollutant may state that so many parts per million are considered safe. However, when we come to discuss an alternative medical practice called homeopathy and the therapeutic use of microdoses, it will become clear that even such tiny amounts of toxic substances can have a subtle but far-reaching effect on human physiology.

The Hidden Dangers of Environmental-Energy Stress

Aside from the better-recognized pollutants, we are also surrounded by an often overlooked form of pollution that comes from the technologies most of us just can't live without. This devil in disguise is electromagnetic pollution. Because of the prolific spread of computers, color TVs, microwave ovens, radio towers, and cellular phones, we now live in a sea of unseen electromagnetic frequencies that did not exist some seventy to eighty years ago. Many electrical appliances generate fairly intense electromagnetic fields that pulse at 60 cycles per second, a frequency that may subtly influence the health of human cells. Although electromagnetic pollution remains a controversial topic, there are a number of epidemiological studies that point toward an increased incidence of certain cancers in individuals who are chronically exposed to high-intensity electromagnetic fields. The relationship between electromagnetic fields and cancer will be reviewed in a later chapter that addresses the use of magnetic fields for healing.

Nutritional, chemical, and even electromagnetic stress are just a few of the recognized sources for potential blockages to the normal flow of energy in the human energy system. In addition to these more obvious chemical and electromagnetic stress factors, there is a subtle type of energy disturbance associated with certain geographic regions that also has the potential to exacerbate human dis-ease. Abnormal magnetic fields produced by underground streams, tectonic-plate stresses along fault lines in the earth, underground crystalline deposits, and other less understood factors may all contribute to the phenomenon of "geopathic stress." Most people are unaware of this type of invisible energy pollution. The term refers to a type of "energetic stress" that affects living systems and seems to come from abnormal energy fields produced by the earth itself. It appears that where a person lives, works, and sleeps may be a more powerful factor in illness causation than previously suspected. In fact, the reality of geopathic stress can be stated as follows: "Where" you live and work can make you sick even if you follow good nutritional and exercise habits.

In Germany, geobiology researchers have concluded that people living in geopathic-stress zones seem to have a higher than normal incidence of cancer. Furthermore, German scientists have successfully applied geopathic-stress theories to explain why there are certain stretches of the famous Autobahn highway that appear to have more traffic accidents and motor-vehicle fatalities than other, similar roads. These innovative scientists sought to find a correlation between geopathic-stress zones and high-accident regions. Using specialized detectors capable of measuring certain aspects of "geopathic energies," the researchers had no trouble verifying this strange connection. Eventually the data collected was used to develop special devices to help neutralize the abnormal geopathic-field zones of the highway. Then several of the devices were placed in strategic locations on the Autobahn and were shown to actually reduce the number of traffic fatalities and collisions in high-accident areas along the highway.

Geopathic stress is perhaps one of the more unusual forms of environmental stress that can lead to adverse health changes. Just as radon can build up in the basement of a building, so can the toxic energy of geopathic stress permeate a person's house and sleeping quarters virtually without detection by our "familiar five senses." However, this type of energy stress can be identified and neutralized to varying degrees using certain types of detectors and special devices. And some individuals who are adept at the art of dowsing can detect geopathic stress with great success. Interestingly enough, the effects of geopathic stress can easily be seen in plant life growing in energetically stressed areas. Dowsers often note that trees growing in geopathic-stress zones tend to develop recurrent plant tumors. Trees growing in geopathic-stress zones are often repeatedly struck by lightning during thunderstorms. This observation suggests that an unusual environmental-energy effect, possibly some type of ionizing radiation, is associated with geopathic stress. It also disproves the popular notion that "lightning never strikes twice in the same spot."

While many people might be surprised by the link between hidden environmental-energy stress and disease, the idea that a person's environment could be capable of producing energetic effects on health and well-being is actually a very old concept. The ancient Chinese art of geomancy, also known as feng shui, is based on an appreciation of the natural flow of ch'i and vital energy through the environment. For many hundreds of years, feng shui masters have been consulted for their specialized geomantic knowledge because they possess an esoteric understanding of specialized architectural and design principles that can enhance the natural flow of environmental ch'i throughout homes and business places. In China, such enhanced ch'i flow is said to promote health and prosperity. In essence, feng shui masters have learned ways to enhance "good earth energy" as opposed to the energetic stagnation and illness produced by "bad earth energy" or geopathic stress. Some American businesses now regularly consult feng shui specialists in designing their homes and offices. While many skeptics consider the practice of feng shui to be nothing more than mere superstition and folklore, the concept of geopathic stress or "bad earth energy" is not really that preposterous.

Subtle-Energy Factors and Health: Traditional Chinese Medicine and the Role of Ch'i

In addition to "external" energy disturbances contributing to energy blockages and illness in human beings, there is also a variety of potential "internal" subtle-energy imbalances (within the human body itself) that can either enhance or impair our health. We have already mentioned that certain environmental and design factors may enhance the flow of ch'i energy through living quarters and business offices. But various forms of ch'i or life energy are also be absorbed from food as well as from the environment. We possess a specialized form of ch'i-energy metabolism that extracts, processes, and distributes ch'i throughout our bodies. According to Chinese medicine, the ch'i that is extracted from the food aids in the formation of our blood. Our blood, which also carries a form of ch'i, is said to serve the function of nourishing and moistening our body's organs and tissues. Ch'i may be the underlying activating force that causes our blood to circulate throughout our bodies, aided by the pumping action and hydraulic-pressure waves created by our beating heart. But ch'i can be used appropriately only if our acupuncture-meridian system is free from blockages or energy imbalances. This system distributes ch'i energy to the different organs of our bodies as part of a subtle-energy network that helps to energize and support our normal bodily functions.

The Many Varieties of Ch'i Energy (According to Traditional Chinese Medicine)

TYPE OF CH'I	ENERGETIC FUNCTIONS	PHYSIOLOGIC FUNCTIONS	LOCATION IN THE BODY
Jing Ch'i (ancestral ch'i)	**Affects transformation and utilization of other types of ch'i**	**Supports fetal development** **Inherited vital energy that supports normal overall bodily functioning**	**Stored in the kidneys**
Rong Ch'i (nourishing ch'i)	**Energetically nourishes organs and tissues**	**Supports organs and tissues of the body**	**Meridians**
Wei Ch'i (defensive ch'i)	**Energetically shields entire body from trauma or attack from external noxious influences**	**Warms the skin** **Stimulates perspiration** **Nourishes body tissues** **Controls opening and closing of pores** **Makes muscles and tendons resilient and resistant to noxious external influences**	**Meridians** **Also flows through skin, muscles, tendons, and fatty tissues of chest and abdomen**
Evil Ch'i (noxious external energy disturbances)	**Environmental-energy "stress" extremes of cold, heat, wind, excessive moisture, extreme dryness**	**Disrupts functioning of bodily organs and tissues** **Predisposes to illness (in combination with other factors)**	**External environment**

According to classical Chinese-medicine theory, illness can result from meridian blockages or imbalances in the supply and flow of ch'i energy to the different organ systems of our bodies. Traditional Chinese medicine further teaches that there are different types of ch'i energy that circulate throughout our bodies and that a deficiency or imbalance in any single type of ch'i energy can also contribute to ill health. Another specific tenet of Chinese medicine states that all individuals receive a type of ch'i energy directly from their parents. The first type of ch'i energy available to us in the womb is known as ancestral ch'i (or jing ch'i). Ancestral ch'i is a kind of hereditary energy charge imparted at the time of conception when a sperm fertilizes an egg to create a new human being. Ancestral ch'i circulates internally throughout our bodies and is said to have both direct and indirect effects upon the transformation and utilization of other types of ch'i used by our bodies. During the nine months of development in the womb, a fetus lives primarily off its inherited ch'i, along with nutrients, oxygen, and other types of subtle life energies supplied by the umbilical cord. The fetus's ability to create and process other types of vital ch'i energy does not actually begin until the infant takes its first breath. Problems may arise if a child is born prematurely, before the natural ch'i-processing systems of the body have become fully activated. This situation parallels the incomplete maturation of the fetal lungs in premature infants who suffer from an inability to absorb and circulate oxygen efficiently throughout their bloodstream.

Other problems with the health of an infant can also arise if during the pregnancy a mother's ch'i energy becomes depleted or the natural flow of her maternal ch'i energy is impeded by drug use or inadequate diet. Following birth, ancestral ch'i energy is said to be stored in our kidneys and must last us for an entire lifetime. Although ancestral ch'i never increases in quantity, it can be weakened by poor health habits or improper acupuncture treatments. Traditional Chinese medicine states that when a person's ancestral ch'i becomes critically weakened, life span may be significantly shortened.

One of the more commonly recognized forms of ch'i is known as rong ch'i or nourishing ch'i, which is said to directly support the organs and tissues of the body. Yet another less well-known form of ch'i, called wei ch'i, is a defensive type of energy used to protect the body from external causes of illness. This defensive ch'i is the type of bioenergy martial artists focus on activating and strengthening in their hands and bodies prior to attempting any brick-smashing, board-splitting, or similar feats of "kung fu power." Unlike nourishing ch'i, which circulates primarily in the meridians, defensive ch'i flows through the skin, muscles, tendons, and the fatty tissues of the chest and abdomen. According to Chinese medicine, wei ch'i defends the body against noxious external influences by warming the skin, causing perspiration, nourishing the tissues, and controlling the opening and closing of the pores. It is said that when wei ch'i is circulating effectively, the skin is soft and the muscles and tendons become resilient to harmful external influences.

A number of different approaches within traditional Chinese medicine allow a practitioner to determine the source of ch'i imbalance that may be related to a patient's illness. One such approach is known as the five-phase or five-element theory. Five-element theory tends to classify specific organ systems and their meridian energies according to their metaphorical associations with the five elements of nature: fire, earth, wood, water, and metal. Five-element theory is a reflection of the ancient principle of correspondences between microcosm and macrocosm as expressed in the aphorism "As above, so below." The workings of the microcosm, the body (below), was thought to follow the same principles governing the macrocosm of the universe (above). Ancient Chinese thinkers felt that relationships between the key elements of nature were mirrored in the relationships between the organ systems of the body. In five-element acupuncture, patients are classified according to their preferences for specific flavors, colors, seasons, and emotions. In a later chapter, we will describe how an acupuncturist—guided by a person's tastes, likes and dislikes, and past medical history—uses the five-element model to identify and correct imbalances in the person's meridian system in order to bring about healing. Determining which of the five elements is out of balance helps the acupuncturist to develop a plan of acupuncture treatments that will rebalance and support the organ systems that are most distressed.

By now you probably realize that five-element Chinese medicine views disturbances in the body's ch'i as the "true" cause of organ pathology and disease. Based on this assumption, a five-element-oriented practitioner of Chinese medicine pays close attention to both "inner" and "outer" factors that might help determine where the patient's ch'i-energy imbalance really lies. According to traditional Chinese-medicine theory, meridian imbalances and disturbances in the body's ch'i energy are thought to be caused by exposure to extremes of environmental conditions, as well as by imbalances in certain internal conditions. The environmental stresses thought to predispose a person to illness (sometimes referred to as "evil ch'i" from outside causes) include exposure to excessive wind, severely cold conditions, extreme heat, excessive moisture, abnormal dryness, and fire. Along this same line of reasoning, chronic imbalances in a person's emotional nature are also thought to contribute to the eventual development of illness. The emotional internal factors said to predispose an individual to illness include excessive anger, unrestrained joy (or manic states), persistent worry, extreme sadness, and long-standing fears of any kind. Chinese medicine was among the first of the mind/body systems of healing to recognize the importance of the influence of the emotions upon health. Today an entire field of research known as "psychoneuroimmunology" is dedicated to uncovering the physiological links between the mind, the body, and the immune system. Other internal causes of illness proposed by practitioners of traditional Chinese medicine include excessive eating and drinking, starvation and dehydration, complete exhaustion, and even total boredom! All internal and external predisposing factors to illness, either alone or in combination, are thought to affect adversely the quantity and quality of the body's ch'i energy, especially over extended periods of time. The flow of ch'i through the meridians can sometimes become disturbed by simple physical trauma to the body that causes scars to form at an injury site. If the scar interrupts the flow of chi' energy through meridians that run through the injury site, health problems may arise. According to acupuncturists, the cutting of meridians and their subsequent blockage by scar tissue may be one of the unforeseen side effects of modern surgery. A therapeutic approach known as "neural therapy," developed by Dr. Dietrich Klinghardt, attempts to unblock the scar-affected meridian by injecting the old scar with the anesthetic drug procaine. The injection of an anesthetic agent into the scar seems to enhance the flow of ch'i through the scar tissue while frequently relieving the chronic pain that sometimes accompanies postsurgical (or post-traumatic) scarring. The preceding example indicates one of the many ways in which a disturbance in ch'i flow can predispose an individual to both pain and ill health.

When the flow of ch'i energy to an organ becomes blocked from either meridian problems or physical organ disease, the acupoints along the diseased organ's meridian actually become painful and tender to the touch. Acupuncture-meridian theory provides an interesting explanation for why people may suddenly develop aches and pains in specific areas of their bodies. In a later chapter, we will see why certain people with painful knees, for example, might experience pain not only from arthritis or bursitis affecting the knee joint but also from stomach problems affecting the flow of ch'i through the stomach meridian that passes through the knee joint. It is not unusual for people to experience pain along a meridian pathway if there are physical problems affecting the corresponding meridian-linked organ. Inflammation in the stomach can cause a person to experience both the typical abdominal pains and indigestion associated with gastritis as well as one-sided or even bilateral knee pains because of the organ-meridian link. The knee pains can sometimes be related to sore acupoints in the region of the knee joint, which in turn have become sensitive because of a ch'i-energy imbalance in the person's stomach meridian.

Chakra Dysfunction and Health: A Critical Energy Link Between Body, Mind, and Spirit

We have discussed how a balanced flow of ch'i energy through the acupuncture-meridian system is crucial to achieving health. But ch'i flow through meridians is only one of the many human subtle-energy systems that must be in balance in order to maintain health. Blockages in the flow of subtle energy through the chakra system can be an equally important contributor to dis-ease. If you recall, in the yogic tradition the chakras are said to function as emotional- and spiritual-energy transducers. They take in prana, the subtle, nutritive life-energy component of sunlight. Although both ch'i and prana are life energies, prana is a distinct and different form of life energy from ch'i. The seven major chakras (and the smaller channels known as nadis) absorb and distribute prana to the various organs and tissues of the body. It seems anything that causes a blockage or disturbance in the flow of subtle energy through one or more chakras can also lead to the development of illness in the body. One of the most important causes of chakra blockage is chronic emotional stress and emotional-energy imbalance. Just as physical toxins can produce illness, emotional toxins can poison the spiritual as well as the physical bodies in various subtle ways. Chronic anger, hatred, bitterness, greed, hopelessness, loneliness, and depression may lead to illness, because these negative emotions create chakra imbalances that shut down the flow of pranic life energy to our body's vital organs.

Yogic tradition further sees each of the seven major chakras resonating with and being influenced by a different type of emotional energy. If our thoughts and our emotions become imbalanced because of chronic emotional stress, a constricted flow of life energy through one or more of the seven major chakras can result. This constricted life-energy flow may then produce a weakness or predisposition to illness in different areas of the body. Such energetic weaknesses in the body tend to act like the weakest link in a chain. When our minds and bodies encounter various stress-producing emotional events or even physical stressors, the weakest links in our chakra chain will be the areas where our bodies will tend to break down and develop problems.

Furthermore, each major chakra has the capacity to function like a computer hard drive for storing the memories of specific emotional and spiritual experiences. Unfortunately, this storage capacity can indirectly set human beings up for illness, because the experiences that make the most impact upon the chakras are those associated with strong emotional reactions to past events in our lives, both negative and positive. While our brains record memories of our significant life events, our years of education, and our professional training, our chakras may be recording the emotional-energy patterns of the way we react and respond to daily life. It is said the very tissues of the physical body remember different types of emotional and physical traumas that we experience during a lifetime. Indeed, chiropractors and massage therapists have witnessed a phenomenon in which clients remember long-forgotten memories associated with an old physical injury when massage or physical manipulation is directed toward the traumatized area of the body. Part of the body's capacity to remember old traumas may be a function of our chakras' implied ability to store emotional and physical memory patterns. This phenomenon may relate to the chakras' intimate relationship with the nervous system. Specifically, each of the seven major chakras is associated with a collection of nerve cells known as ganglia. The ganglia could function like miniature brains that actually "remember" different emotional memories. More about these types of chakra phenomena will be presented later.

In recent years there has been increased recognition by the medical profession of the role emotional stress plays in health and illness. We live in extremely stressful times. Poverty, limited resources, and lack of social support have always been sources of stress to people living within the lower socioeconomic levels of society. But now people at the middle and higher levels of society are experiencing stress as well. Corporate downsizing has many people in the workforce worrying about their long-term job stability. Some early retirees who were once promised ongoing health-care coverage are now facing the reality of their former employer's reneging on this commitment. Military conflicts in Eastern Europe and the presence of nuclear weapons in the former Soviet Union, India, and the Middle East remain a continuing source of concern for all people throughout the world. Simply raising a family in today's fast-paced world can prove to be extremely demanding and stressful.

Such chronic stress can take a toll on our bodies, depleting the vital energy reserves needed to keep illness at bay. Depending upon our level of emotional reactivity to the turbulent world around us, we may find ourselves becoming overly stressed and plagued with a variety of subtle (or not so subtle) illnesses. Emotional stress can affect the body's biochemistry in very specific ways, depending upon how we react to each new stressful event. As you will soon learn, each of our seven chakras responds to different types of emotional and spiritual stresses and issues. Our chakras, like a spiritual biofeedback system, subtly tell us where the strongest emotional and spiritual imbalances exist in our lives, as well as where we need to direct our greatest healing efforts in order to re-achieve health and inner balance.

Clairvoyant diagnostician Carolyn Myss has done a great deal of research to substantiate the energy connection between the chakra system and the physical body in health and illness. Her perceptions of the chakra system and the patterns of chakra imbalances behind physical illness are based upon her clairvoyant diagnostic evaluations on hundreds of patients. Dr. Norman Shealy, a prominent neurosurgeon and pioneer in the field of energy medicine, found Myss to be accurate in her clairvoyant medical diagnosis more than 90 percent of the time! Much of what Myss perceives as dysfunction in the different chakras relates to painful or stressful life experiences and emotions that people hold on to. Myss believes that psychic wounds create damage to our spiritual as well as our physical bodies at the level of our chakras. If we continually focus upon our past hurts and psychic wounds, we never allow ourselves to heal in the present, thus limiting what we can make of our futures. As Myss suggests, it is as if we are borrowing from the energy currency allocated for our body's current health needs in order to support and "finance" our past traumas and old psychic wounds.

Extremely traumatic events in our lives do take time to heal. But unless we move beyond them, they can preoccupy the rest of our lives and can eventually manifest in the form of illness in the physical body. If we keep energizing those emotional- and thought-energy patterns of past traumatic experiences, we take away from the subtle energy needed to sustain our lives in the present. With Myss's rationale, focusing on past hurts causes us to continually reinvest energy into maintaining old wounds at the expense of our "bioenergy bank-account balance." When we continue to focus upon past injuries, even at an unconscious level, we divert energy away from our bioenergy bank reserves. In essence, we finance our past at the expense of our present health, ultimately leading to a kind of bioenergy "debt." If our bioenergy bank reserves become depeleted, the energy needed to support the tissues of the body becomes diminished, eventually resulting in illness. Energy imbalances caused by misdirected thoughts and emotional energies can impair health via a shutting-down of energy flow through our chakras, especially our heart chakras. As such, a primary disturbance originating in our emotional and mental bodies can lead to patterns of chakra imbalance and subtle-energy starvation to specific organ systems of our physical body, ultimately resulting in dis-ease.

The energies of our old dysfunctional thoughts and emotions are investments that we're making on a spiritual level. Such energy investments keep alive the negative emotions linked to old traumatic experiences, never allowing us to process them and move on. Because human beings have only a certain amount of bioenergy to work with in their subtle-energy bank accounts, we must learn to manage this limited resource wisely in order to maintain the health of our physical bodies. If too much time and energy are spent fixating upon past hurts, associated pain or revenge emotions take away from the limited biological-energy resources we have available for nourishing our bodies and keeping us healthy. Our energy is literally wasted when we focus upon an immutable past. The key to moving beyond traumatic memories is in learning to perceive them from a higher spiritual perspective. While it is difficult to perceive the lesson from a painful life experience in the midst of that pain, it eventually becomes easier to grasp the spiritual significance of traumatic events by interpreting them from the symbolic perspective of the soul.

When we can learn to shift our focus to that of the eternal soul, by using what Myss refers to as "symbolic sight," it becomes possible to see the spiritual lessons contained within painful experiences. Viewing events from this higher, symbolic level of perception is what Myss calls "seeing from the eighth chakra level." Sometimes called the transpersonal point, the eighth chakra is located above the head. It is the chakra Myss sees as the connecting link between the personality and the higher self. Perceiving life from the symbolic perspective of the higher self allows for an appreciation of how our soul chooses to place us into situations of potential stress, adversity, and illness in order for us to learn more about our true power and our higher spiritual nature.

Even mainstream psychologists can accept the belief that the human soul attempts to communicate with the ego (the conscious mind) through the use of symbols, the language of the right brain, and the dream state. Unfortunately, most of us operate on a very literal left-brain level, especially when we seek to find meaning in the events of our daily lives. We tend to view events in the physical world in terms of their personal effects upon ourselves and our families. For this reason, when people are told about a serious health challenge, they often react with shock and denial. First there are thoughts about how the illness will affect loved ones, personal earning power, or the quality of life. Often people will search for "concrete" reasons as to why such a horrible thing could have happened to them. Increasingly, depression, despondence, and anger may set in. Some people even interpret the diagnosis of serious illness as a punishment for some secret sin or source of guilt. But by looking beyond the literal level, we can view the diagnosis of illness not as punishment but as a spiritual challenge requiring all our physical, emotional, mental, and spiritual resources. Perhaps when we begin to understand the life lessons from the symbolic perspective of our higher selves and our souls, we will move closer to discovering our inner healing power, a part of our divine spiritual nature.

Each of the seven chakras is viewed as being tied to specific life lessons stemming from the need to adapt to change or to new life experiences. How we choose to react to each new lesson determines the emotional charge with which we imbue each experience. For example, suppose two students go to take their final exam for a science course in college. Both students have studied equally hard, but each enters the testing facility with a different attitude. One student says, "Hey, I studied as much as I can. What's the use of getting stressed out and anxious at this point?" She calmly works her way through the exam. The other student, preoccupied by fears that he'll fail the test and ruin his chances for getting into graduate school, freezes up and actually inhibits his performance, scoring lower than he otherwise might have. Although it's the exact same test, the two people approach it with different attitudes, resulting in two different ways of emotionally processing the same experience.

Learning how to approach new life lessons, relationships, or emotional issues with a calm, nonreactive, and optimistic attitude allows us to arrive at the greater spiritual lessons our soul seeks to teach us. Furthermore, remember that each of our seven chakras acts as a memory bank for specific categories of experiences our soul presents to us in order to teach us important spiritual lessons. As you will see, the energy flow to our vital organs from each of the seven major chakras is influenced by how we deal with and react to these major and minor life lessons of the soul. Whether we respond to life with a negative emotion or an optimistic feeling may significantly influence our state of health. By understanding the physical, emotional, mental, and spiritual dynamics of each of the seven chakras, we may be better able to make changes in our thinking, feeling, and reaction patterns which ultimately contributes to either illness or wellness.

The following information, from medical intuitives like Carolyn Myss, from ancient doctrines, yogic chakra theory, and Chinese-medicine teachings, helps to provide an interpretation of each main chakra's correlation with specific illnesses, emotions, and relationships. It will become easier to see the possible links between chakra imbalance and disease as you begin to understand certain chakric associations and life lessons. But take heed: Emotional imbalances and chakra-energy disturbances are only two of many factors to be considered in terms of health-risk factors that can predispose people to develop specific illnesses. Whether a particular organ system breaks down or develops problems at any one point in a person's life will also depend upon that person's current levels of stress, inherited tendencies toward specific illnesses or organ problems, level of exposure to various environmental toxins and viruses, recent and long-term nutritional

The Major Chakras and Their Physiologic and Psychospiritual Associations

CHAKRA CENTER	ASSOCIATED ORGANS	ASSOCIATED ENDOCRINE GLAND
First Chakra (Root Center)	**anus and rectum, lower pelvic tissues, bones and bone marrow**	**gonads**
Second Chakra (Sacral Center)	**bladder, prostate, uterus, fallopian tubes, ovaries, genitals, colon, intestines**	**Leydig cells gonads**
Third Chakra (Solar Plexus Center)	**stomach, liver, gallbladder, pancreas, kidneys**	**adrenals**
Fourth Chakra (Heart Center)	**heart and lungs bronchial tubes**	**thymus**
Fifth Chakra (Throat Center)	**larynx trachea carotid arteries**	**thyroid**
Sixth Chakra (Brow center/Third eye)	**part of cerebral cortex eyes and ears midbrain**	**hypothalamus pituitary gland**
Seventh Chakra (Crown Center)	**cerebral cortex**	**pineal**
Eighth Chakra (Transpersonal Point)	**higher spiritual bodies**	

habits, as well as many other potentially important risk factors. This is an important point to understand when we try to make sense of chakra imbalance and disease tendencies. While an imbalanced chakra can eventually contribute to the development of an illness, that subtle-energy imbalance is only one of many factors acting in concert to bring about varying states of illness versus wellness. We must always take into account the whole person without placing too much

ASSOCIATED PHYSIOLOGIC FUNCTIONS	ASSOCIATED EMOTIONAL AND SPIRITUAL ISSUES	LEVEL OF RELATIONSHIP
excretory reproductive immunity	grounding fear and safety issues basic survival instincts	tribal
genitourinary	sexuality financial issues creativity codes of honor and ethics	one-on-one relationships
digestive	self-esteem self-image issues of responsibility	basic relationship with inner self
circulatory immunity	issues of love depression and loneliness matters of the heart unconditional love forgiveness wisdom of the heart	loving relationships with significant others, with oneself, and with all life
respiratory	right use of will communication self-expression speaking one's truth self-discipline	our relationship with our personal "truths"
autonomic nervous system and intellectual functioning	ability to see one's life clearly intuition inner vision use of the mind/intellect	our relationship to knowledge, wisdom and vision
central nervous system control	spiritual search for meaning issues of karma and grace grace bank account spiritual awakening divine discontent	our relationship to our life's true purpose relationship between the ego and the higher self relationship to God and spirit
coordination between physical and spiritual dimensions	archetypal/symbolic ways of perceiving hidden spiritual significance behind daily "synchronistic" events	our relationship with our soul

emphasis on any single factor. We cannot, however, ignore the concept that changes in the subtle energy-field caused by our attitudes, our emotions, and our mental outlook on life can result in powerful physiological changes, both positive and negative, that strongly affect our physical bodies. It is therefore important to start recognizing the state of our emotional and mental balance as powerful contributing factors in the ongoing creation of health or illness.

First Chakra

According to clairvoyants and healers who can directly observe the chakras, the first or root chakra, located directly over the anal region, supplies vital pranic energy to support the rectum, the lower pelvic organs and tissues, the skeletal system, and the bone marrow. Since the bone marrow is the place where many of the body's red and white blood cells are manufactured, dysfunction of the root chakra can contribute to immune-deficiency problems, various types of anemias, and states of chronic fatigue associated with a general loss of vitality. Chakra theory and clairvoyant observation also suggest that the root chakra can become distorted from severe pain and trauma, such as the trauma of a mother's extremely painful childbirth experience. More specifically, intuitives who can see the chakras sometimes describe a "blown-out" root chakra in women who have undergone traumatic birthing experiences. Such distorted root chakras may lead to a variety of recurrent health problems unless therapeutic measures are taken to heal and repair the damaged root center.

While physical traumas to the body have been clairvoyantly observed to adversely affect the health of the chakras, chronic emotional and spiritual imbalances are thought to be a more common cause of chakra dysfunction. The first chakra can become impaired by specific types of subtle emotional imbalances, which if left uncorrected may lead to long-term health consequences. The first chakra is associated with the emotional issues surrounding basic physical survival. People who have been traumatized by the violence of war sometimes continue to suffer painful emotional memories that can create an impairment in the way the root chakra absorbs and distributes life energy to its associated organs. Chronic fear can strongly influence a person's consciousness and lead to a paranoia of, or suspicion toward, the world in general. Over time, long-standing fear can impair the functioning of the immune system, making an individual more susceptible to illness from a variety of causes, including chronic infections, cancer, and even autoimmune diseases.

In the five-element theory of Chinese medicine, excessive fear is thought to impair the functioning of the kidneys (connected with the third chakra). Oddly enough, the kidneys are hormonally linked (albeit indirectly) with the bones and skeletal system, whose vital energy comes from the root chakra. The kidneys produce a hormone known as or erythropoetin, which stimulates the marrow-precursor cells within the bones of our body to increase their production of red blood cells. One way in which root-chakra dysfunction (produced by chronic fear) might physically damage the kidneys is in the form of chronic pyelonephritis or recurrent kidney infections. Since the first chakra is indirectly linked with immune function through the bone-marrow connection, dysfunction of the first chakra might contribute to a depressed immune response, which in turn could transform a mild case of cystitis into a chronic kidney infection that could damage the kidneys over time.

According to observations made by Carolyn Myss, the first chakra is also associated with a type of group consciousness that relates to the survival of one's tribe. In ancient days, people obeyed tribal laws to ensure the safety and continuance of their tribe. However, today the concept of a tribe can refer to different social groupings, ranging from one's family, place of employment, society of professional affiliation (such as the AMA for doctors), and even the nation of allegience. The emotional-energy structure of the first chakra is such that whenever we go against the laws of our tribe, we may suffer inner strife on some level because we have chosen to go against the wishes of the tribal majority. That tribal majority might be the elders of a person's family or even the "moral majority" of a particular ethnic group. Whenever people act in a way that makes them appear as if they have "gone out on a limb," so to speak, against the tribal norms, there is always the potential for first-chakra imbalances to occur. Even though a person consciously believes that the decision to go against tribal wishes "feels right," the root chakra just might react instinctively to the breaking of "tribal codes of ethics."

Carolyn Myss is perhaps the first medical intuitive to recognize the important link between our emotional reactions to tribal laws governing social behavior and the health of the root chakra. Myss has found that our first chakra contains the belief patterns that have become established by experiences with our biological family and our early social environment. In addition, the spiritual challenge of the first chakra relates primarily to how well we manage our physical world. Through her clairvoyant diagnostic work, Myss has found that the physical problems associated with root-chakra dysfunction can range from chronic low-back pain and sciatica to varicose veins, depression, rectal tumors, and even immune disorders such as HIV-related illness.

There also seems to be a definite correlation between the underlying blockages that can occur in each of the seven chakras and seven different types of dysfunctional relationships that can impair energy flow through our chakras. At a very primary level, first-chakra imbalances tend to occur in individuals who are "ungrounded" and suffer from a kind of dysfunctional relationship with the earth itself. They lack a sense of connection to the earth, often feeling that the world is a threatening place in which to live. In other types of root-chakra imbalances, the dysfunctional relationship is at the tribal level. There may be intolerable friction between an individual and his or her family, religious group, or business and professional associates. This type of root-chakra problem might affect those who choose to go against the wishes of their parents and their extended family by marrying someone outside of their religious or ethnic group. In a slightly different scenario, where the tribe involved is the AMA, a physician who chooses to practice alternative medicine sometimes risks going against the mainstream thinking of medical peers. Here, too, an imbalance in the root chakra could be waiting to happen. In either case, individuals make a decision to travel a path that follows their heart but goes against tribal beliefs and tribal rules of appropriate behavior. While the dysfunctional relationship might not produce conflict at a conscious level, the root chakra is often strongly affected by such issues, resulting in a subtle form of energy imbalance that might eventually lead to health problems in the abdominal and pelvic regions of the body.

Second Chakra

The second chakra is located directly over the genitals and the gonads (the ovaries in women and the testicles in men). Our reproductive organs are energetically nourished by the second chakra. Psychoenergetically speaking, the second chakra is the center of creativity, procreation, and new ideas. Continuing with information gleaned from Far Eastern doctrine, yogic chakra theory, and medical-intuitive investigations, the second or sacral chakra is connected with emotional issues that concern our one-to-one relationships. Unlike first-chakra problems, which arise from being out of sync in our tribal relationships, second-chakra energy blockages are often due to dysfunctional relationships of a more personal nature. These include sexual, business-related, or even simple friendship relationships. Second-chakra emotional-energy dynamics also involve aspects of honor and ethics with regard to the way we relate to our life partners, our family members, and our professional and business associates, since our morals and ethics always come into play when we deal with others. Consider the example of a person who ruthlessly steals colleagues' business accounts while outwardly denying such unethical behavior. Problems in the prostate or bladder, two organs that are energetically nourished by the second chakra, could eventually develop. Another potential cause of problems in the second chakra involves the level of morality and ethics we display in dealing with our life partners. For example, the temptation for married individuals to have affairs is a test of their personal code of ethics. Is a person capable of truly honoring professed marital vows of monogamy? While many people routinely have extramarital affairs without thinking twice, at an unconscious energetic level they may be developing problems in the functioning of their second chakras.

Two other issues that influence the activity of the second chakra have to do with different aspects of power and control over others. Power issues can be of either a positive or a negative nature, depending upon how one exerts power over another individual in a relationship. In situations where an individual is using or abusing other people, sexually or in business relationships, there can be a greater susceptibility to developing blockages in the second chakra.

The sacral chakra is strongly linked with the issues surrounding both sex and money, too. The link with sex is obvious. The second chakra energizes and supports our sexual organs. So at a very basic level, the second chakra is affected by our level of comfort with our own sexuality as well as by the way that we sexually relate to others. People who have chronic difficulties in their sexual relationships or who are ambiguous in their sexual preferences may be more likely to develop second-chakra problems. Second-chakra difficulties can also be triggered by sexually traumatic events such as sexual abuse, incest, or rape. Traumatic sexual experiences often leave deep psychological scars as well as an energetic recording of fear and terror produced by the sexual violation, which remains stored in the memory banks of the sacral chakra. It is not uncommon for women who have been sexually abused or raped to experience recurrent pelvic infections such as chronic cystitis and pelvic inflammatory disease. The problem may lie not only in the microbes that inflame the bladder or fallopian tubes but also in the energetic starvation of the second-chakra-fed pelvic tissues, thus creating a greater susceptibility to recurrent pelvic infections.

The second-chakra link with money relates to our sense of financial security in the world as well as with whether we feel that we are earning our true monetary worth in our current occupations. Because the anxiety of personal financial difficulties can sometimes impair energy flow through the second chakra (which feeds energy to the region of the lower back), it is not uncommon for people with money problems to experience recurrent back pains, sciatica, or sexual-potency problems. Dysfunction in the second chakra can also contribute to problems with the urinary tract, ranging from chronic bladder infections to bladder tumors. It is entirely possible that the rising rates of prostate cancer in men could be partly related to an upsurge in second-chakra blockages caused by dysfunctional relationships concerning issues of power, ethics, and sexuality. This may relate to an increase in ruthless business practices occurring over the last few decades. After all, we now live in an era in which corporate takeovers of small businesses and even larger companies have become commonplace. One must ask, "What are the energetic consequences of making money at the expense of eliminating people's jobs and livelihoods?" In order to heal second-chakra-related health problems, and to prevent their eventual recurrence, attention must be addressed toward healing the emotional and psychic trauma that may have energetically damaged this vital-energy center. A comprehensive healing approach would use both counseling and energetic work, as well as therapies directed to healing the specific physical problems.

Third Chakra

The third chakra, known as the solar plexus chakra, has traditionally been linked with our sense of personal power in the world. This is the energy center associated with our sense of self-esteem and self-worth. People who have a low sense of self-esteem often develop chronic energy imbalances in the solar plexus center. The third-chakra center is where we are said to hold on to those concepts about how we look, our confidence in our physical abilities, our sense of how smart we are compared with others, and how we rate our work skills compared with those of our co-workers. If we believe that we are somehow inferior to those around us, our insecurities may leave us with a poorer sense of self-esteem and diminished self-confidence. Whether or not these feelings of inferiority might have originated in critical comments made by parents or teachers or from our own self-doubts, they limit what we can achieve in life.

The third chakra is also a place where we tend to feel strongly the emotions of fear and intimidation. For instance, it is fairly common for people who frequently experience these emotions to describe feeling "all knotted up in their gut," when they are intimidated and made to feel afraid. This makes sense, because the solar plexus center is physiologically connected to the stomach and the intestinal tract.

The solar plexus area—"the gut," essentially—is where most people tend to experience pain and discomfort after being criticized for their actions or their beliefs. It is also the region of the body linked with responsibility for decision-making. People who feel overwhelmed by responsibility may experience health problems stemming from energy imbalances in their solar plexus centers. For example, an employee who constantly fears making mistakes, lest he be fired by upper management, or who dreads the wrath of an overbearing supervisor or office manager, is especially susceptible to third-chakra weaknesses. Since the third chakra feeds subtle energy to the stomach and the digestive tract, people who feel chronically stressed by responsibility often complain of indigestion, heartburn, and abdominal pain caused by recurrent ulcers, gastritis, or esophagitis. However, a question arises as to why one organ might be affected over another when a chakra imbalance exists, since a number of different organs receive energy from the same chakra. One way of understanding how a solar plexus chakra imbalance might lead to ulcers versus liver problems versus kidney problems can be found in Chinese medicine's five-element theory. According to five-element theory, each of the organs of the body is connected with a specific emotion or emotional state. For instance, the liver is the organ associated with the emotion of anger, while the emotion of excessive fear is thought to be connected with the kidneys. As such, in Chinese medicine, someone with a great deal of repressed anger, due to a sense of overwhelming responsibility, might be expected to develop liver problems due to a solar plexus blockage. Conversely, someone who instead of being angry is consumed by fear and panic as an emotional reaction to having too many responsibilities, might develop kidney problems. When observing from the energetic viewpoint of chakras and meridian energies, one can see how different emotional states associated with solar plexus chakra imbalances could easily contribute to a variety of different illnesses.

As the reader may have noticed, both Chinese medicine and chakra theory have been mentioned separately to describe specific pathways by which imbalanced emotions can lead to different diseases of the body. But these two viewpoints, though different, will often be complementary in explaining how certain emotional imbalances may contribute to the development of specific illnesses. Let's next blend Chinese medicine and chakra theory together to see how these two different yet compatible approaches might explain how the energy imbalances in the solar plexus chakra could contribute to the expression of a complicated disease like diabetes. We know that the solar plexus center feeds nutritive subtle energy to the pancreas. The pancreas is a major digestive and endocrine gland that produces both digestive enzymes, to break down the foods we eat, and insulin, to help us absorb the ingested sugars and digested carbohyrdrates. Insulin is a key hormone in the process of assimilating and processing glucose, or sugar, the major fuel for the body and the brain. From the chakra perspective, emotional problems such as chronic low self-esteem can sometimes impair the flow of subtle energy to the pancreas. Such an energetic imbalance may contribute to abnormal functioning of the pancreas gland, resulting in diseases as varied as hypoglycemia, diabetes, and pancreatitis. According to Chinese five-element theory, the pancreas is also strongly associated with the emotions of worry and obsession. By following a five-element model, we see that individuals with a solar plexus imbalance, caused by both low self-esteem and a sense of overwhelming responsibility, might begin to worry and obsess constantly about their problems. The excessive worry or obsession can contribute to a kind of pancreatic burnout, which might initially show up as the medical condition of hypoglycemia. In time, hypoglycemia could eventually turn into diabetes. Diabetes, when viewed from the perspectives of both Chinese medicine and chakra theory, provides us with an example of how the spectrum of different emotional imbalances connected with chakra dysfunction may contribute to specific types of health problems. Of course, diabetes is a multifactorial disease that may develop not only because of subtle-energetic and spiritual factors but also because these factors frequently work in concert with other more physical causes of diabetes, which include inherited tendencies, dietary indiscretions associated with weight gain and insulin resistance, toxic damage to the pancreas by alcohol consumption, autoimmune disorders, and viral infections of the pancreas. Nevertheless, an examination of diseases such as diabetes from a chakra perspective may help us to better understand the various physiological and bioenergetic pathways by which our emotions may contribute to the development of specific illnesses. The example of pancreatic dysfunction in diabetes also shows us how ancient Chinese medicine and yogic teachings about energy flow in the body can sometimes provide complementary clues as to how different emotional and personality expressions may contribute to specific health problems associated with imbalances in the third or solar plexus chakra. As an aside, the symbolic significance of diabetes, a disease that causes those afflicted with it to eliminate sugar from their diets, is also quite interesting. Louise Hay, author of *You Can Heal Your Life,* has studied the symbolic meaning behind various diseases. Hay views diabetes an as illness in which "all of the sweetness has gone out of life." By avoiding sugary foods, diabetics have literally been told to eliminate all the "sweetness" from their dietary lifestyles. Then having to give onself daily insulin injections while constantly monitoring blood-sugar levels may also contribute to a feeling of "no sweetness in one's life." Since the pancreas is the organ of sweet, sugary food metabolism, one can see an interesting symbolic correlation between diabetes and a decreased ability to appreciate "the sweetness of life."

Fourth Chakra

The fourth chakra sits directly over the heart and lungs. In addition to energizing and nurturing the heart, the fourth chakra provides nutritive subtle energy to the lungs as well. Thus, if the flow of subtle energy through the heart chakra becomes chronically constricted, the resulting energy blockage may contribute to the development of physical heart and/or lung problems.

Clairvoyant research and yogic chakra theory associates the heart chakra with the emotional issues, expressions, and relationships involving love. The broad spectrum of loving emotions connected with the heart chakra include romantic love (between lovers), the familial love of a parent for a child, and self-love. As most people will probably agree, the ability to love ourselves in spite of all of our faults and past mistakes is frequently the hardest form of love for us to develop. We often have higher expectations of ourselves than we have of the other people in our lives. The necessity for love in our lives cannot be understated. For some people, the absence of love may mean the difference between suffering from a minor case of the flu versus succumbing to a a life-threatening illness. Such a scenario might actually occur, because a "lack of love" in one's life can eventually lead to heart-chakra dysfunction.

Perhaps the most universal love relationship is one that involves romantic love. Being in love with someone often produces a sense of expansive euphoria in the heart center, especially when we are thinking of the individual who loves us and whom we love in return. Conversely, the heartbreak of a lost love is also something we can literally feel in our heart chakras. According to esoteric wisdom, this occurs because there is a constriction of subtle energy in the heart chakra whenever any profound sense of emotional loss is experienced. Even the popular music of our culture speaks of the ache of a broken heart as being synonymous with the ending of a romantic relationship.

We first begin to learn about loving relationships in our early home lives. Our first and most basic form of love is the love that we, as children, feel for our parents. Then, too, the love parents feel for their children can be a powerful emotional bond. However, not all early parent-child relationships are filled with love and affection. Sometimes family relationships become dysfunctional, such as when a parent is unable or unwilling to show love toward a child. Each of us needs loving reinforcement from our parents when we are young to help us validate our growing sense of self-worth. A lack of nurturing parental love in early home life can impair our ability to express love for others or even to feel love for ourselves during adulthood. Sometimes, there can be "too much" parental love. "Smothering love" from a parent can stifle or impair the health and psychological development of a child. Some psychosomatic therapists believe that "smothering" love from an overprotective parent can be a contributing factor to asthma attacks in a number of cases of childhood asthma. In such childhood asthma cases, the dysfunctional love relationship between parent and child may cause an energy impairment or constriction within the heart chakra, the energy center that feeds subtle energy to the lungs, leading to recurrent airway constriction from asthma attacks. Of course, allergies and other medical factors are frequently involved, but the subtle-energy component might function as the underlying trigger in some cases of childhood asthma exacerbations.

Besides being influenced by the emotional issues surrounding love, the heart chakra is also affected by hatred, resentment, and bitterness toward other people—the exact opposite of love. When an individual holds on to bitterness or hatred, it is sometimes said that the individual has a "hardened heart." Medical researchers have noted that chronic anger and hostility are significant risk factors that may be linked with a higher incidence of heart disease. Even when the flow of love, and thus the flow of heart chakra energy, becomes blocked by negative emotional expressions, there still exists a powerful inner source of healing for the heart. That source of healing can often be found in the expression of forgiveness toward those individuals whose actions generated the original bitterness and resentment. Learning to forgive another person for perceived "wrongful acts" committed against us can often be more healing to the heart than any drug or dietary program. Although the major spiritual traditions teach us to "turn the other cheek" and to love and forgive, too often such advice goes ignored except on Sundays or religious holidays.

In contrast to the healing forces of love and forgiveness, many in the health-care field are finding increasing evidence that the emotional issues of grief and loneliness can also powerfully affect the heart center. Look at some of our common "heart-oriented" sayings. For example, we frequently refer to someone who is overwhelmed by grief or loneliness as having a "heavy heart." Dr. Lawrence LeShan is a psychologist who has researched the mind/body connection between certain emotional states and the development of cancer.

LeShan conducted extensive personality surveys of hundreds of cancer patients. He found many cancer patients to be pleasant, agreeable individuals who tended to bottle up and suppress their true feelings. But LeShan's investigation also found another thought-provoking commonality in many cancer patients. Several years prior to their diagnosis of cancer, quite a few of LeShan's cancer patients suffered from some form of depression. The depression was frequently due to the loss of a loved one or to the loss of a career. To LeShan, it seemed that persistent grief and chronic depression following a major loss were significant predisposing factors to the eventual development of cancer, possibly because of depression's ability to weaken the immune system.

Recent health research has also shown a significant correlation between loneliness and higher mortality rates from heart attacks. Individuals who feel loved and supported by friends and family or even by a loving pet (such as a dog or a cat) tend to have higher survival rates following heart attacks than other cardiac patients who experience loneliness and a sense of social isolation. It appears that filling the loneliness in a person's life with a meaningful, caring relationship, no matter how small (e.g., the unconditional love expressed by many dogs and cats for their owners) can sometimes, quite literally, "mend a broken heart."

Dr. Dean Ornish, a pioneering cardiologist who has demonstrated the ability of cardiac patients to reverse coronary-artery disease using healthy diets, meditation, and psychological support, notes that there are now many medical studies demonstrating the importance love plays in surviving heart attacks as well as other forms of serious heart disease. Love is truly the great healer, for when it is absent, the heart chakra and eventually the physical heart tend to suffer from dysfunction and dis-ease. Perhaps one of the reasons for the increased survival from heart attacks in certain individuals is a healing of the emotional block that led to the energy imbalance of the heart chakra in the first place. The sudden realization that you might die from a heart attack can be a powerful motivating force for a lifestyle change. It can also provide a stimulus for people to shift their attitudes, their emotional reactivity, and their thinking patterns. Potentially life-threatening illnesses may serve as a wake-up call to remind us of our physical mortality while simultaneously providing us with fresh insights into how a renewed spirit may reenergize and heal our physical bodies. For some people it takes the threat of losing everything before they will stop behaving in old dysfunctional ways and finally allow the healing power of love to trickle back into their lives.

Heart attacks from coronary-artery disease are only one health challenge that chakra-oriented researchers associate with an imbalanced flow of energy through the heart center. Other cardiac problems, including enlargement of the heart, congestive heart failure, and even mitral-valve prolapse (a somewhat common leaky-valve condition), may also be somehow connected to dysfunction of the heart chakra.

In addition to energizing and nurturing the heart, the heart-chakra also provides nutritive subtle energy to the lungs as well. When heart-chakra blockage causes energetic impairment of the lungs, clairvoyant diagnosticians like Carolyn Myss and Mona Lisa Schultz have noted that such energy blocks may manifest as chronic bronchitis, pneumonia, asthma, emphysema, or even lung cancer. We already know that cigarette-smoking and air pollution contribute to various lung diseases. But smoking seems to affect each person differently. Some people develop emphysema from smoking in their early forties, while others are puffing away into their eighties with only a minor smoker's cough. Is it possible one of the reasons for the variable response to smoking (aside from genetics) may be the level of functioning or degree of impairment of each smoker's heart chakra?

Besides contributing to heart and lung diseases, heart-chakra blockage may have widespread effects upon the entire body because of the heart center's link with the immune system. Since the heart center also feeds energy to the thymus gland, the master immune gland of the body, heart-chakra blockage can sometimes result in immune-system dysfunction due to impaired thymic-energy flow. As most people know, a weakened immune system leaves an individual open to many different forms of disease. Recently, microscopic analysis of diseased coronary arteries has suggested that even atherosclerotic heart disease may have an unsuspected immunological component. The degree of an individual's immune impairment could be related not only to the severity of heart-chakra imbalance but also to the effect of lifestyle-related, genetic, and environmental predisposing factors that may be present as well. Perhaps the rising rates of lung cancer, breast cancer, heart disease, and immune-related problems in our society has to do with an increasing incidence of heart-chakra dysfunction within the population. Although risk factors such as high-fat diets, smoking, nutritional deficiencies, infectious agents, environmental carcinogens, and cancer-promoting genes have all been implicated as potential causes of cancer and coronary-artery disease, no medical research has studied the critical relationship between heart-chakra dysfunction and these major diseases of our modern industrialized society. We already know that an abnormal immune functioning may result in a wide range of health problems, including allergies, chronic fatigue syndrome, increased susceptibility to colds and viruses, autoimmune disorders such as lupus and rheumatoid arthritis, and even certain forms of cancer. Autoimmune disorders are caused by the body's literally attacking itself via autoantibodies that target various normal cell constituents. Symbolically and literally, individuals with an autoimmune disorder are attacking themselves at a cellular level. While there are those who would consider this in the realm of pure speculation, I suggest that on a subtle-energy level, some autoimmune disorders might be occurring in people who are (unconsciously) more self-critical than self-loving or accepting of themselves. It is often easier to love someone else than it is to love yourself without judgment or self-criticism. Over time, a chronic lack of self-love could eventually lead to a distorted flow of subtle energy through the heart chakra, and thus an abnormally functioning immune system.

The wide spectrum of manifestations of heart-chakra dysfunction, from simple colds and respiratory infections to heart attacks and cancer, demonstrates how incredibly important the heart center is. The lessons of the heart chakra teach us about the necessity for having love in our lives if we want to maintain our health and well-being. It is not an exaggeration or a mere metaphor to say that "love makes the world go round." Without it, we would all become chronically ill. The need for love is openly discussed in many of the world's great spiritual teachings. For centuries, our world's greatest esoteric and spiritual teachers have known and written about the healing power of love and its ability to stimulate rejuvention of body, mind, and spirit. Love can be a powerful motivating force in human relationships on many different levels. Unfortunately, some individuals tend to use love as a controlling influence over others, threatening to withhold love unless a spouse or child behaves in a certain manner. This type of love could be considered "conditional love."

However, the highest form of love is conscious, unconditional love and acceptance of others, as well as of ourselves, completely and without judgment. Unconditional love is a form of love to which many aspire but few may actually achieve during their lifetime. Few people really appreciate how this aspect of the heart chakra, the full expression of the energy of unconditional love, can be such a powerful healing force. But when our energy systems become "clogged up" because of our inability to love ourselves or the other people in our lives, the result is often an imbalance in the heart center. In all forms of laying-on-of-hands healing, from Therapeutic Touch to spiritual healing, the energetic trigger that initiates the healing interaction between healer and client is often the healer's focus on unconditional love for the client and a sincere desire or intention to help and to heal. The key lesson of the heart chakra is to accept and love ourselves unconditionally. We must forgive ourselves for our past mistakes, our faults, and our failures, while also giving ourselves credit for the good things we have accomplished in our lives. We must learn to move through life with a more open attitude toward what life has given us to work with, lovingly accepting ourselves with all of our little flaws. Love and forgiveness are the healing forces that release the energy blocks to the heart chakra created by bitterness, resentment, anger, and even grief. Each of us needs to learn to process and accept the mistakes and failures of our past as well as our perceived injuries and wounds. We need to move beyond them so that we can continue to grow spiritually as well as to heal physically. As long as we continue to hold on to and "energize" past feelings of having been embarrassed, injured, or attacked by people we have encountered throughout our lives, we become overly focused in emotional patterns that block the perception of love in our lives. This makes it more difficult for love to penetrate the emotional-energy armoring that we may unconsciously surround ourselves with. Walking around with hatred, bitterness, and resentment in our hearts because of some unfortunate past circumstance may keep us from the happiness and contentment we seek. It is easier to blame someone else when bad things happen to us than to look deeper for causes. By chronically feeding energy to those emotional thoughtforms of hatred and resentment, bitter people ultimately hurt themselves by constricting the flow of energy through their own heart chakra.

Ironically, some people may actually develop heart-chakra problems because they hold back from loving another individual for fear of being hurt and suffering a "broken heart." Although no one consciously wants to experience the emotional pain of lost love or the loneliness that often follows, sometimes you have to take chances in life, risking heartbreak and disappointment in order to gain from new experiences, even if the experiences might be potentially stressful to the heart chakra. We have all heard the saying "It is better to have loved and lost than never to have loved at all." Love is a necessity of human existence and a powerful healing force for both self and others. In fact, a memory of love can sometimes be life-sustaining when one is dealing with major life challenges, obstacles, and illnesses because, you see, love is the primary energy of the heart chakra, and perhaps it may even be the "binding force of the universe." Though we may have conflicts with our loved ones, we can always come back to those moments of love that we have shared, even if they are only in memory.

It is important that we extend an attitude of love and caring toward all the people in our tiny part of the universe, whether they be strangers or family. And what better way to extend an attitude of caring than by random acts of kindness? While small acts of kindness and compassion may not immediately benefit the donor, they help to create a more fluid and stable flow of heart-chakra energy for the "doer of kind deeds," no matter how small the deed might be. Sometimes sitting down on a park bench, sharing a kind word, and lending a sympathetic ear may avert a potential act of negativity or desperation in another person. We all know that it is often difficult to "overtly" demonstrate love toward people who are creating conflict in our lives. But please understand, on some level these irksome people are really providing a life experience for us that will force us to grow spiritually. We must learn to love on many levels, to experience forgiveness for ourselves and others, and to move through life with greater joy. Only then can we grasp the higher perspective of the soul while we begin to heal the heart chakra and the vital organs it helps to nourish and sustain.

Fifth Chakra

The fifth or throat chakra is located directly over the thyroid gland and the larynx. Emotional issues that influence the throat chakra involve the way we choose to exert our personal will. There is a "right use of will," which follows the golden rule of "Do unto others as you would have them do unto you." In a sense, the right use of will is really the expression of personal will tempered by the wisdom of the heart. The strength of our personal will is measured not by how strongly we can exert our will over others but by how effectively we can muster up our own willpower to control and discipline ourselves. This is perhaps one of the key lessons of the throat chakra. Self-discipline is really the true test of one's willpower. Conscious self-control is not merely the discipline of eating properly or exercising regularly or following the accepted rules of society. It is a discipline extending to the mind, the body, and the spirit. For example, a dysfunctional relationship between our pleasure-seeking behaviors and our drive to better ourselves through self-discipline can sometimes result in an energy imbalance in the throat chakra. For many, chronic throat-chakra blockages have the potential to develop into physical disorders of the thyroid gland such as hypothyroidism and autoimmune thyroiditis (an immune-related inflammation of the thryoid gland). Interestingly, one of the chief features of hypothyroidism is fatigue and lack of motivation (the very opposite of self-discipline). Our need for self-discipline and introspective thought becomes even more important when we seek to consciously heal ourselves of illness.

The throat chakra is also the center from which we speak our "truth" about what we believe or hold to be true about ourselves and the world around us. By chronically suppressing our real thoughts and feelings, holding back from saying what we really believe or know to be true, we may eventually suppress the natural flow of subtle energy through the throat chakra. Of course, there is no such thing as an exact time to reveal your true feelings about a particular emotionally charged subject. And there are also times when your truth is not necessarily another person's truth. But always holding back from saying what you feel to be true can have the negative-energy consequence of blocking the normal flow of subtle energy through your throat chakra. Temper this thought with a reminder, though. What we say when we express ourselves can have uplifting as well as negative influences upon others, depending upon our choice of words. Just as "speaking one's truth" can have healing influences upon the throat center, the use of the voice to continually express criticism or negative gossip or to express harsh judgments of other people can sometimes produce deleterious effects upon the throat chakra.

Chakra dysfunction of the throat center does not necessarily produce illness directly. But the malfunctioning of fifth-center energy flow is thought to create a weakness in the anatomical structures that are energetically nourished by this chakra. Chronic subtle-energy starvation creates a weakness in the cells and tissues of the body, synergistically adding to other disease-producing factors to create diseases within the stressed chakra-fed region of the body. Constriction of energy flow through the throat center affects the thyroid gland, the larynx, and other anatomical structures within the neck, sometimes contributing to chronic hoarseness and recurrent sore throats, especially when other predisposing factors such as cigarette-smoking are present. As previously mentioned, throat-chakra blockage is a hidden energetic factor that may predispose an individual to developing hypothyroidism (an underactive thyroid gland). Conversely, excessive energy buildup in the throat center could potentially lead to an overactive thyroid condition known as hyperthyroidism, in which there is a continual overproduction of thyroid hormone. Other physical illnesses that might result from throat-chakra dysfunction include recurrent cold sores and sore throats, laryngitis, gum disease, and even TMJ or temperomandibular-joint syndrome, which can produce jaw pain and recurrent headaches. Since the throat center is the center of will, a "weakness of will" may lead to various types of substance-abuse problems, including addictions to drugs, alcohol, food, and even gambling. The key to balancing the energies of the throat center is in learning to make more conscious choices in our lives and to live according to the right use of will. A disciplined will can help us make better dietary choices, follow daily exercise regimens, practice regular meditation, and direct our thinking in ways that will attract greater positive energies into our lives.

Sixth Chakra

The sixth chakra or brow center is said to be the energy center where we process issues primarily involving the mind. The brow or "third-eye" chakra is the seat of the intellect as well as the energy center having the most to do with the power of the mind. The challenge of the sixth chakra is to become aware of the power of consciousness itself and of the reality we each create for ourselves through our thoughts and belief systems. That we each have the power to create and influence our own reality is a physical as well as a spiritual truth. Our thoughts and our physical actions have the power to shape our lives and the lives of those around us. The power of our thoughts and beliefs can also be a potent trigger for various self-healing systems of the body. The ability to use thought to change physical reality is demonstrated by the healing power of the so-called placebo effect. If people believe that a pill they ingest is imbued with powerful healing properties, healing is more likely to occur, whether or not there is any actual medicine in the pill. The simple belief in the healing effects of a substance has enough therapeutic value to produce significant healing in as much as 30 percent of the general population. You see, the power of the placebo really rests with our thoughts and our beliefs and their ability to trigger the self-healing mechanisms inherent within our minds, which are in turn intimately linked with our physical bodies. Because of the powerful mind/body connection, the mere belief that a pill will cure you can actually bring about healing changes within your physical body.

Unfortunately, most of us have belief systems that limit our perceptions in certain ways. Our belief systems act as a filter to our consciousness and allow us to perceive only those things that fit into the narrow range of what we believe to be possible and acceptable phenomena in the external world. But fixed belief systems and thought patterns sometimes need to be modified when we are confronted with changes and challenges in our personal and professional lives. Throughout life, we will find ourselves in situations where our belief systems are strongly challenged. Confrontational situations sometimes force us to reexamine whether our beliefs need to be modified or updated in order to accommodate new information and new experiences. In reality, it is the energy of the soul and the actions of the higher self that place us in difficult situations in order to stimulate our spiritual growth and development. Even though we may not always be conscious of the hidden spiritual agendas behind challenging learning experiences, they contain at their core symbolically coded messages from our higher selves that await our conscious decoding so that we may understand their hidden meaning. As we learn to embrace new ideas and experiences with greater openness and awareness, without preconceptions, the resulting shift in consciousness can eventually lead to a clearing and opening-up of energy flow through the sixth or third eye chakra.

According to medical intuitive Carolyn Myss, the sixth chakra may become blocked when we lack a sense of "vision" or when we hold ourselves back from seeing certain aspects of our lives. It is as if there are certain times in our lives when we do not allow ourselves to see, either because things are too painful or awkward or because they do not fit into our family, cultural, or scientific belief systems. A key issue of the sixth chakra is our ability to look truthfully at our lives, to evaluate our flaws and failures, and to acknowledge our strengths and accomplishments. Many people see the world only from their self-centered perspective of how events affect them personally. They often become blind to the effects of their own actions upon others, as long as such actions serve to further specific accomplishments or to help them attain personal goals. In personal and business relationships, some people may actually be keeping themselves from really "seeing" how their actions manipulate other people and create negative effects for those around them. Spiritual blindness to the consequences of one's actions is an important cause of energy blockages in the sixth chakra. Anything that keeps us from closely examining our lives with truthful introspection can lead to dysfunction of the sixth chakra.

The sixth chakra is thought to have an energetic connection to the eyes and ears. Energy blockages in this chakra may lead to diseases of the eyes that produce progressive blindness, as well as deafness caused by deterioration of the nerves of the ear. The metaphoric message behind sixth-chakra dysfunction is "What is it that I am keeping myself from seeing or hearing?" Since the sixth chakra also feeds subtle energy to the brain and central nervous system, energy dysfunction here can sometimes contribute to neurological disorders. As such, the sixth chakra is considered the energy center that is most influenced by the mind and the intellect. One of the key relationships of the sixth chakra is the relationship between the mind and the emotions. It is said that our sixth chakra may become impaired when we relate to others and the world around us only through our emotions while entirely ignoring what our minds have to say. While following our hearts can usually lead us to better solutions for problems and challenges, we cannot always be swayed by our sexual passions (second chakra) instead of our thinking selves (sixth chakra), or we may occasionally find ourselves in trouble. Our emotions can also override our intellect when we allow ourselves to act out of anger, rage, or jealousy. If we have a tendency toward rash behavior as part of a habitual pattern, continually blocking the mind and following only our emotions can eventually lead to problems with energy flow in our sixth chakra. Yet when we do learn the lessons of our sixth chakra, we may become privy to the world that exists beyond the physical plane, the multicolored universe that clairvoyants can see and directly experience. When our sixth chakra fully opens, then auras, subtle energies, and the invisible spiritual world surrounding and interpenetrating our physical world are said to become visible to us.

The sixth chakra is also known more familiarly as the third eye. The term appropriately fits this chakra, which is considered the seat of intuition, inner vision, and clairvoyance (from the French, which literally means "clear seeing"). Chakra theory suggests that as we start to mature spiritually and allow ourselves to become more open to new ideas and information about ourselves and the world we live in, we will also begin to make a spiritual shift in consciousness. This shift may eventually allow us to see beyond the visible physical world into the higher vibrational world of normally invisible subtle energies that surround and interpentrate our physical bodies (such as the energies of our chakras and our auric fields). Clairvoyant medical intuitives like Carolyn Myss and Mona Lisa Schultz have demonstrated high degrees of accuracy in diagnosing physical and emotional problems (even at great distances) through just such third-eye-chakra insightfulness.

Seventh Chakra

Finally, the seventh or crown chakra has been esoterically linked with issues relating to our faith, our sense of spirituality, and our search for personal meaning in the universe. Chakra theory states that the seventh chakra emerges from the top of the head. It is the energy center that connects us with our search for the divine essence in each of us and with our personal exploration of our relationship with God. At the seventh-chakra level, we begin to move into the eternal relationship between our conscious personality and our soul. Each human being has an innate need to connect with the divine creative energy of God, whether or not the connection is of a conscious or unconscious nature. Even atheists make up a part of the great universal-energy matrix that is the sum total of creation, which in turn is really formed from the "body of God" (sometimes referred to as "All That Is"). As such, we are not composed simply of molecules born in the solar furnaces of stars, so-called starstuff. We are also composed of "Godstuff." Energetically speaking, all matter and all beings that make up the universe are "holographically connected" to God, regardless of their conscious acknowledgment of this divine relationship. There is a part of us, our divine essence if you will, that at times seeks to be reunited with the energy of the Creator. When we block that divine urge, we also block the natural flow of subtle energy through the seventh chakra.

Sooner or later, depending upon the circumstances, all people ask questions about the meaning of their lives. Many times, such questions of higher meaning come only at the end of life or during a potentially life-threatening illness. Those who rely solely on science as their religion, choosing to look only at hard data and physical measurements of the universe, might deny that God plays any role in their lives. But in spite of following a scientific creed of life, all individuals still need to acknowledge their relationship with God at the spiritual fabric of their being sometime during their earthly existence. It is a basic human need, regardless of what some may think. All humans need to experience sacred moments during which they can directly connect with the divine energies of creation, whether that connection is expressed through religious ceremonies or merely in personal prayers. As we begin to ask questions about why we were born and what the meaning of our lives is, our seventh chakras begin to activate the connection to our higher selves or souls. Sometimes the higher self seeks to provide us with answers to our questions by producing dreams that are symbolically connected to the deeper issues of our life's meaning. At other times, the spiritual search for meaning triggers a response from the higher self that forces us to focus intensively upon these spiritual truths and deeper meanings we often desperately seek. The seventh chakra is also considered the seat of our spiritual conscience. In other words, here is where we find that small inner voice that is too often ignored. If we totally block out thoughts about the meaning of life, intense feelings of anxiety and depression, as well as an overwhelming sense of heaviness and fatigue, can sometimes manifest. We in a sense can "come down with" a kind of spiritually triggered malaise caused by consciously tuning out the inner wisdom of the higher self, refusing to look at the deeper meaning of our lives. In seventh-chakra-blocked individuals, there seems to be a schism between the ego (the conscious personality) and the true soul essence trying to communicate with the ego. In extreme cases, some people may experience a "mystical depression" that is associated with a sense of overpowering emptiness and lack of fulfillment in life. As clairvoyant Carolyn Myss has noted, mystical depression is often characterized by a psychological state referred to as the "long dark night of the soul." During this long dark night, an individual becomes incessantly driven by burning, soul-searching questions about life's meaning, which keeps that person in a cloud of temporary spiritual madness. Such spiritual madness can literally force people into an inner journey to ponder the deeper meaning of existence and try to make sense of the jumble of their life's path. This deep inner search for meaning becomes activated as an inner test of one's faith in the energies of the divine and is often a time of personal isolation and inner searching that can result in true spiritual transformation.

Another milder form of seventh-chakra imbalance causes people to experience a form of what is known as "divine discontent." They might feel that their chosen profession is no longer fulfilling or challenging, no longer causing their "soul to sing." Mystical depression and divine discontent may have as their underlying basis a feeling of being "disconnected from God" as well as a sense of being on a path that is at odds with one's true "soul purpose" in life. This sense of disconnection is a common thread in many seventh-chakra-based disorders, which may serve as a cosmic trigger to help people to waken from a "spiritual sleep" and reconnect with the soul, as well as to develop a greater awareness of their connectedness with all life.

The seventh chakra is also the site of what Carolyn Myss refers to as our "grace bank account" or "cellular bank account," where we store the energy generated by our prayers and our positive acts of spiritual service and kindness toward others. As we begin to "practice what we preach" in terms of acting out of love instead of fear, and as we start to follow the core tenets of the world's major spiritual teachings, performing little random acts of kindness in our day-to-day lives, we gradually generate positive karmic energy. This karmic energy is deposited into our seventh-chakra energy bank account, where both positive and negative karmic energies are stored. The seventh chakra is also connected with emotional and spiritual issues of service to others, humanitarianism, courage, faith, and devotion, as well as issues relating to personal ethics. When we donate our time for volunteerism and perform acts of kindness and service that benefit others, more positive karmic energy is generated. This positive karma is actually an energy of "grace," which some consider "karmic brownie points." The energy of grace can help us to grow and to thrive with fewer obstacles in our path, assisting us in the greater spiritual evolutionary process of learning shared by all human souls who attend the Earth School of Life.

The positive energy charge created by our loving acts of kindness and grace can actually attract to us wonderful situations and opportunities that almost seem too good to be true. This is partly because the higher self uses the energy from our grace bank account to "finance" little miracles and incredible opportunities that often seem to come out of nowhere. The higher self frequently works through the synchronicity of strange coincidences that place us in the right place at the right time. As we give of ourselves in love and service toward others, acting with morals, ethics, truth, and honor in our personal and business relationships, we generate more of this positive energy of grace stored in our seventh chakra. Conversely, negative or harmful actions toward people, animals, or even the environment can generate negative karma that may ultimately depelete the energy level in our grace bank account.

When significant levels of energy depletion occur within the seventh chakra, the spiritual and biological energy reserves needed to maintain the cells of the body may drop to critical levels and result in a variety of physical and emotional problems. While blockages in the lower chakras tend to cause localized problems in specific organ systems, seventh-chakra energy problems frequently affect the entire body. Seventh-chakra blockages can sometimes be purely energetic in nature and thus do not result in a specific physiologic imbalance. But at times a seventh-chakra dysfunction may lead to a form of chronic exhaustion and overwhelming fatigue that seems to lack any apparent physical basis. Certain cases of fibromyalgia and chronic fatigue syndrome might actually be due to problems with the seventh chakra.

The "Eighth" Chakra

The question of life's deeper meaning frequently goes unanswered when we view the events of our lives from a literal perspective. The key to perceiving the hidden meaning behind life's events and in understanding our soul's journey requires a leap in perception to the symbolic level of the eighth chakra. According to clairvoyant investigations, the eighth or transpersonal chakra sits directly above the head and acts as a connecting link between the soul and the ego or conscious personality. The soul or higher self views life from the causal-plane perspective. The causal plane is a higher-dimensional level of reality where cause and effect from the spiritual viewpoint of the soul can be readily perceived. From this higher-dimensional level, the soul attempts to communicate with the ego through the use of meaningful symbols. Symbols are the native language of the right cerebral hemisphere. Our right brain represents our intuitive, artistic side. Normally, our right brain functions quietly in the background, while our left brain's linear, analytical perspective of reality defines the "superficial meaning" of events to us. This typical linear interpretation of everyday events often leaves us searching for deeper meaning. However, the deeper spiritual meaning behind the events in our lives can be fully comprehended only if we drop our analytical, left-brain view of the world and shift into a more right-brain, symbolic perspective on our lives. When we start to develop what Carolyn Myss calls "symbolic vision," we can more easily comprehend the "hidden agenda of the soul." This spiritual agenda holds the key to understanding why various positive and negative events seem to occur "randomly" at certain critical moments in our lives. If we choose to see life from only the literal perspective of the logical, analytical mind, these strange, random events and coincidences that happen to us throughout our lives seem to occur without rhyme or reason. But when these same events are examined from the symbolic perspective of the eighth chakra, the soul's view of the world, we begin to understand that there are reasons and purposes for everything that happens to us. We begin to understand how various situations and relationships may actually be a series of learning experiences or "spiritual tests" that our soul carefully "arranges" for us to undergo during specific times in our lives.

In Summary: The Energetics of Illness and Wellness in the Multidimensional Human

We humans are unique energetic beings. Ours lives are sustained by a delicate balance of different forms of energy that we assimilate in order to sustain our physical bodies. The most primary form of energy is the metabolic energy extracted from the foods we eat and the fluids we drink. If our food or water supply is nutritionally lacking or is chemically contaminated, our ability to make use of this energetic fuel may be diminished. Similarly, we need unpolluted air to provide the oxygen necessary for the transformation and utilization of metabolic energy extracted from the foods we consume. Increasing care and discrimination must be exercised when selecting quality water and nutritionally balanced foods for our diets in order to maintain proper health. Besides the potential contamination of our metabolic-energy supplies, there are also other energies we human beings can be affected by. We are constantly bathed in an invisible sea of different frequencies of energy. Our energetic environment includes the electromagnetic fields produced by our TV sets, microwave ovens, and electrical power-distribution systems, as well as the magnetic and subtle-energy fields coming from the earth itself. All these electromagnetic fields have the potential to influence our health either positively or negatively. Aside from the health risks from man-made energy fields, researchers may have to start paying closer attention to the negative health consequences of geopathic stress, too. European scientists are finding mounting evidence of a possible connection between geopathic stress and cancer, especially in regional clusterings of cancer cases that cannot be explained by exposure to environmental carcinogens. It is becoming increasingly clear that future architects and designers will have to look more closely not only at the research on geopathic stress but also at the benefits of feng shui architectural consulting in home and business planning, in order to promote greater health and productivity in energetically supportive environments.

The earth-energy connection is a part of the larger spectrum of environmental energies that can contribute to either health or illness. While strong electromagnetic fields and "noxious earth energies" can sometimes play a role in human illness, there are other environmental energies that we need just as much as food and water in order to maintain health. The vital life energies of prana, possibly carried by the sunlight that bathes our planet, and of ch'i are required by all human beings in order to function properly. When the flow of these vital life energies throughout the body becomes impaired, disease can be the ultimate result. While nutrient-rich food, clean water, and unpolluted air may be important physical-energy inputs needed to sustain life, a strong and balanced flow of vital life energies flowing through the meridians, the chakras, and the etheric and spiritual bodies are just as crucial for achieving and maintaining optimal health and wellness. Blockages in the meridians and chakras can create imbalances in the flow of vital energy to the organs of the body, contributing to the development of diseases in specific organ systems. By recognizing the importance of ch'i-energy flow and meridian imbalances as hidden causes of illness, vibrational-medicine researchers have gradually developed new ways of diagnosing illness as well as a new respect for "older ways" of treating diseases in the body. Once an area of chakra imbalance or meridian dysfunction is identified, the treatments of acupuncture, homeopathy, or even flower essences can be prescribed to restore balance in the affected energy systems. The vibrational medicine of the future will rely on technologies and devices that health-care professionals will use to objectively measure the body's meridian-energy balance as a means of diagnosing potential diseases while they are still at an early energetic stage. It may eventually become possible to create the ultimate preventive medicine, which will correct energy imbalances in the body before overt illness ever develops.

Along with a proper balance of ch'i flowing through the body's acupuncture-meridian system, an adequate supply of pranic energy through the chakras is just as important. The chakras also form a major part of our link with the soul; so when an individual's chakras are blocked, it may reflect difficulties with key psychological and spiritual issues to which the individual may not be giving adequate attention. Chakra-linked issues are intimately tied to the functionality of our relationships with ourselves, our family members, our loved ones, our co-workers, our social and ethnic groups, and even with the planet on which we live. Our attitudes and emotional reactions toward life's events play an important role in whether we will be vibrantly healthy or chronically ill. The chakras might be considered a kind of emotional and spiritual feedback system between the soul and the conscious personality. Exactly "where" an illness occurs in the body is often a symbolic indicator of which particular psychospiritual issues the individual needs to direct attention to in order to achieve inner rebalancing and a recovery of health.

Vibrational medicine looks at illness and wellness from these many different levels of energy dysfunction that can affect the multidimensional human being. The road to wellness is about more than proper nutrition and exercise. It is about our total energy environment, which includes the emotional-energy environment we create through our consciousness and our attitudes toward people and events in our lives. That energy environment also includes the spiritual energies of the soul that sustain us through the high and low points of our lives. In the eyes of the immortal soul, there are no absolutely good or bad events. All experiences are relative to the journey of the soul and its search to know itself better by partaking of life upon the physical plane. To the soul, all of life's situations, positive or negative, are merely different types of learning experiences chosen by the higher self to provide challenges to the conscious personality or ego as it travels the path to spiritual awakening and enlightenment, the true goal of life. Vibrational medicine is one of the few approaches that addresses not only the physical dimension of stress and illness but also the higher spiritual aspects of illness as they apply to the soul and spiritual bodies.

Within vibrational medicine there are many different therapeutic approaches that can help us to achieve physical healing, positive shifts in our conscious awareness, and a rebalancing of our whole being at the higher spiritual levels of our existence. One of the keys to finding balance in a stressful world may be found in the potential of our consciousness to steer a calm course through troubled and turbulent waters. Our filter of consciousness determines whether we perceive the world around us as a threatening place or as a spiritual universe with hidden soul agendas and learning experiences waiting to be discovered. Our consciousness and our attitudes affect how we respond to each new stressful event and can strongly influence whether we are able to achieve and maintain a state of wellness. Through a variety of biochemical, neurological, and energetic connections, our consciousness and our emotions can powerfully affect our health in both positive and negative ways.

In the next chapter, we will begin to examine some of the ways we can use our consciousness to relieve stress, to perceive the world differently, and to assist the body in achieving an optimal state of wellness. We will look at a variety of physical, emotional, and spiritual techniques everyone can use to stay healthy and to find balance, peace, and harmony in an increasingly stressful world.

Three

THE ROAD TO HEALTH AND WELLNESS: A MULTIDIMENSIONAL APPROACH

IN THE LAST chapter we explored different internal and external chemical and energetic stresses that can potentially throw us out of balance, shifting us from health into states of disease. In view of these many stresses that can negatively affect us, we need to consider the process of achieving optimal health from the larger multidimensional perspective of the extended human being. This means we need to pay attention not only to the needs of the physical body but also to the energy "requirements" of the chakras, the meridians, and the human spiritual bodies as well. There is a diverse variety of techniques anyone can use to achieve optimal wellness. While most people are familiar with physical exercise programs that help to tone the muscles of our bodies, they may not be aware that there are also spiritual exercise programs that can improve the functioning of the meridians and chakras. We will examine a number of these different approaches according to how each can positively influence the physical and subtle-energy systems of the multidimensional human being.

Taking Care of the Physical Body

Let's first address a person's nutritional needs. As alluded to in the last chapter, our bodies are dependent upon a high-quality source of metabolic fuel to power our cells, our organs, and our nervous system. Our basic nutritional needs must be met on a daily basis. Many in the health profession claim that eating a balanced proportion of foods from each major food group is enough to supply all the vital nutrients needed for optimal health. However, even if this were partly true at one time in our history, we now eat different types of foods than our great grandparents ate. Older farming techniques are being replaced by large agribusinesses that grow crops in mass quantities. Many chemical insecticides are routinely applied for crop pest control. Also, much of the food we consume is prepackaged. And many "fast" foods are overly refined. Nutrients that exist naturally in raw foods are being removed by progressive refinement and processing techniques. Even the morning breakfast cereal now has to be "fortified with vitamins and minerals" to make up for the nutrients that have largely been removed from natural grains during the manufacturing process.

The question of what constitutes adequate versus optimal nutrition remains a matter of open debate. However, a number of things have to be factored into the nutrition equation. The current medical establishment has created a list, known as the RDA (Recommended Daily Allowance), that suggests how much of each vitamin and nutrient is necessary for maintaining health. But the quantities of vitamins listed in the RDA are not the amounts necessary to achieve optimal health. The original RDA dosages for vitamins were based upon how much of each individual vitamin was needed to prevent a vitamin-deficiency disorder like scurvy, pellagra, or rickets from occurring. What the original RDA did not take into account was the influence of an increasingly stressful and polluted world.

Adding to the debate about the adequacy of the RDA is research into the biomolecular nature of aging. Many studies suggest that the continual formation of so-called free radicals in the body—highly unstable chemicals generated by a biochemical process known as oxidation—play a large part in premature cell death and rapid aging. Much of the same biochemical research on aging has also demonstrated that the use of antioxidant nutrients like vitamin C, vitamin E, and selenium is a must to help minimize free-radical-related cell damage. Recent investigations into the use of B vitamins during pregnancy has revealed a lower incidence of neurological birth defects in infants born to women taking folic acid compared with babies whose mothers did not take the vitamin supplement. In fact, research evidence was so strong that the FDA recently readjusted upward the RDA dose of folic acid for pregnant women. Additional nutritional research into heart disease and B vitamins has discovered a higher incidence of coronary-artery disease in individuals whose blood had lower levels of vitamin B6, B12, and folic acid. What this new research shows us is that our cutting-edge knowledge about nutrition is continually changing. There are growing numbers of "nutritionally aware" health-care professionals who now advocate treating illnesses by using not only medications and supplements but also diets containing high amounts of various "healing foods" like fruits, vegetables, and "live" sprouts.

We live in an age in which vitamin supplements are readily available at the health-food store, the drugstore, and even the local supermarket. People are bombarded with television advertisements for various herbal and nutritional supplements. In the midst of all this information and easy access, it is often difficult for people to know what is the right combination and dosage of nutrients they should take. On the one hand is the policy of the orthodox medical establishment, which suggests that you don't need to take any vitamins or supplements if you eat a balanced diet. On the other hand there is emerging nutritional research on heart disease, birth defects, early osteoporosis, and other disorders suggesting that vitamin supplementation may indeed be beneficial in preventing the eventual development of various illnesses. For many people, a daily intake of a good high-potency multivitamin containing minerals and trace elements as well as 1,000 to 2,000 milligrams of vitamin C may be adequate if they do eat a so-called balanced diet. For individuals with a family history of heart disease or cancer, or for those people seeking to retard premature aging caused by free radicals, a slightly larger list of daily supplements may be beneficial. For instance, an additional 400 to 800 International Units (IU) of supplementary vitamin E, 200 micrograms of selenium, 25 to 50 milligrams of vitamin B6, 1 milligram of folic acid, and 25,000 IU of beta-carotene provide increased antioxidant protection.

With regard to eating a balanced diet of fruits, vegetables, whole grains, meat, and dairy products, it is always advisable to purchase food from high-quality sources. Organically grown meats and grains are preferred when available. Washing fruits and vegetables to remove pesticides and other chemicals is a must. Also, eating fruits and vegetables in an uncooked state may help to preserve the natural enzymes and vital energy contained within the food. There is evidence to suggest that heating and microwaving of foods can destroy many of the naturally occurring vitamins. Cooking can also produce various chemical by-products that may be harmful to human health. Charring burgers on the backyard barbecue grill, for example, has been shown to create small amounts of potentially toxic carcinogens in the meat. The heating of oils, especially those high in saturated fats, can also produce toxic by-products that could be unhealthy to consume over a period of many years. High-fat diets in general have been associated with higher rates of colon cancer, breast cancer, and heart disease. Most doctors recommend low-fat diets to maintain optimal health and promote increased longevity.

Presuming that one is eating a healthy diet, another critical factor in maintaining good health of the physical body is daily exercise. We live in a fast-paced world with many time demands. Physical exercise is difficult for most people to fit into their busy schedules. However, physical conditioning has many important health benefits. Regular vigorous exercise has been shown to raise the so-called good HDL cholesterol in the bloodstream (which is said to protect against heart disease). Also, regular exercise increases bone density and is an important factor in preventing or minimizing premature osteoporosis. In elderly individuals, exercise programs have been associated with modest improvements in health and longevity. Aerobic exercise has obvious benefits to the heart by helping to improve collateral circulation in the heart muscle and raising one's level of cardiovascular fitness. Regular physical exercise also helps to combat stress and depression by boosting levels of brain neurochemicals that elevate mood. Exercise tends to improve the quality of restful sleep as well. At a more basic level, regular exercise stimulates the body's fat metabolism to aid in preventing obesity.

There is no single "best" exercise activity. In general, it is good to participate in activities that not only tone muscle but also temporarily raise the heart rate (i.e., aerobic exercise). Running, tennis, swimming, and competitive sports are pleasurable social exercises. Also in recent years there has been a proliferation of "fitness machines," from treadmills to specialized whole-body workout systems, which people can also use in their efforts to achieve physical fitness and optimal health.

Taking Care of the Spiritual Bodies Along with the Physical

In addition to conventional physical exercises, there are a variety of other exercise programs that benefit the physical body as well as the subtle-energy systems supporting the physical body. Yoga is a discipline of the physical body, the mind, and the spiritual bodies. There are actually many different forms of yoga. To most people, yoga means sitting in a cross-legged position while meditating. The most commonly practiced form of yoga, known as hatha yoga, consists of certain body postures—known as asanas—breathing techniques, and specific forms of concentration and meditation. These body postures and breathing exercises produce muscle toning and benefical physiological changes in the physical body. Scientific research in India and in the United States has suggested that the regular practice of certain yoga postures may actually help to stimulate the body's endocrine glands and metabolism. In fact, a number of diabetic patients in yoga research projects have practiced specialized forms of yoga designed to help with pancreatic functioning. It worked so well that some patients were actually able to reduce their daily dosage of diabetic medication. Furthermore, the regular practice of yoga can increase physical endurance, strengthen resistance to illness, ease stress and tension, improve control over the body, help to regulate weight, and aid in training and focusing the mind.

Certain yoga techniques also help release trapped life-force energies in the body and move subtle energies through the chakras as well. There are specific forms of yoga, especially kundalini yoga, developed primarily to move energy through the chakras (from the lowest chakra all the way up to the crown center). In yoga, one is taught about a form of "sleeping" energy known as the "kundalini" that remains stored in the root chakra at the base of the spine. "Kundalini" is a Sanskrit word meaning "the coiled serpent." Yogic texts describe kundalini energy as the sleeping serpent fire that when awakened can lead to enlightenment and "higher powers." In fact, the medical symbol of the caduceus, represented by twin serpents coiled around a winged staff, might be an esoteric symbolic representation of the serpentlike kundalini energy as it moves up the seven chakras. The consistent practice of kundalini yoga, along with other spiritual and meditative practices, is said to awaken the "sleeping serpent." To experience an awakening of kundalini energy is to feel an unbelievable surge of energy move from the root chakra up the spinal column to the crown chakra. When kundalini energy reaches the crown center, it is said to activate higher mental abilities and to bring about a state of enlightened awareness. The practice of kundalini yoga might be suitable for spiritual aspirants who are prepared for a very rigorous journey of self-exploration. But the regular practice of hatha yoga (the more familiar form of yoga known to Westerners) is likely to be better suited for most people as a form of daily exercise.

Unlike typical calisthenics, hatha yoga begins with quieting the mind and emotions. Yoga exercises are performed slowly, deliberately, with poise, rhythm, and concentration. There are few repetitions of yoga exercises in a single session. Yoga exercises involve holding a body position for a brief interval while the mind is fully focused on what the body is feeling. People who practice yoga may develop a regimen of basic yogic postures that they will practice sequentially on a daily basis. Besides the postures, there are a variety of yogic breathing techniques such as alternate-nostril breathing and "pranayama" or deep-breathing exercises. Such controlled rhythmic breathing may help to balance the flow of prana or life-force energy throughout the body, while improving oxygen delivery to the tissues of the body as well. Many excellent books, videotapes, and classes on yoga are available. Some easy-to-follow videotapes are listed at the end of the book. Suffice it to say, hatha yoga, as a daily practice, is an excellent regimen for promoting the health of the mind, the body, and the spiritual-energy systems of the body, especially the chakras. Generally speaking, it is often beneficial to end a session of hatha yoga (or any other form of exercise) with a brief cooling-down period using either meditation or a progressive relaxation technique. It should be added that serious students of yoga also subscribe to specific dietary practices, a spiritual philosophy, and a unique lifestyle and mindstyle. For those people who wish to pursue yoga beyond the classic hatha-yoga asanas, a long-term study of the various forms of yoga can be extremely helpful in achieving and maintaining optimal physical and spiritual health.

Another form of daily exercise beneficial to the flow of subtle energy throughout the body is tai ch'i chuan, also known simply as tai ch'i. Like yoga, tai ch'i is centuries old, originating in the ancient Chinese Taoist philosophy. It is a form of meditation that tends to emphasize slow, gentle movement, relaxation, and a calmness of the mind. The movements are continuous, light, rhythmic and graceful. The regular practice of tai ch'i movements can help people develop a greater sense of inner and outer balance, assist in moving subtle energy through the body, and may possibly even slow down the aging process. Recent studies on the benefits of teaching senior citizens tai ch'i exercises found that elderly tai ch'i practitioners developed a steadier sense of balance and movement and had fewer falls than other, "nonpracticing" seniors.

Chi gong, a form of physical and spiritual discipline originating in China, has also seen increasing popularity in the West over the last twenty years. As the "ch'i" in tai ch'i and chi gong implies, these exercises are intended to help move ch'i energy throughout the body in healthful ways. Like yoga, the practice of chi gong (sometimes spelled as qigong and ch'i kung) is an entire system of self-healing exercises and meditation that includes healing postures, special movement exercises, breathing techniques, specific forms of meditation, and self-massage. While yoga is oriented toward encouraging the healthy flow of pranic life energies through the physical and spiritual bodies and the chakras, chi gong focuses on the healthy flow of ch'i energy through the physical body and the meridians. Through the various exercises and methods of chi gong, ch'i energy is accumulated and stored in the body as in a reservoir or charging the body's "bioelectrical batteries." In ch'i gong philosophy, "impure" or "polluted ch'i" (considered the essence of disease) can be cleansed and refined into pure, healing ch'i energy in order to restore health to the body. Some of the breathing exercises of chi gong are intended to discharge and eliminate impure ch'i from the body, just as one would exhale carbon dioxide, one of the end products of normal metabolism.

There are a variety of chi gong categories or applications. "Medical" chi gong is primarily oriented toward self-healing as well as healing others. Individuals highly skilled in chi gong are known as "chi gong masters." Chi gong masters can consciously circulate the flow of healthy chi energy through their meridians and send it out their hands to a sick individual, even at a distance of several feet away. They can supply ch'i energy to areas of a sick person's body that may be low in ch'i energy. Conversely, they can also draw excess ch'i energy away from body regions that are "oversupplied." Such healing is known as "external ch'i healing" and is considered a branch of medical chi gong. Other forms of chi gong include "meditative or spiritual chi gong," which is used to achieve inner tranquillity of thought and deeper awareness of self, and "martial ch'i gong," a kung fu–like application, considered one of the Chinese martial arts. There is even a category sometimes referred to as "business chi gong" in which management encourages employees to practice daily chi gong exercises at the beginning of the workday so that they can achieve greater focus in their work, improve productivity, relieve stress, and maintain better health.

Chi gong is designed to direct the flow of ch'i energy throughout the body in a more balanced and healthful manner. The regular practice of chi gong exercises and self-massage techniques may be helpful in strengthening the immune system and in fortifying the body's vital life-energy reserves to combat many different types of illnesses. Certain types of chi gong exercises can be especially useful in preventing or lessening the pain and stiffness of many different forms of arthritis. While a daily twenty-minute regimen of chi gong may be useful for maintaining bodily health, the practice of various chi gong exercises, breathing techniques, and meditative approaches can also be of help in achieving greater mental and spiritual focus as well.

In chi gong, yoga, and a number of other physical and spiritual disciplines, there is an emphasis on specialized breathing exercises that have been developed to help move energy through the body and to clear the mind. Many people tend to "hypoventilate," or take very shallow breaths, never fully inflating their lungs. Breathing is important to life because it allows us to take in life-giving oxygen and prana. Breathing is also a method by which we excrete toxic waste products such as carbon dioxide. When we are under great physical and emotional stress, we tend to unconsciously take shallow breaths. Regular deep breathing helps to relieve stress and focus the mind toward a state of deep relaxation. Unless breathing is slow and rhythmic, it becomes very difficult to relax either physically or mentally. Just sitting quietly and taking slow deep breaths can be extremely relaxing. A simple deep-breathing exercise can be done while lying on the floor, with knees bent, hands resting on either side of the rib cage. Inhale slowly to a slow count of three, feeling the hands at the sides of the rib cage move outward as the lungs expand. When the lungs are filled, exhale slowly to a count of four, applying slight pressure through the hands to the rib cage at the end of the breath to fully empty the lungs. Alternately, the hands can be placed over the abdomen. Breathe in slowly to a count of three, the hands moving upward as the abdomen rises. Following inhalation, exhale to a count of four, pressing the hands on the abdomen gently down toward the floor to complete emptying of the lungs. Practicing this simple exercise once or twice a day for five minutes each time can be quite relaxing. After completing this breathing exercise, it is important to rest a few minutes before rising to avoid minor dizziness that can sometimes happen after deep breathing.

The different exercises, from simple physical exercise to yoga to chi gong, all deliver benefits to the physical and to the subtle or spiritual bodies. While there are no specific exercises to "tone up" the etheric body, following a regimen of good nutrition, proper breathing, and adequate consumption of clean and uncontaminated water, along with the judicious use of vitamins and supplements, can help to strengthen the etheric body. The astral and mental bodies are less affected by what we do physically than by what we think and feel. In other words, our emotional and mental lifestyles and our very consciousness have a powerful effect upon the health of our higher spiritual bodies. Therefore, we shall now focus on how one can shift or train one's consciousness to achieve mental and emotional states associated with deep relaxation, improved emotional balance, greater focus, and increased mental clarity.

The Role of Consciousness in Health and Illness: A Guide to Tools for Transformation

There are a variety of techniques people can use to achieve states of consciousness associated with better health, well-being, and creative expression. This is of great importance because our consciousness strongly influences our health in many subtle ways. In the last chapter we discussed how our chakras are affected by our consciousness and our emotions in ways that may ultimately lead to illness. Dysfunctional emotions and erroneous patterns of perceiving the world can lead to blockages in the flow of subtle energy through our chakras. The resulting areas of "life-energy starvation" and weakness can make our bodies more susceptible to illness from a variety of biological, chemical, and energetic stresses.

Our perceptions of the world and our emotional reactions to events are influenced by our experiences and by the way our brains are organized. People talk about seeing the world from a left-brain or a right-brain perspective. The concept of left-brain/right-brain orientation is based upon brain research studying the differences between the two cerebral hemispheres that make up the largest part of the human brain. Each hemisphere seems to have its own unique perceptual and cognitive style. The left cerebral hemisphere, or left brain, perceives the world through the language of logical, deductive reasoning while analyzing the world in terms of linear sequences of cause and effect. Our left brain is the hemisphere that participates in speech, language, and writing. The loss of such skills can be readily seen in individuals who have suffered a stroke that causes damage to the left side of the brain. Left-hemispheric stroke patients will often lose the ability to communicate verbally. The left brain is also the side that uses numbers and mathematics. With regard to music, the left hemisphere possesses the skill to organize musical notes on paper and analyze musical rhythms and cadences.

Conversely, the right cerebral hemisphere, or right brain, is the side where music appreciation takes place. Whereas the left brain analyzes musical rhythms and instruments used in the performance, the right brain is more concerned with the pure enjoyment of music. While the right brain takes in the whole picture as a single visual gestalt, the left brain looks at a picture and analyzes it in terms of individual elements and details. In some ways this situation is analogous to the old saying "You can't see the forest for the trees." As the right brain sees the forest, the left brain often misses the forest because it gets lost in the details of the individual trees and shrubs. The right hemisphere represents the artist who is concerned more with overall aesthetics than with minuscule details. In contrast, the left hemisphere might be considered the site of the analytical scientist in each of us. The analytical left hemisphere tends to be verbal, while the right hemisphere is primarily visual. Another fascinating aspect of the dual-hemisphere model is that the right brain perceives the world in terms of symbols and symbolic meanings. Dreams, which are often symbolic in nature, are constructed using the symbolic language of the right brain. But the left brain interprets the world quite literally. The left cerebral hemisphere is also the part of our brain that seems to know that it is thinking, as if it were our conscious self. Conversely, the right brain seems to be more connected to our unconscious self. Interestingly enough, there is evidence to suggest that our higher selves are more directly connected or "hardwired" to our right brains than to our left brains. This observation relates to the fact that the higher self often tries to communicate with the conscious personality or ego using the language of symbols, whether dream symbols or symbolic events in our waking lives. Paradoxically, the analytical skills of the left brain are needed to help translate the symbolic language of the right-brain messages received in dream state so that the conscious ego can understand the hidden message. In a sense, dreams are a way for the unconscious aspects of the right brain to communicate with the conscious awareness of the left brain.

For most people, one hemisphere of the brain tends to be dominant over the other. While the two hemispheres do communicate directly with each other via the collection of nerve bundles called the corpus callosum, one hemispheric cognitive style tends to predominate over the other. Because the nerve connections between the brain and body crisscross from one side to the other, the left brain controls the right side of the body and vice versa. Thus the left brain controls the right hand, while the right brain controls the left hand. The fact that our culture is predominantly right-handed suggests that most of Western civilization tends to be "left-brain dominant." Some researchers have discovered important connections between hemispheric dominance and immune-system susceptibility to various diseases. Dr. Norman Geshwind of Harvard University has documented that left-handers (who are typically right-brain dominant) tend to have more general problems with their immune systems. But right-handers (left-brain dominant) seem to be more susceptible to autoimmune disorders (illnesses in which people's dysfunctional immune systems actually attack their own bodies). Research suggests that certain areas of our left brain may be critical to stimulating or modulating our immune functioning. For instance, animal studies on the effects of brain lesions have shown that damage to large areas of the left brain, especially the frontal lobe, can sometimes result in reduced numbers of certain types of immune cells. Some immune cells, known as T suppressor cells, actually keep other immune cells from attacking the body. Fewer T suppressor cells could potentially lead to a higher incidence of autoimmune problems. So it is possible that certain problems with left-brain functioning might somehow contribute to so-called autoimmune disorders. Whether altering a person's cognitive style can directly affect immune functioning is still a matter of debate. But preliminary evidence suggests that right-brain visualization and imagery techniques may actually boost cellular activity in certain parts of the immune system.

Although it varies, depending upon the type of education a given person receives, as well as any specialized training in the arts and sciences, most people in our culture tend to be left-brain dominant. This is primarily because our educational system is dedicated to teaching children the three R's of reading, 'riting, and 'rithmetic—in other words, the skills of the left brain. While many schools try to encourage children to develop an interest in right-brain-centered activities like art and music, the major emphasis is still directed toward increasing children's proficiency in skills considered crucial for surviving in a left-brain-oriented culture. However, without the right-brain creative, artistic, "big picture"-perceiving skills, we would all become overly dry, analytical, and reductionistic. By becoming stuck in either left-brain or right-brain modes of seeing, we create perceptual filters that act like blinders to us, limiting what we are able to see in the world around us.

Whether we become overly focused in a left-brain, analytical viewpoint of daily life or whether we explore the mysteries of the right-brain-oriented, symbolic synchronicities of daily events, either one can affect not only our consciousness but our health as well. Our cognitive styles can strongly influence our perceptions. And our perceptions of events in our lives, and the emotional significance we give to those events, can determine whether or not we are overwhelmed by life stresses to the degree that we begin to develop illnesses in our bodies. In becoming more open to the symbolic, spiritual dimension of life—a perspective that can still coexist alongside our rational, linear, left-brain view of the world—we open a window through which we may glimpse the hidden perspective of our soul. A right-brain viewpoint can give positive meaning to seemingly negative life events that make sense only when the bigger picture of the soul's perspective is truly understood. Specifically, a negative life experience that forces us to adapt and develop new coping strategies may actually be providing us with the knowledge and experience that will eventually prepare us to survive a future crisis.

All of life's experiences have the capacity to teach us valuable lessons if only we remain open enough to perceiving their "hidden" value. The symbolic perspective of the soul allows us to find the hidden meaning behind events occurring in our lives that would otherwise be interpreted as negative, random, or meaningless. Our perception of negative or stressful events in our lives can also influence whether we approach such events optimistically or pessimistically. The ability to comprehend the positive value behind potentially negative learning experiences by seeing them from a more symbolic, spiritual interpretation might be considered a kind of "spiritual optimism." Surprisingly, an optimistic versus a pessimistic attitude in responding to stressful events can also play a role in health and illness by influencing our immune system. A recent study[1] by Dr. Suzanne Segerstrom at the University of Kentucky on the role of stress and immune function in first-year law students found that students who remained optimistic in spite of the stress of classes had more helper T cells and more effective killer T cells (both important immune cells) than did pessimistic students. Dr. Segerstom's study suggested that students' outlooks and attitudes may play a role in how their immune systems would respond to common immune challenges such as exposure to cold viruses. Being able to see the world from a symbolic and spiritually optimistic perspective, or—metaphorically—to glimpse the silver lining behind the dark clouds, can affect not only our moods and our attitudes toward life events but our health as well. While such a higher perspective may be easy to talk about, it is sometimes difficult to appreciate at the time one is actually undergoing a negative, stressful experience. It is only when we force ourselves to shift to a different level of consciousness, to "reframe" the dynamics of the event in terms of its symbolic meaning, that we may begin to appreciate any "teaching value" derived from our stress reaction to the negative event.

Some of the seemingly random events happening to us in daily life are often orchestrated by our higher selves so as to unconsciously act as a "mirror" to our inner emotional processes. Being confronted by a very angry person in our job, for example, might seem like a chance meeting with a mentally unbalanced individual. Most people would dismiss the encounter as something having no personal relevence, perhaps even asking themselves, "What the heck happened to make that guy so crazy?" However, when seen through the symbolic perspective of the higher self, the person who seems to be directing his anger toward us might actually be symbolically functioning as a mirror, reflecting back to us a reminder of our own unexpressed anger. In other words, the external event of being confronted by an angry person may remind us about certain unresolved issues or angry emotions that we have chosen not to deal with. If we view the encounter with the angry person as something that exists entirely outside of ourselves, we tend to dismiss his rantings as merely those of a crazy person. But by shifting to a more symbolic mode of perceiving reality, we may begin to ask ourselves, "What might this person be showing me that has relevance to my own personality issues?" or "What is the hidden message about how I am dealing with my own unexpressed anger?" Sometimes our higher self puts difficulties in our way that are intended as subtle, symbolic reminders of areas of personality development we still need to work on. By learning to examine our lives from the symbolic viewpoint of the right brain and the higher self, we may eventually find a higher, more spiritual meaning from even the simplest random events of everyday life.

In addition to the role that consciousness plays in our health, new research is pointing to the influence of the "heart/brain connection" upon how we think and feel and experience stress. Researchers at the HeartMath Institute in Boulder City, California, have researched the unique connection between the heart and the brain to provide evidence that supports the concept of a heart-based or "heartful" consciousness. Many people are familiar with the expressions "acting from the heart, not from the head" and "following your heart." There may be much more to this than just metaphor. Focusing the consciousness in the heart region while reexperiencing a positive emotional memory can help shift a person into a heart-oriented intutitive state that can assist in decision-making and coping with life's little struggles in ways that are healthier and lead to fewer and less severe stress reactions within the body. Researchers at the HeartMath Institute are pursuing different ways this insight might be applied to dealing with work-related stress and health in general. They have accumulated considerable scientific evidence supporting the idea that the heart center possesses a unique form of consciousness. There appear to be nerve connections that, when stimulated, send messages from the heart to higher brain centers involved in learning. Conversely, there are nerve connections from the brain to the heart (as part of the autonomic nervous system) that send messages to the heart to speed up or slow down depending upon physical demands and emotional reactions to the outside world. Our emotions strongly influence the activity of our heart. Anger, depression, and hostility are now recognized to be possible contributing factors to the development of heart disease. It even appears that during an anger-producing or frustrating experience, the heart is actually affected on a beat-to-beat basis.

Computerized analysis of the heart's rhythm has provided subtle clues that demonstrate how feelings of anger and frustration, as well as love and compassion, can create measurable changes in the electrical functioning of the heart. Most people are unaware of how their heartbeat changes from moment to moment, because the beating of the heart is usually below the level of conscious awareness. Rather than beating at a fixed rate all the time, the heart speeds up and slows down in small, imperceptible ways. The degree to which the heart rate changes over time is referred to as heart-rate variability or HRV. There is medical evidence to suggest that a decrease in HRV as people age may be associated with the onset of a variety of illnesses, including heart disease. Conversely, a high degree of HRV is often associated with efficient cardiovascular conditioning. When scientists at the HeartMath Institute analyzed the moment-to-moment variability in heart rate using special computer programs, they found unique patterns that correlated with different states of consciousness and emotional feelings. When people were experiencing discordant, chaotic emotions of anger and frustration, their heartbeat-frequency patterns were also chaotic, incoherent, and disorganized-looking. In contrast, test subjects experiencing feelings of love, compassionate caring, and an appreciation for life produced a coherent heart-frequency pattern that looked like a mathematically regular oscillating sine wave. In other words, the presence of loving, compassionate states of feeling in the heart area actually produced measurable changes in the electrical activity of the heart.

Based upon this finding and similar research, Doc Lew Childre, president of the HeartMath Institute, developed a powerful, simple technique of focusing consciousness in the heart center in order to shift quickly out of stressful emotional patterns. The technique, known as "Freeze-Frame," involves calling a kind of "time-out" from the stressful activity of the moment in order to shift from a head-based or mental focus to a more heart-based focus of awareness. The "Freeze-Frame" terminology refers to the metaphor of viewing life as a movie in which one attempts to freeze the action temporarily. A movie is made up of many individual still images running past the projector lens and light source at a rapid rate of speed. Similarly, our lives are made up of a continual series of moments and experiences strung together like the frames of a movie. Sometimes our lives can feel as if they're moving so rapidly that it seems like we're watching a high-speed movie, moving faster and faster without any apparent way to slow it all down. How you perceive and respond to each moment of the movie frames of your life determines how fast or slow the pace of life seems to you. Your perceptions will also determine the emotional tone of the way you will react to each succeeding frame of your life's movie. When you continually react to life events with distress, anxiety, frustration, or anger, an alteration takes place in the way you perceive your life's movie. Simultaneously, you create negative changes in your body's physiology that can eventually lead to burnout and illness. The concept of Freeze-Framing is similar to watching the movie of your life on a VCR. Suddenly you hit the pause button on the remote control in order to freeze the moment, call for a brief time-out, and shift your focus of awareness to a state that is more conducive to relaxed and introspective decision-making.

The five steps of the Freeze-Frame technique are simple but powerful in their effects. For the purposes of practice, think of a current situation that is making you feel anxious, angry, or just plain frustrated. Go through each of the five steps one at a time in order to attempt to shift to a new perspective that will help you decrease your stress level. In the first step, recognize that you are reacting stressfully to the events of the situation and call for a quick time-out. In the second step, make a concerted effort to shift your focus away from your racing mind and scattered emotional feelings to the region around your heart. It is often helpful to imagine breathing in and out through your heart center or to place your hand over your heart in order to focus your energy and attention to this area of your body. You should try to keep the heart focus for at least ten seconds or more. This allows you to alter your perspective to a viewpoint where you can more clearly perceive the "bigger picture" of what is really going on in your stressful situation. By shifting into a mode of heart consciousness for guidance, you are able to move to a perspective that allows you to make quality decisions based on your true heart's feelings, and not just on the immediate constraints of the situation at hand. This shift in awareness allows you to disengage temporarily from your stressful perceptions of the current situation in order to consider other, more effective possibilities or solutions that you might not have initially considered. As you shift your focus to your "inner" heart center, you help to relieve the physiological strain placed on your heart by the stressful emotional state. You allow yourself to disengage from the discordant emotions of the current dilemma while assisting your heart in operating with greater efficiency. Try holding a smile on your face in step two and beyond. It's difficult to feel bad when you're smiling.

In the third step of Freeze-Framing, think back and recall a positive, enjoyable experience you've had in the past and attempt to reexperience that same pleasurable feeling again. This could be a loving relationship from your past or present, a fun experience, or merely an enjoyable outing with nature. Research carried out at the HeartMath Institute has demonstrated that experiencing the emotional feelings of love, compassion, caring, and appreciation can produce immediate benefits to the immune system as well as shifts in awareness. By shifting your emotional state into a past memory of joy or appreciation, you disengage from the stressful feeling that might have been overwhelming to you before starting the Freeze-Frame technique. You may only be able to shift into a state of "neutral." However, by moving into a neutral state, you obtain a more objective, less emotional viewpoint of the situation at hand. In the fourth step, using both your common sense and your intuition, ask your heart what would be a more efficient response to the stressful situation you just shifted out of. Ask your heart for a solution or response that would also minimize future stress in the same situation. Finally, in the fifth step, listen to what your heart says to you in answer to the question you have posed. By listening to the advice of your heart and not your head, you are able to make decisions that reflect what is best for you. You can move beyond impulsive mental reactions and choices that might temporarily resolve the current situation but could also lead to future stressful consequences. The Freeze-Frame technique involves practice at first but becomes easier each time you try it.

Techniques such as Freeze-Framing are simple but effective approaches that allow us to shift consciousness into states more conducive to health and wellness. Our thoughts and our emotions can produce powerful effects upon the physiology of the body through a variety of biochemical and energetic pathways. Much of what we have discussed in this chapter in terms of mind/body effects is related to basic physiological changes in the body produced by stress. That is, we have described the effects of emotional stress in depressing our immune system's ability to respond to potential threats to our health.

We've touched on the effects of discordant thoughts and emotions upon the electrical activity of the heart. However, from a vibrational-medicine perspective, the effects of human thought and emotions often go far beyond the influence of brain-wave patterns and neurochemical interactions with the glands and organs of the body. The human mind is not just a hardwired, patterned, chemical organ. The astral and mental bodies also operate in concert with our brain's inherent processes, and thus contribute additional influences to our thoughts and feelings. Many clairvoyant researchers, who can directly perceive the higher-dimensional levels of our higher spiritual bodies, believe that our thoughts possess a distinct energy and form. In fact, at the astral and mental levels, it has been suggested that our thoughts can actually exist as individualized thoughtforms or patterns of subtle energy. What an interesting concept! Clairvoyants believe that such patterns manifest in the energy field known as the human aura. The aura is described as a distinct higher-dimensional energy field that surrounds an individual. One of the theories associated with auric thoughtform creation is that the stronger the thought or the emotional charge behind a thought, the stronger the thoughtform will be. Healer and teacher Dael Walker proposes a general rule of thoughtform behavior based upon his own experiments and observations: "The strongest thoughtform wins." This means that strong emotional patterns of reaction to life events, especially traumatic life events, can produce thoughtforms that can exist and persist in the auric field. Just as there are negative emotional thoughtforms, there can also be positive thoughtforms created by strong positive and loving emotional thoughts and feelings. But then again, if negative emotional-energy patterns can produce strong negative emotional thoughtforms, could they not overpower weaker, positive thoughtforms and thus create havoc in our lives and our physical bodies? While we once considered thoughts to be merely the province of neurochemical and brain activity, it seems our thoughts may actually possess a life of their own, independent of the gray matter of the brain.

Another proposed theoretical aspect of thoughtforms is that they are made of a subtle magnetic energy generated by our astral and mental bodies. If thoughtforms are composed of a kind of subtle magnetic energy, then couldn't they influence and perturb our etheric body over time? Perhaps such an influence could eventually produce cellular distortion and organ imbalance at the level of the physical body. Also, if thoughtforms are characterized by a subtle magnetic property, they might tend to magnetically attract other thought-energy patterns to themselves. Most people have heard the adage that "misery loves company." That is, people who are miserable or have problems in their lives tend to gravitate toward social groups made up of other individuals who have a similar fixation on their problems. Many of us know individuals who can relate to the world only from the perspective of the long suffering they have endured because of various health problems or personal misfortunes. In fact, the dynamics of how our culture has become fixated upon negativity are wonderfully explained in clairvoyant Carolyn Myss's writings on what she refers to as "woundology." People with past traumatic emotional wounds often communicate with each other using the language of "wound-speak," in which almost every social interaction with another individual is a chance to share the pain and trauma of how they have been wounded through failed relationships, physical attacks, accidents, illnesses, or other negative life events in their past. While woundology partially explains why people with similar "wound histories" tend to cluster together, the dynamics of people's energy fields may also explain how negatively fixated individuals tend to attract like-minded individuals to themselves. Perhaps a "subtle magnetic action" of thoughtforms in an auric field plays a part. It could be that the subtle magnetic field of the thoughtform resonates sympathetically with the astral and mental bodies of other, similarly predisposed individuals who are "emotionally stuck" in negative patterns. Often "wound-fixated" people are both emotionally and physically draining to be around. Maybe their subtle-energy fields "suck up" life energy from the healthy auric fields of sympathetic individuals almost like unconscious psychic vampires or "energy leeches." One of the reasons people find themselves in destructive relationships that repeat the same negative patterns may relate to the as yet undocumented magnetic tendency of thoughtforms to attract negative people and events to an individual. In order to shift toward a healthier mind and body and to create more positive relationships with others, maybe we need to be able to change our own emotional and mental patterns and sweep away old thoughtforms still hanging around like cobwebs within our auric fields. Even though we may not be able to easily observe such vibrational-energy patterns around ourselves, they could still continue to influence us through our chakras, our spiritual bodies, and through the people we continue to attract.

Although the idea of thoughtforms is very foreign to the Western concept of the way things work, there have been many clairvoyant individuals who fully ascribe to the concept of thoughtform energy. The Reverend Charles Leadbeater was a famous clairvoyant of the late 1800s with deep beliefs regarding thoughtform energy. According to Leadbeater, there are three basic classes of thoughtforms. The first category relates to thoughtforms that are not centered on the thinker or aimed at any individual but are left behind as a kind of "thoughtform trail" showing where an individual has been. The second class of thoughtforms are those that center on the thinker and tend to hover around the individual's auric field, following wherever the person goes. The last category of thoughtform represents an energy form that shoots straight out from a thinker and aims for a particular objective or individual who is the focus of the thinker's attention. This last class was of considerable interest to Leadbeater because it represented a part of the person's spiritual energy that was sent out from the auric field and directed toward another individual. Besides Leadbeater, Carolyn Myss also sheds light on these third-category thoughtforms. In her book *Anatomy of the Spirit,* Myss proposes that when people continue to think about old traumatic events in their lives, especially about someone who has done something emotionally hurtful to them, they actually send a part of their spirit to that distant person or event in time. Her explanation is similar to Leadbeater's third class of thoughtforms that migrate from the individual to a distant place or person.

Myss makes this point: When enough of our spiritual energy becomes invested elsewhere and is locked into a continual cycle of thinking about an emotionally traumatic past, bad things tend to happen to our health. If we never allow ourselves to heal or move beyond our wounds, we end up depriving ourselves of the spiritual energy needed in the present in order to maintain the health of the cells of our physical body. If we unconsciously create negative thoughtforms and continue to reenergize them as we relive the hurtful past events in our minds over and over again, we may be negatively affecting our health. From an energetic standpoint, this may be one of the reasons that the expression of forgiveness toward an individual who has hurt us in the past can be important to us both psychologically and spiritually. The cleansing generated through forgiveness helps to release us from the spiritual-energy links we have unconsciously created between ourselves and the people who have hurt us. As we learn to forgive other individuals for their hurtful acts, we call back our spirit from the people and places in whom we have unconsciously invested our spiritual energy.

If "calling back your spirit" rings true to you, here is an exercise to do while using the energy of the heart center, forgiveness, and unconditional love to cleanse and heal your physical and spiritual bodies. Sit in a comfortable position with your hands placed on your lap with palms facing upward. Imagine taking a very long cord made of rainbow colors and tying it around your waist. Imagine inserting the long tail of the rainbow cord into the back of your spine and down through your root center, deep into the earth where the rainbow cord grounds your energies into the earth. Now envision in your heart center a glowing ball of golden-white energy, similar to a miniature sun of brilliant light. Feel the sense of warmth of the golden-white sphere of light within your heart center. As you focus your attention in your heart center, say the following prayer to yourself: "Father/Mother/God, I pray that my heart center is filled and healed with a beautiful sphere of healing white and golden flame of unconditional love in which perfect love casts out all fear. My heart center becomes filled and healed within a spherical white and golden flame of divine healing unconditional love, forgiveness, and acceptance of myself . . . for all of my strengths as well as my weaknesses . . . for all of my achievements as well as my past mistakes. My heart center becomes filled and healed within a beautiful, sphere of the white and golden flame of divine healing, unconditional love, forgiveness, and acceptance of myself in all of my totality." Now imagine that the sphere of white and golden flame begins to pulsate and glow more brightly as it expands infinitely outward from your heart center. As the white and golden sphere expands ever outward, say to yourself, "Father/Mother/God, I pray that the white and golden flame expands infinitely outward from my heart center, and as it does, it completely fills and heals, balances and aligns, blesses and forgives, regenerates and renews, transmutes and tranforms, and heals and encompasses my entire physical body, my etheric body, my astral body, my emotional body, my mental body, my causal body, my higher spiritual bodies, as well as my auric field within a brilliant healing white and golden sphere of divine healing, unconditional love in which perfect love casts out all fear."

If you begin your day practicing this little exercise, it can help to cleanse your physical, emotional, and spiritual bodies and also help you to start the day feeling uplifted and more optimistic about yourself and what you are capable of accomplishing. In addition, this exercise can be used not only for self-healing and rebalancing but for distant healing of another individual as well. When your spritual energies are tied up in conflict over someone you feel has injured you, is your enemy, or has rubbed you the wrong way, part of your spirit remains linked to that other party, robbing you of energy that you need for health maintenance. You can use the previous exercise to send a beam of golden-white light to the individual with whom you are experiencing an uncomfortable relationship. By doing so, you help to free up your own spirit's energies from areas that no longer serve you and may actually end up harming you. The fascinating part of this distant-healing exercise is that after you direct the golden-white light of unconditional love toward someone who has previously been a source of conflict and friction, the irritating individual often seems to soften and may actually change their attitude toward you! To do the exercise of sending healing and forgiveness to another individual, start by forming the sphere of golden-white light in your own heart center and go through the entire exercise for healing yourself. Then envision sending a laserlike beam of the golden-white light of divine healing, unconditional love from your heart chakra directly to the other individual's heart chakra. Repeat the prayer, substituting the other individual's name. For instance, if you were trying to send healing energy to "George," you would repeat the prayer, substituting his name.

Next, imagine that the sphere of white and golden flame expands infinitely outward from George's heart chakra, filling and healing and balancing his physical, etheric, and spiritual bodies just as you did for yourself. Using this technique with others can not only be quite healing to oneself and the person on the other end of the visualization, but it can also be great in resolving conflicts with someone else without having to say anything to that person at all.

Another technique people can use to clear away their mental- and emotional-energy cobwebs is a brief visualization exercise involving a cleansing tornado of white light. In the visualization, imagine that you create a large, spinning tornadolike vortex composed of white light above your head. Picture the tornado as being fairly large, at least ten to twenty feet in diameter. As you picture the tornado of white light spinning above your head, imagine the vortex slowly descending through your body, first passing through your head, neck, and shoulders, then moving down through your abdomen, your pelvis, then down through your legs. Envision the tornado cleaning up and sucking away all the negative energies as well as the disturbing emotional- and mental-energy patterns as it travels downward from head to toe. As the tornado moves down through your body, envision the tornado going deep into the Earth, where it dissipates and releases its absorbed negative energies. Then imagine the energy being positively recycled and reabsorbed by the earth.

To clairvoyants who see auras and thoughtforms, the atmosphere of the earth is marred not only by airborne chemicals and hyrdrocarbon exhaust fumes but also by huge masses of negative thoughtforms and background "thoughtfield noise" generated by centuries of human negativity, perversity, greed, lust, killing, and the by-products of undisciplined thoughts and lower emotions. The thoughtform "debris" littering the earth's field mirrors the first class of thoughtforms described by Leadbeater's early work. Leadbeater's explanation of level-one thoughtforms is fascinating. He believed that such thoughtforms tended to dissipate unless reenergized by thoughtfields of a similar nature generated by other individuals. During the 1970s, another researcher, Robert Monroe, concurred with and extended Leadbeater's work. As an explorer of so-called out-of-body experiences (OOBEs) or astral travel, Monroe described astrally traveling through bands of thought pollution in the earth's atmosphere similiar to those Leadbeater had described. Monroe labeled the bands "H-Band" noise, referring to the background noise created mainly by undisciplined human thought. He documented hundreds of his own personal out-of-body travels and experiments in several diary-type publications. In *Ultimate Journey,* Monroe's last book to detail his observations of the nature of the astral and other higher dimensions, he describes the region of thought pollution that surrounds the earth. To quote Monroe, "The H-Band noise is the peak of uncontrolled thought that emanates from all living forms on earth, particularly humans. Just imagine the magnitude of this disorganized, cacophonous mass of messy energy. The amplitude of each segment of the band is determined by the emotion involved in the thought. Yet our civilization does not even recognize that the H-Band exists."[2]

The preceding explanation of thoughtform energy leads us next to the subject of using the energy of thought to create healthier energy patterns through verbal affirmations. By mentally repeating a specific, positive, verbal affirmation over and over again, it seems that healing benefits on a variety of energetic and physiological levels can be achieved. The simple act of mentally verbalizing (as well as repeating out loud) a wish to heal our bodies and change our lives can often produce powerful effects upon our subtle bodies and in turn our physical bodies, especially when the affirmation is repeated regularly.

The use of structured affirmations in healing goes back nearly a hundred years. Emile Coué, a French pharmacist around the turn of the century, opened a free clinic. Patients used affirmations, or what Coué referred to as "conscious autosuggestions," as part of a mind/body healing approach. The basic affirmation was "Every day in every way, I am getting better and better." With repetition of this simple mantra a total of twenty times every morning upon waking, people who practiced Coué's autosuggestion program actually began to feel better. Of course, Coué developed many other more specific affirmations that were intended to assist with relaxation as well as healing the body of different health problems. His work became the precursor to many progressive-relaxation techniques now widely used for stress reduction and pain relief. Coué's techniques were similar to present-day autosuggestion approaches of self-hypnosis.

Coué's work on autosuggestion stemmed from his belief that people's thoughts eventually materialized as physical reality. As such, he believed that dysfunctional thinking patterns could actually worsen and perpetuate different forms of illness. He felt that imagination, coupled with affirmations, was a powerful pathway to healing illnesses of both mind and body. Coué and many of his successors discovered that a critical ingredient to success was the mental repetition of affirmations "without effort." It seemed autosuggestion worked best when brute willpower was put aside in favor of a more relaxed, effortless repetition of the verbal affirmations. For optimal efficacy, Coué taught that affirmations needed to be expressed calmly, without passion or will, yet with absolute confidence and expectation of success. In the mid-1970s Dr. Norman Shealy, first president of the American Holistic Medical Association, adapted Coué's autosuggestion approach along with imagery exercises, progressive-relaxation techniques, and other modalities into a healing system known as "Biogenics." Biogenics became a highly successful therapeutic program for dealing with chronic pain conditions. Shealy and many other physicians working with chronic pain and stress-related disorders found affirmations to be of great benefit in relieving pain and stress and for promoting healing.

Louise Hay, a popular metaphysical teacher and author, has also developed a series of affirmations for dealing with particular illnesses. Through her own personal healing and her work with others, she has developed a deeper understanding of the symbolic meaning behind various illnesses. In her book, she lists specific affirmations to help people change some of their "erroneous thinking patterns" that may have contributed to the development of specific illnesses. Hay feels that each disease is associated with particular negative mental and emotional patterns that can be healed through the use of specific affirmations. The affirmations are said to help create new healthier patterns of thinking. Many of her affirmations deal with learning to love yourself, to release old fears, to feel safe in the universe, and to accept joy into your life. By the repetition of affirmations, the emotional energies of love and joy can flow more easily into your life. New patterns are created that can attract more positive experiences into your life, as well as help to heal physical illness. By repeating affirmations that cancel out old, faulty beliefs and replacing them with more appropriate ways of perceiving your life, a significant shift in both consciousness and health is possible. And by shifting your consciousness into healthier emotional and mental patterns; your emotional, mental, and spiritual bodies become restructured and realigned, which in turn helps to restructure and repair your physical body.

Another application of affirmations involves a technique known as "programming." The popular Silva Mind Control courses, developed by founder José Silva, have long taught this method. Verbal affirmations are used to bring about positive changes in an individual's health and in the nature of the individual's life in general. In a very real sense, affirmations allow people to literally "reprogram" their mind, body, and spirit, because as people continually repeat their affirmations, a shift occurs in the mental and emotional bodies as well as in the neural patterns of the brain itself. Old, dysfunctional messages that the subconscious mind has been replaying for years are replaced by positive, life-affirming new messages that can subtly shift a person's attitude toward life in general. People who have used the Silva Mind Control method of programming swear that by using structured affirmations to "verbally visualize" the desired outcome of their efforts, they can "program" not only for health changes but also for job opportunities and even instant parking spaces. By repeating affirmations, are people literally creating thoughtforms in their energy fields that can magnetically attract the object of their desire? Is it possible that affirmations can be really more than just autosuggestion and "wishful thinking?"

Another form of verbal affirmation that attempts to influence events in the physical world is prayer. Many people attribute personal healings, as well as changes in the world around them, to prayer. The power of focused consciousness to shape physical reality is much greater than most people have even dreamed. When groups of people focus their attention through prayers in order to bring about positive social changes, the collective energies of group consciousness can often have a tremendous amplification effect.

Prayer is simply one form of spiritual affirmation that can bring about healing changes. However, lasting healing created through any form of affirmation may ultimately depend upon a permanent shift in people's attitudes and belief systems. One spiritually based system of affirmations that can literally reprogram people's basic beliefs about themselves and the world around them can be found in the workbook for *A Course in Miracles.* The source of the information in the Course is purported to be of a "divinely inspired" nature. The Course has found worldwide appeal in its universal message about the healing nature of unconditional love and the divine nature of humanity. Published by the Foundation for Inner Peace in Tiburon, California, and now translated into an increasing number of foreign languages, *A Course in Miracles* consists of a textbook, a workbook, and a manual for teachers. The workbook for the Course contains a different set of affirmations for each day of the year. If the daily affirmations are performed with sincerity, the results can be truly amazing. Radical shifts in spiritual awareness can be one of the many beneficial results. The basic tenet behind *A Course in Miracles* is that miracles are natural, and when they do not happen, something has gone wrong. The something going wrong is most often our dysfunctional belief systems and our ways of reacting to the world more out of fear than out of a loving connection with life and the connectedness of all things. By shifting people's beliefs through affirmations and a kind of "spiritual reprogramming of consciousness," by replacing fear with love, miracles may actually become possible. The power of our consciousness and our spiritual energies to bring about healing, positive changes is so great that we have the power to literally "reengineer" our own personal and collective realities if we would only open ourselves to the idea that healing changes are possible and within our reach.

Whether or not one chooses to create a program of personal affirmations for shifting spiritual awareness, manifesting abundance, improving health, or assisting in stress reduction, there are basic guideliness to follow. With regard to specific affirmations, people should create their own program according to their own needs, their health problems, and what they are trying to achieve in their life. In the beginning, it is best to write out the affirmations and carry them with you. That way the affirmations become your script to read out loud. Later on, you can switch to silent repetition. Eventually the affirmations become incorporated into your consciousness and memory. There are several coaching points needed while using affirmations for programming general health and life outcomes. For example, an expectation of success in programming is important to having a successful outcome. Also, since affirmations work in part by reinforcing new patterns in the subconscious mind as well as in the mental and emotional bodies, it is important to structure affirmations in specific ways. As the subconscious mind operates at a fairly primitive level, almost childlike in nature, rhythms and rhymes incorporated into the phrases of an affirmation are often effective reinforcers. Rhymes are easier to remember and may sometimes have a more potent effect upon the subsconscious mind. Also, the subconscious mind has difficulty dealing with certain types of phrases containing "negatives." In fact, Dr. Shealy found that in teaching patients with chronic pain to manage their pain via his Biogenics-program affirmations, it was best for people to not create affirmation statements saying that they wished to have "no pain." If such a strategy was employed by patients, this seemed only to intensify the pain they were trying to get rid of. It appears that the subconscious mind doesn't know how to create a "no-pain." It would just create more pain when affirmation phrases were structured using negatives. A more effective solution was found in teaching patients programming or affirming for "an absence of pain" or a feeling of calm, pleasant, enjoyable sensations in the body.

Another key to successful programming involves being very specific in what you are trying to create with your consciousness. For instance, Silva Mind Control graduates sometimes report that if someone was trying to program for meeting that special "Mr. Right," she would create an affirmation or visualization or prayer for meeting a person with all the qualities she thought would be important in a "soul mate." After repetition of her programming, she would often meet someone who fit all the requirements in her list of desirable qualities but would occasionally be disappointed because Mr. Right also had bad breath, was penniless, or possessed some other irritating quality she hadn't considered when creating her program. This type of unintended outcome from programming serves as a caution in creating your program or affirmation. Remember to "be careful what you wish for." At first it is probably best to program for broad goals that summarize the end result of what is desired without worrying about every specific detail along the way.

With this kind of autosuggestion, we are often unconsciously programming our minds and our spiritual bodies with simple statements made about ourselves without really thinking, which can turn out to have powerful impacts on our health and our entire lives. To illustrate, one of the psychologists working with Dr. Shealy as a therapist in his early Biogenics programs told a story about a client she had worked with in a previous group of patients. This particular client related to the psychologist a story about how he was always so overconfident of his abilities that he would jokingly tell people, "Hey, I could do that standing on one leg!" The client had been enrolled in Dr. Shealy's program because he was suffering from severe chronic pain at an amputation site where his leg had been surgically resected to remove a malignant bone tumor that had developed several years earlier. It was the opinion of the psychologist that the individual might have programmed his body with repeated autosuggestion statements that eventually led him to create an illness that quite literally left him "standing on one leg." While such stories are only anecdotal, they reinforce the need for caution in not only composing positive verbal affirmations but in paying attention to how powerful our consciousness can be in creating our reality and in maintaining the health of our physical and spiritual bodies. Emile Coué, one of the pioneers of autosuggestion, had long believed that a person's health could be easily influenced by autosuggestive statements and that children were particularly susceptible. Coué felt that illnesses should not be discussed in front of young children and that they should be taught that good health is the natural state for people to be in.

One additional key to success in using affirmations for effective change is repetition. A person may start the day repeating affirmations a minimum of three times upon waking. Three repetitions seems to be a kind of magic number in reinforcing or amplifying thoughtforms, although more than three repetitions can be even more effective. The more often you repeat your affirmations, the stronger your thoughtform becomes. You can do sets of three repetitions of your basic affirmations at different times throughout the day, then again at bedtime. As mentioned earlier, writing down the affirmations on paper as well as reading them aloud can help to reinforce their effectiveness, since writing, seeing, saying, and hearing the phrases involve four different perceptual and kinesthetic senses.

For added reinforcement, it is especially helpful to create a mental image or "inner picture" of the desired outcome from your affirmations. In recent years, the use of mental imagery or visualization for self-healing has been shown to be of great value in many different illnesses, from relieving stress to recovering from cancer. While we have focused mainly on verbal affirmations here, there is also a wealth of literature on various healing visualizations that can provide additional help in dealing with specific illnesses. However, not everyone is a good visualizer (myself included).

I will, however, describe one visualization technique that can perhaps help to energize or reinforce the creations of our consciousness. The interesting proposition that thoughtforms are actually of a subtle magnetic nature was mentioned early on in this chapter. This magnetic theory comes into play now. You can use mental imagery to create a powerful magnetic field to strengthen positive thoughtform generation by envisioning a tightly coiled wire wrapped around a central piece of steel, similar to an electromagnet, situated along the vertical axis of your spine. During the process of programming with either imagery or verbal affirmations, envision the electromagnet turned on with current flowing though it. Hear the electrical hum from the coil as the energy flows through the magnet. Imagine that a powerful magnetic field is pouring out through the top of your head. By imagining the generation of a powerful magnetic field during the time one is practicing affirmations, perhaps the thoughtforms produced can become magnetically energized and strengthened.

As a basic summary of early-level programming and affirmation-making, start by writing out a simple script to follow for doing daily affirmations. One of the easiest affirmations to do regularly is Emile Coué's simple phrase "Every day in every way I am getting better and better." The phrase can be modified to "Every day in every way I am getting healthier and healthier," as well as other variations on a theme. Some people find that affirmations work better for them if they are constructed in the form of a prayer to a higher power such as God or the universe itself. One affirmation that covers many different aspects of life is a prayer that goes "Father/Mother/God, I pray that my life is filled with boundless joy, endless wealth, infinite abundance, daily miracles, perfect health, and perfect balance in body, mind, and spirit." Such prayer/programs cover all the bases, from enhancing health to increasing the physical and spiritual abundance in one's life as well. Another affirmation that can simply help the day to flow more smoothly is "Father/Mother/God, I pray that everything will flow incredibly, miraculously, and divinely smoothly, swiftly, efficiently, and easily for me in everything that I do today." Affirmation and visualization techniques help to reprogram your consciousness and belief systems to expect abundance and perfect health in life as opposed to operating from an attitude of fear, frustration, stress, and the paranoid expectation that everyone in the world is out to get you or to give you a hard time.

Thoughtforms, Probable Futures, and the Fine Art of Reality Engineering

When using affirmations and imagery for "reality engineering," I believe that you are able to change not only the reality of your own physical and spiritual anatomy but also physical events outside of yourself. How might this be possible if we are "separate" individualized beings, disconnected from one another and from the world around us? The concept that we are separate from one another, that we end at the boundaries of our skin, is actually a reflection of the pervasive Newtonian viewpoint of the universe as a mechanistic clockwork in which the whole is thought to be merely the sum of its component parts and that our consciousness is limited to the confines of our physical brain. In the vibrational perspective of the multidimensional universe, we are actually less separated from each other than our physical senses would seem to be telling us. Because we have energetic components that extend beyond our physical form, human beings are more than just the sum of their physical parts. Our physical bodies are, in a sense, containers for our consciousness that allow us to interact strongly with objects and people in the physical world. In reality, our consciousness reaches beyond the brain and the physical body to the level of our spiritual anatomy and our extended multidimensional energy fields. Consciousness has the capacity to go within and direct physiologic events in our physical bodies, as well as the power to reach out through the subtle-energy fields of our thoughtforms.

Biofeedback research has proven we can consciously influence body physiology if given the proper feedback. For example, by using instruments and sensors to give feedback on the bodily processes of blood flow and skin temperature, most people are easily able to increase blood flow through their fingers in order to consciously raise hand temperature. However, long before biofeedback instruments were even developed, early autosuggestion researchers showed that people could accomplish the same thing by repeating the affirmation "My hands are warm and heavy." Specially constructed verbal affirmations allow us to tap the hidden powers of consciousness and to access any potential for consciously regulating our bodies and minds. The ability to consciously create our own personal and collective realities may be limited only by the power of our imagination.

Researchers Robert Jahn and Brenda Dunne at the Princeton School of Engineering have shown that a large number of test subjects were able to influence microscopic physical events at a distance, such as the output of random-number generators, merely by focusing their consciousness to "will" events to happen in a specific direction. Their work on psychokinesis (PK), the power of the mind to move objects, has been replicated in various forms by other researchers. Jahn and Dunne's research demonstrates that microscopic or quantum events may be more easily affected by the power of the mind than macro-PK events such as "spoon bending." The ability of consciousness to directly influence micro-events has been repeatedly demonstrated at statistically significant levels in a variety of laboratories. But is it possible to influence events larger than the quantum processes governing the output of a random-number generator? Can we consciously influence micro-events in such a way as to produce larger macro-effects upon the world around us?

I believe that when using consciousness techniques such as visualization or programming for reality engineering, an individual is consciously influencing reality at a micro-level. By selectively influencing certain micro-events, it is possible to create a much larger effect because of the fluid, malleable texture of real-world events. The world today is a mosaic of many simultaneous ongoing events that are extremely nonlinear, fluid, and turbulent in nature. Look at something as complex as the presidential elections. While both political parties rely on the use of polls to make predictions of how voters will choose candidates, sometimes an unpredicted event such as a scandal or a response to a hostage situation can have unforeseen effects in skewing the election one way or another. When the Internet was first created by the Defense Department, no one dreamed that it would become fertile ground for new businesses as well as the basis for an evolving global electronic village. Because of this fluidity of reality, predictions of future events are often approximations to reality or merely educated guesses. Very few things in the world are as linear or predictable as our high-school physics textbooks would have us believe. An entire science of chaology or chaos theory has been developed over the last twenty years in order to study how natural, nonlinear systems like weather patterns and other turbulent processes could be better understood, anticipated, and predicted. When random events are analyzed using specialized chaos computer programs, repeating "patterns of order" can often be found. One of the tenets of chaos theory is the so-called "butterfly effect." In essence, the butterfly effect suggests that under the right circumstances, very small events can be amplified in such a way as to create very big effects. The quintessential example of this principle is the mathematical demonstration that a gently beating butterfly's wings in Tokyo eventually can be amplified into the hurricane winds of a major storm system that could inundate the Eastern seaboard. The bottom line is this: Under the right circumstances, micro-events can eventually become amplified into macro-events of significant consequence. Reality is actually so fluid, so filled with seemingly random events, that a little push one way or another is sometimes capable of shifting probable futures toward a particular direction.

The relationship between programming and conscious reality engineering is that human consciousness is capable of creating micro-events that under the appropriate conditions can become amplified into major shifts in physical reality. Every day we are confronted with big and little decisions. Do I turn left or right at the street corner ahead? Do I take my usual expressway route to work or listen to the traffic report on the radio and take an alternate street that will get me to work faster by missing that big accident on the expressway? Do I go to school to become a doctor or become an apprentice and join the family plumbing business? Each branch of the decision tree of our life consists of a shift in what are known as "probable futures." Every time we make a decision to zig right or to zag left, there is a probable future left behind down a path we never followed. The path our lives take and the direction of events in the world around us are partly determined by our choices and our reactions. Whether you decide to turn right or left at the next corner might determine whether you are stuck in traffic for an extra half hour on the way to work or have smooth sailing all the way. The choices we make are a reflection of our perception, our mental and emotional bias at the time we have to make these choices, and our belief systems about the way the world works. If we truly believed we had a greater ability to influence ourselves and also the world around us, we might make very different choices.

I believe that by focusing our consciousness in very specific ways, as in the case of programming for a particular event or outcome to take place, our consciousness produces an energetic effect in shaping our reality by creating micro-PK-influenced quantum events that act as little course corrections in our lives. Imagine rowing a canoe along a rapidly flowing river. You must constantly make such corrections as you encounter rocks, turbulent waterways, and fallen branches in order to successfully navigate the watery path. In a similar way, making little course corrections in our daily lives helps us to make our day move along more smoothly, with fewer obstacles and greater ease and enjoyment along the way. By actively visualizing or verbally affirming how you wish the day to go or how your body should respond to stress, you're able to tap in to the abilities of the mind and the higher-dimensional energies of your spiritual body to actively tip the balance of inner or outer reality in one direction or another. It is as though programming allows us to make subtle nudges to fluid, nonlinear, unstable events in reality to either zig or zag in a particular desired direction. These gentle nudges are a product of the thoughtforms we consciously create. If properly directed with focused intention and repetition, an energy-amplification effect can occur. To borrow from the old adage of "the straw that broke the camel's back," if the camel was already carrying too much weight, and was poised to collapse if any more weight were added to its load, a small effect could become amplified into a much larger event. The ultimate effect of successful programming is an energetic shifting of reality, causing it to unfold in a way that more closely approximates the probable future that has been visualized or affirmed in the program. That probable future you are programming for may be one in which your body completely heals itself of a serious illness, or it may be a probable future in which you are finally working in a job you really enjoy, surrounded by positive, supportive people.

Probable futures are unique aspects of multidimensional reality that have been discussed at great length in the book *The Nature of Personal Reality*. The origin of the material in this book was Seth, a higher-dimensional being who frequently "spoke through" the late Jane Roberts. Seth alluded to the concept that, of all the different frequency dimensions of energy making up the layers of the multidimensional universe, there are not only etheric, astral, and higher spiritual dimensions of reality but also many different probable futures or "parallel universes" that coexist simultaneously but are separated from each other by different vibrational states of being. If this were true, it could mean that there exists somewhere in an alternative universe "another you" following a different path. For instance, let us say that at the point in your life when you chose one career over another, you decided to go to engineering school to become an electrical engineer just like your father. You did this instead of studying art and graphic design, as you once dreamed you would. While in this life, you might be employed as an engineer for a large car company. In another simultaneous yet alternate probable future that went forward in a different direction from that critical life-decision point, an "alternate you" could very well be a successful graphic artist employed by the advertising department of the same car company. In other words, your life along the "path not chosen" might actually continue to exist as another probable future in the multiple parallel dimensions of multidimensional reality. This concept of exploring these different probable futures is the theme of the popular science-fiction television program *Sliders.* Each week a group of "travelers" uses a unique technology known as "sliding" to travel to different alternate probable futures. There are probable futures where dinosaurs never became extinct, or alternate futures in which the Nazi regime runs the government of the United States. In that particular parallel universe Germany never lost World War II. According to Seth, people may occasionally intersect with their alternate selves in these different probable futures while astrally traveling during the sleep state. But most of the time each of our alternate selves remains unaware of the others.

The importance of the concept of probable futures and parallel universes is that many different probable futures can exist simultaneously and that a slight shift in one or two critical elements can change how the future unfolds in radical ways. If, for example, you alter your travel plans because you listen to the warnings of a precognitive dream depicting an impending air disaster, it is possible to change your fate and thereby shift to a different probable future in which you don't die in a fatal plane crash. Similarly, you might have an intuitive feeling about a potential health problem that sends you to your doctor's office where a potentially serious illness is detected early enough to avoid a potentially fatal outcome. By acting on intuition and following our gut instincts, we may be able to shift our probable futures to those with happier, less stressful endings. This malleability of multidimensional reality is one of the reasons that psychically predicting the future can be so fraught with difficulty. As people act on precognitive or intutive information, the flow of reality can be changed in such a way that events are shifted toward an alternate probable future where, for instance, the predicted disaster never happens because a potentially tragic technical problem is identified and repaired in time.

We've just explored how we can program for specific outcomes and events to happen in our lives by using the power of visualization, affirmations, and prayer. However, we may also be capable of using programming to shift not only our own personal reality but also small portions of the greater "collective reality." Perhaps a strong enough thoughtform created by an individual produces a kind of micro-PK effect outside that person, which in turn is able to gently nudge small events in the larger collective reality closer to the direction of the "desired" (and perhaps more positive) probable future. Remember, the strongest thoughtform usually wins. However, in considering the strength of personal thoughtforms required to accomplish a particular shift in external reality, it is best to remember Robert Monroe's observation of the H-band noise layer of human thought that permeates our planetary atmosphere. One must still contend with the energetic effects of the vast number of thoughtforms already being generated by the rest of humanity. But with a sufficiently powerful focused thoughtform, a person might possibly create subtle positive shifts in "local events" that would help to manifest "programmed for" or visualized "optimal" probable futures for themselves and for their small portion of the world.

One example of how a small shift produced by a low-energy event can sometimes create a much bigger effect is the so-called domino effect. Many people have seen the large geometric arrangements of dominoes that creative domino hobbyists have designed and set up for the beauty and intricacy of their complex patterning. Pushing over a single "key" domino in the pattern sets the entire display into vivid motion, almost as if lighting a match to ignite the fuse of a fireworks display. By gently pushing the first domino in the larger pattern to fall over against the second domino, a cascade effect is created. One domino tumbles over into the next, and then the next, and so on, amplifying the first tiny movement into a massive display of cascading dominoes. Metaphorically speaking, we can also see the domino effect in real life, such as in the case of the increasing problem of "road rage." Consider an individual who is preparing to drive to work during morning rush hour. It is the middle of the summer and the weather is extremely hot and humid. Let us say the individual is married and has just had an argument with his wife, so he leaves home in a huff. While muttering to himself about the argument, he gets into the car and starts the ignition, only to find that his air-conditioner is on the fritz. He grumblingly drives off to work, only to find himself in a typical traffic jam caused by ongoing road construction. He calls work on his cellular phone to let his boss know he'll be later than expected, only to be chewed out over paperwork he'd been struggling to finish. Suddenly someone pulls over into his lane without signaling and nearly cuts him off. If this had been any other morning, this little incident might not have bothered him. However, this new irritation now becomes the "straw that broke the camel's back." He reaches into his glove compartment and pulls out a handgun he keeps in the car for "security purposes" and fires a shot into the car ahead.

The aggravated driver, poised to explode because of a series of many cumulative little irritations, is triggered into violent action by the mere event of another car's moving over into his lane. Similar to knocking over the first domino in a larger pattern, a series of events is set in motion. While the trigger itself is only a very small action, the subsequent consequences can sometimes be quite powerful. The relevance of this example to programming is that by focusing a minimum of energy with maximum intention into a system that is potentially unstable and extremely sensitive to change, a tiny energy event can often become translated into a much larger shift in outcome. By focusing our consciousness to do mental programming, we may be energetically affecting tiny changes in the flow of real-world events that help to shift reality more in the direction of the desired probable future we are affirming or imaging in our mind's eye. By properly focusing our awareness to consciously create thought-forms that program for the life we seek to live and for the body we wish our spirit to inhabit, we begin to activate a series of energy events on many multi-dimensional levels.

As we attempt to program or use specific affirmations to heal our bodies and also the patterns of our lives, we also begin to make closer contact with our higher selves. Such a course of action may enlist the aid of the true self, the soul, which exists in a higher spiritual reality that overshadows and subtly influences the events of physical reality. As we engage the higher self in activities more consistent with finding love, seeking peace, banishing fear, acting with courage and kindness, and dissolving resistance to following the soul's true path, we frequently get "help from upstairs" in subtle ways that go beyond what the conscious mind can sometimes conceive as possible. As *A Course in Miracles* states, miracles are natural. They can become everyday occurrences if you can shift your consciousness and your belief systems to accept miracles as natural, commonplace events. But unless you create the space for such miraculous events to exist, even if only as a probable future, you will never see them in your own life. We each need to break through a different kind of unconscious mental programming we have all been made to undergo during the process of growing up. The typical programming of our minds by our parents as well as by our society leaves many of us conditioned into believing we can never change our lives because our fate is somehow fixed and immutable. Silently we might even suspect that our diseases and accidents might be some type of divine retribution against us for somehow being "bad." Such programming makes us think that miracles are only for the few fortunate, religiously devout individuals who really deserve it, but certainly not for us.

Unfortunately, patterns of limited, superstitious, fatalistic thinking are often far too common. Most of us tend to look to the "priesthood of science" to heal us of our ills, to free us from our perceived punishments for some forgotten sin or misdeed. Many people never suspect that they may already hold the key to healing within their own minds and spirits. Part of that healing is dependent upon a shift in consciousness that allows people to lovingly forgive themselves for the perceived misdeeds of a past that cannot be changed. But what is also needed is a belief in the unlimited capacity we all have for healing and positive change. Positive changes in our lives may be facilitated by tools such as prayer, affirmations, visualization, and the active enlisting of our spirit as an ally, all of which can provide the critical catalyst necessary to shift things in a healing direction. Unlike conventional science, which often tends to dismiss things spiritual, vibrational medicine acknowledges the human energetic connection to the spiritual dimension. Indeed, the spiritual connection is often the hidden driving force behind why certain life events unfold in particular patterns here in the physical world. Contrary to popular belief, I believe that most physical events and so-called accidents in the world of human affairs are not merely random acts. Instead there may be an underlying spiritual order that directs and subtly orchestrates events throughout the physical universe from the very big to the very small. When one begins to study one's life for synchronistic events and meaningful accidents, an awareness develops of a subtle trail of bread crumbs our higher self tries to leave us to provide feedback on whether or not we are following the correct path out of illness, chaos, stress, and confusion. If we would only attempt to tap in to our limitless potential for growth and change, we would find that the healing changes we seek for ourselves may actually be within easy reach. By expanding our consciousness to search out the helping hand of our higher self, and then enlisting its aid and support through prayer and programming, we move a little closer to living the life we always dreamed about. And isn't that a life of love, peace, spiritual growth, and healing?

As stated earlier, most individuals are unaware of the easy access they have to existing spiritual support systems. And while many spiritual philosophies speak of the human connection to a higher divine nature and of a connection we all have to one another, our society often conditions us into believing we are separate, disconnected beings. The illusion that we are separate is the fallacy of the Newtonian perspective. However, at the quantum level, we seem to be closer to being less a physical machine and more like a structured energy field with boundaries a little fuzzier than our physical senses would indicate. Our consciousness has been shown to extend beyond the boundaries of the physical body, so we are constantly interacting, albeit at an unconscious level, with the many beings around us in ways that are difficult for many to fathom. Remember the HeartMath Institute's research into heart-rate variability and illness discussed earlier? Another researcher with the HeartMath Institute, biologist Glen Rein, conducted a slightly different experiment. Rein's research indirectly supports the theory that consciousness extends beyond the level of our physical bodies. While working at the HeartMath Institue, biologist Rein discovered that individuals who could enter a state of "heart consciousness" such as heartfelt appreciation or unconditional love could actually alter the winding and unwinding of DNA in solution! They could do this whether they were holding the DNA in a test tube or merely if the test tube were in the same room as the test subject. When test subjects held the loving feeling in their hearts, their heart rhythms became extremely coherent. As their electrocardiograms were analyzed by sophisticated frequency-analysis software, the feeling of love in the heart center was associated with a heartbeat-frequency distribution pattern resembling a smooth, regular sine wave. Fellow HeartMath researcher Dan Winter also noted that when individuals were holding loving, appreciative consciousness in their heart centers, the same unique, coherent rhythm pattern found in test subjects' electrocardiograms could also be measured in the trees growing outside the HeartMath facility some distance away from where the test subjects were being monitored! This research tends to support the idea that certain qualities of human consciousness are capable of influencing events at a distance, and that we are not as separate from one another as the Newtonian thinkers would have us believe.

While some might consider these ideas to be in the realm of scientific heresy or perhaps the product of "magical thinking," I have seen numerous cases in which people have been able to influence distant events in physical reality in very specific ways through the power of their focused consciousness. After all, if such an influence of consciousness on reality were not even possible, why would so many people waste their time praying for world peace, believing that their efforts were only so much wishful thinking? I believe that the people who are successful in programming for specific outcomes of events in external reality are actually "shifting probable futures" toward a future more consistent with the goal of their affirmation or program. Yet another example of how consciousness can literally restructure physical reality is the "Maharishi effect" documented by transcendental meditation (TM) researchers. In one study, crime rates in certain cities in Africa were studied during the time periods just before, during, and after large groups of advanced TM instructors were conducting large meditation sessions. It was noted that crime rates in these cities fell to very low levels at these times. Crime rates had been high in these cities just prior to the arrival of the TM groups, and they returned to baseline levels following the TM groups' departure. This drop in crime rate demonstrated that there was a statistically significant effect at work during the periods of group meditation. Rather than praying for world peace, the meditators were merely trying to create a peaceful consciousness within themselves. While the meditators didn't necessarily focus on reducing crime or violence, their consciousness seemed to have a pronounced effect in decreasing the incidence of crime throughout the cities being studied. Of course, the fact that many people were focusing their consciousness in unison to create inner peace produced a much greater amplification effect than if just a single individual had been meditating. Who knows what the effect might have been had the meditators been united in a conscious intention to actually eliminate crime and violence in the cities, or if they had literally prayed for world peace? The fact that a single individual or a group of individuals working together with conscious, loving, peaceful intention could bring about a real change supports the notion that our consciousness possesses genuine power to manifest positive changes, not only in ourselves but in the world around us as well.

The Higher Self and Our Connection to the Divine Source

We have explored how various techniques—including affirmations, mental programming, visualization, Freeze-Framing, and prayer—can help us create positive changes within ourselves. We have also examined ways to bring about healing changes in our life circumstances, helping our lives to go more smoothly and with less stress. Healing shifts include a subtle restructuring of our astral, emotional, and mental bodies as we reinforce new and better patterns of responding to everyday life and new ways of relating to the people in our lives. But the preceding explanations should not lead you to believe that just "wishing away" illness, stress, and dysfunctional relationships works effectively. And all the mentioned paths of self-help must be used correctly, with the proper intention and focus, in order for true change to occur. While visualization, prayer, and programming may be capable of producing great changes, they are usually only a partial answer to healing. Often, other medical and vibrational-healing approaches may be needed in conjunction with any chosen self-help technique. The important thing to realize is that our capacity to bring about inner change is much greater than most people tend to think. However, the ability to recognize the reasons behind the illness may be just as important as its vibrationally assisted cures.

When we are confronted by physical illness and frustrating life events, it is frequently useful to consider our circumstances from the perspective of our higher selves. The higher self or soul is able to see the "bigger picture" behind the events that shape our lives. It often seems many of us struggle from day to day, searching for reasons to explain why we are suddenly confronted with illness or tragic life events. But our soul is always aware of the hidden meaning behind such circumstances. After all, aren't we on the physical plane for the main purpose of learning and spiritual growth? Many people, though, are already aware of a pecularity associated with inner growth. It seems growth is sometimes difficult if not impossible unless life provides little challenges as well as major obstacles to be navigated around or overcome. Even the human body requires stress to maintain proper health. Just think about astronauts deprived of the stress of earth's gravity for prolonged periods of time. Unless the space travelers maintain regular exercise to stress and challenge their bodies, their bones become weaker. The manner in which we overcome the early obstacles of our lives often determines how we will respond to future challenges, as well as which patterns of spiritual growth we might tend to follow. Some individuals who were teased and taunted for being a little different from other kids often strive to higher levels of achievement and growth as adults. In the long run, such prodding may stimulate individuals to become stronger, smarter, and strive harder to achieve success than the unthinking children who had originally done the teasing. In other words, on a soul level, sometimes your worst critics may ironically turn out to be your greatest allies by prodding you on to achieve greater levels of success and personal spiritual growth than you would have had you not been "challenged." While the superficial motivation may be to prove competence and true worth, external challenges may actually drive us to higher levels of spiritual growth, too. Under certain circumstances, physical disabilities in childhood or adulthood may provide the stimulus to overcompensate as well. One can find examples in which the challenge of a physical disability has driven a person to overachieve, as in the case of the gifted yet physically challenged physicist Stephen Hawking, whose theories on black holes and the nature of the universe are so fascinating.

On a conscious level, few of us are ever aware of the "larger agenda" of the higher self. Yet at an inner psychic level, possibly, we are all capable of accessing information from our higher selves. We truly do have guidance through the quagmires and pitfalls of everyday life. Help is always available from higher levels of the spiritual universe. One needs only to ask inwardly for assistance or to pray for help and guidance. However, help may not always come in the form we expect. It is said that "God works in strange and mysterious ways." The same might be said of our own higher selves. This should not seem so unusual, since the higher self represents the part of us that is more directly linked to the divine energies of God. In fact, the higher self, sometimes referred to as "the true self," is the aspect of our total being that seeks to reunite the conscious personality or ego with the sacred energies of God the Creator. It is at the level of the higher self that we become fully aware of our connection to each other, not only person to person but also to all forms of life on our planet. Within each of us isn't there a secret longing to reconnect with our higher spiritual nature as well as with the divine essence of God? It is a hunger we each have at some level to be reunited with the divine energies of creation and of the Creator. This deep inner longing is an important and integral aspect of all human beings.

A way to attune to the energies of God and to our divine higher selves is by creating a sacred space within our home. This sacred space functions as a kind of makeshift altar for meditation, worship, and attunement to the energies of the divine. It is a place where we can go to be reminded of the fact that in spite of outward appearances, we are really beings of light occupying physical bodies. As such, we need to reconnect with that divine-light energy from time to time in order to recharge our "spiritual batteries." One need not be too elaborate in creating a holy or religious shrine in the home. All that is needed is a conscious intention to create and then consecrate a sacred space. You might create such a sacred space in a corner of the bedroom where you could place a small table with a candle and a statue of the Buddha, or a picture of Jesus or a favorite saint, or merely a small object with personal spiritual significance. The most important thing is that this space be consciously designated as a consecrated place with the sole purpose of "attunement" via meditation and prayer with the divine energies of God as well as a dialogue with the higher self. Just beginning the day with a brief meditation or an attuning prayer can set the outlook for the whole day in a positive light. By attuning to the energies of God and giving thanks for each day of life, we are briefly reminded of our divine roots as multi-dimensional beings of light. By starting the day with a prayer and a "program" for a smooth, flowing day, we make a brief attitudinal adjustment before tackling the tasks at hand. Attunement and spiritual optimism are not only healthy, but also practical. Through them, we begin the day by focusing the energies of our thoughts and by drawing to us those experiences and minor course corrections needed to make our day flow smoothly.

We each have access to the wisdom of our higher self all the time. Yet few of us are aware that it is so readily available. You can use your higher-self connection to determine a health-filled path through life. When deciding on what to eat for a meal, simply ask inwardly about each food, "Is this food I am considering eating for my highest health?" and then listen for a simple yes or no. The answer may come in the form of hearing your own voice telling you yes or no, or it may come in the form of an intuitive feeling. Many of us know the foods that would be better dietary choices for us, as well as the food items that are less than ideal for our long-term health. We often make food choices based on issues of convenience, time constraints, and what we think will taste the best at that particular moment. While Kentucky Fried Chicken and french-fried potatoes may taste great, living solely on KFC and french fries would likely result in clogged coronary arteries before too long. Similarly, chocolate and candy may taste wonderful, but living on a sugar-rich diet can ultimately lead to exhaustion of the pancreas and a variety of other health problems over time. Of course, asking for advice from the higher self and following that advice are often two completely different things. Many of us tend to ignore what we think is the best dietary choice for us because stress, aggravation, and the frustrations of the day often make us reach for foods that will provide instant gratification. The key to using the wisdom of the higher self is not merely in the asking for advice but also in listening to the message received.

A similar approach may be used in accessing the wisdom of the higher self for advice in other day-to-day choices. Often, the wisdom of the heart is synonymous with that of the higher self. By learning to listen to that small inner voice of the heart, we become closer to that inner wisdom we all possess. The brain may hold information and knowledge, but the key to applying that knowledge is true wisdom, a wisdom that sometimes transcends the obvious facts of the situation. As we begin to pay attention to the small inner voice of our hearts, we begin to develop a more loving and intuitive approach to life in general. Even though we may need facts and figures to make educated choices about how to live our lives, we still need room to follow our heart-based intuition and the wisdom of the higher self, even when the advice we receive might sometimes fly in the face of logic. Remember, when we stop for a moment and view our lives from the higher-dimensional perspective of the higher self, the bigger picture can be more easily appreciated and understood.

Another important way to become more in tune with your higher self is by keeping a dream diary. Because your higher self is more directly "hardwired" to the circuitry of your right cerebral hemisphere, it attempts to communicate with your conscious personality or ego through the symbolic language of dreams. By keeping a dream diary and attempting to analyze the metaphorical language of your dreaming mind, you can begin to create a stronger link with your higher self while expanding your psychic and intuitive abilities to greater levels of proficiency. A number of gifted psychics claim that their psychic abilities and precognitive skills really began to blossom when they started keeping dream diaries and analyzing their dreams. Dreams are extremely nonlinear, highly symbolic forms of communication. As such, most dreams are rarely understood through a literal interpretation. Once a person becomes more proficient in understanding the symbolic language of dreams, it becomes possible to begin an examination of the symbolic significance of everyday life events. This higher-self perspective of our lives, what expert intuitive Carolyn Myss refers to as "symbolic vision," can bring clearer insight into daily events that often seem random, yet are laden with "hidden" meaning.

Of course, as Freud said, "Sometimes a cigar is just a cigar," referring to his students' penchants for seeing phallic symbols in everything. In other words, we can become so focused in searching for the symbolic meaning of life events that we may become too ungrounded, looking for deeper meanings in everything to the extent that the literal interpretations of events may actually be ignored. This goes along with many of the objections made by critics of New Age thought who feel that people are already reading too much into things. Some suggest that the idea of people creating their our own reality is not a functional psychological concept because it can actually make people feel overcome by "spiritual guilt." The belief that people create their own illnesses or are somehow not "strong enough" to heal themselves can sometimes do more harm than good because of the accompanying guilt it may produce if they fail to "heal themselves." It is important to recognize that while affirmations may have healing benefits, affirmations and visualizations alone will not easily heal a potentially serious disease like cancer. The key is in learning to take advantage of all our physical resources while still acknowledging the higher guidance and healing potential of our inner spiritual resources.

There is no question that there is often a deeper meaning behind illness, relationships, and the simple events of everyday life. We must learn to walk that fine line between understanding the meaning of our lives on the physical plane while still paying attention to our spiritual roots in the multidimensional worlds that underlie and energize physical reality. However, we need to be active participants in life, not merely mystical observers. This means we must become practical mystics, so to speak. As we stretch our consciousness beyond the limiting filters of emotional reactivity (created by old wounds and past conditioning), we begin to see the bigger picture of our life's meaning. And when we start to use prayer, affirmations, and programming to shift our personal reality in positive ways, we move closer to living lives filled with greater joy, inner growth, abundant health, and spiritual enlightenment, the true goals of life as seen from the soul's perspective.

FOOTNOTES

1. "Optimistic outlook may protect immune system," *Medical Tribune,* vol. 39, no. 14, July 16, 1998, p. 1.
2. Robert Monroe, *Ultimate Journey* (New York: Doubleday, 1994), p. 17.

F o u r

~~

HEALING WITH HOMEOPATHY

What Is Homeopathy?

Homeopathy is one of the most basic forms of vibrational medicine in use today. It has evolved considerably since its early days of medical usage over a hundred and thirty years ago. The development and popularization of homeopathy has been credited to Samuel Hahnemann, a German physician who is said to have come upon the healing principle of "like cures like" in the mid-1800s. This principle refers to giving a sick patient a specific homeopathic medication that in larger doses could actually produce the patient's symptoms in a normal, healthy person. While this type of healing logic might seem strange to most Western-trained physicians, homeopathy is still considered to be a valuable treatment for many illnesses by a large number of European and American health practitioners. To this day, the physician to Queen Elizabeth II and the royal family of England remains a homeopathic practitioner.

Homeopathy was once one of the most popular forms of medicine used in this country. Into the late 1800s and the early 1900s, a large percentage of the hospitals throughout the United States were actually homeopathic hospitals. Few laypeople are aware that the burgeoning homeopathic movement was one of the driving forces behind the formation of the American Medical Association (AMA). The nonhomeopathic or "allopathic" physicians who used drugs and surgery as their only form of treatment believed that there was an encroachment of homeopaths upon what was considered the allopathic physician's "economic turf." Feeling the economic pinch in the battle for health-care dollars back in the mid-1800s, allopathic physicians formed the AMA a year or two after the creation of the American Homeopathic Association (AHA), its main competition. The political pressure exerted by MDs to fight homeopathy back in the 1800s was so vehement that one of the early bylaws of the AMA included a section strictly prohibiting fraternization with homeopathic practitioners or medicinal use of homeopathic remedies. Gradually the allopathic philosophy became the dominant path to health in America. Homeopathy lost its practitioners and its power base.

Homeopathy as a system of healing has its own peculiar rules and principles, one of which is the principle just elucidated, the concept of like cures like. In some ways, this principle is analogous to treating a dog bite by vaccinating with the "hair of the dog that bit you." In a way, the practice of homeopathy has certain similarities to the medical approach of giving vaccinations against various disease-inducing agents. In a typical vaccination, small amounts of a weakened virus are administered in order to build up the body's immunity to a potent virus. In this manner, a potential illness is warded off. Essentially, one is giving tiny amounts of the disease agent to the patient in order to build up an immunity to the disease and to the illnesses it causes. Similarly, allergy-densensitization therapy, which uses weak dilutions of the offending allergens in an intramuscular injection, also follows a homeopathic-type principle whereby "like cures like." When Hahnemann first came across the healing principle of like cures like, he sought to discover which natural substances might be useful for treating specific illnesses. He did this by carefully determining the toxic side effects of hundreds of naturally occurring substances. Hahnemann based this prescribing approach upon his early experience with cinchona bark and the disease malaria. He had been impressed by the fact that cinchona, the treatment of the day for malaria, was found to produce the same intermittent fevers as those seen in malaria when cinchona was given to a normal, healthy individual. Using this same reasoning, Hahnemann theorized that the side effects produced in a healthy person after ingesting a specific medicinal substance could provide prescribing guidelines for that particular substance's healing properties and applications.

In order to carry out his research, Hahnemann enlisted the aid of medical students, who would first ingest large quantities of specific medicinal substances and then meticulously document their physical, emotional, and mental reactions to those substances in journals that would be collected and compared for commonalities. The technique of pooling information about medicinal side effects in order to determine a remedy's medical indications is sometimes referred to as homeopathic drug "proving." During the drug provings, those symptoms that are most consistently and strongly experienced by the majority of drug testers are interpreted to be the guiding or "keynote" symptoms for a particular drug (when it is given in an extremely diluted form). Such symptoms come to be regarded as useful indications for prescribing a particular homeopathic remedy. For instance, being in close proximity to recently cut onions will often cause immediate irritation to a person's eyes, leading to watering and tearing from the eyes, a runny nose, and copious clear nasal drainage. Now, if a homeopathic practitioner has a patient who presents with early cold symptoms such as a watery, runny nose and irritated eyes, the healer will administer allium cepa. And allium cepa is a specially prepared homeopathic remedy made from onions! Allium cepa offers a perfect example of curing an illness by giving a homeopathic remedy made from a substance in nature that reproduces the main symptoms of the illness. In order to characterize all the different uses for single remedies, homeopaths frequently teach students what a single patient with all the different symptoms or indications for a particular remedy would look like. The "homeopathic drug picture," as it is known, is a complete portrait of a single individual suffering from all of the many toxically induced symptoms produced by a medicinal substance, literally the entire list of toxic side effects associated with ingesting large amounts of the raw substance. Before becoming alarmed, though, remember that in homeopathy only minuscule amounts of the original medicinal substance are therapeutically used, thus avoiding the actual production of any toxic side effects. In order to treat an illness, the homeopathic physician searches homeopathic reference books to find a remedy with a homeopathic drug picture matching the totality of symptoms experienced by the sick patient. If the correct remedy is chosen, a healing of the illness often occurs. But if the incorrect remedy is prescribed, nothing happens, and the illness continues unabated. Healing will occur only if the symptom complex of the patient exactly matches the homeopathic drug picture of the remedy given.

While many of Hahnemann's original homeopathic drug provings with medical students often used raw concentrates of medicinal substances for their testing procedure, homeopathic remedies in Hahnemann's day as well as today are still prepared from extremely weak dilutions of the same medicinal substances. Most homeopathic remedies are so diluted they rarely contain more than a few molecules of the original healing substance. This is the paradox behind homeopathy that is so unfathomable and frustrating to most Western-trained, logical physicians. The idea that extremely dilute remedies can produce strong healing effects seems to make no sense from a logical, Newtonian viewpoint. Western medical practitioners expect stronger effects

The Preparation of a (Plant-Based) Homeopathic Remedy

healing plant, herb, etc.
+
alcohol-water mixture
→ **plant is thoroughly ground up and solution is filtered**
→ **mother tincture of plant / herb**
→ **1 drop of mother tincture + 99 drops water (1:100 dilution)**
→ **mixture is shaken violently (succussion)**
↓
1C (1:100) potency homeopathic remedy
↓
1 drop of 1C potency remedy
+
99 drops water
↓
succussion (shaking)
↓
2C potency (1:10,000) remedy

process of homeopathic potentization

← **process is repeated (dilution and succussion)**
← **progressively higher-potency remedies made**

The higher the dilution, the more "potent/potentized" the homeopathic remedy
↑
Increased potencies mean less gross physical plant substance and more plant life-force energy in a purer form

← **homeopathic solution taken directly (sublingual)**
← **homeopathic solution added to milk-sugar tablets for ingestion (sublingual)**
← **homeopathic solution added to neutral cream base for making salves (topical usage)**

from increasing, not decreasing, dosages. Yet the clinical effectiveness of homeopathic remedies is finding growing validation in both laboratory and clinical studies. The unique homeopathic method of progressive dilution seems paradoxically to transform a potentially toxic substance from a physically based medicine into a more energetic type of vibrational-healing modality. The specific mechanism behind the effectiveness of dilute concentrations of herbs and other healing substances remains something of a mystery as well as a matter of ongoing debate and speculation in various medical and scientific circles. However, there does seem to be something special about homeopathic preparation techniques that appear to transfer some kind of healing memory pattern or vibrational aspect of the healing plant into the water used to make the extremely diluted remedies.

Homeopathic remedies are commonly prepared using a specialized technique known as succussion and potentization. Succussion refers to the process of violently shaking a diluted homeopathic remedy in solution. When using a plant-based herbal substance to prepare a typical homeopathic remedy, a concentrated solution of alcohol, herbs, and water is ground up and mixed together to form a "mother tincture." From this mother tincture, 1 drop of plant-extract solution is added to 99 drops of water and the combined solution is shaken together vigorously. In the old school of preparation, the sealed tube of solution was hit against the palm of the practitioner's hand over and over again. This technique of succussion, of shaking and hitting a sealed vial of homeopathic dilution repeatedly against a hard surface, is felt to be critical to the homeopathic preparation process. Today, modern homeopathic pharmaceutical companies have developed mechanical equipment that duplicates this rather time-consuming process. After a 1-to-99 parts dilution of plant extract has been shaken, a single drop of that solution is added to another 99 drops of water, and the succussion process is repeated. In homeopathic vernacular, a 1-to-100 dilution is referred to as a 1C potency, "C" referring to a centessimal scale of 1 to 100. When that 1C potency is again diluted, 1 to 100, it is now called a 2C potency, and so on. The number that describes the homeopathic potency refers to the number of times the 1-to-100 process of dilution has been repeated. For instance, in a 100C-potency remedy, the process of 1-to-100 dilution and succussion has been repeated 100 times. Another common potency of homeopathic remedies, based on a 1-to-10 dilution or decimal scale, is referred to as "X" potencies, i.e., 1X, 2X, 100X potencies, etc. The real paradox of the homeopathic dilution process is that the more times a remedy is diluted and succussed, the more potent its effect is considered to be by homeopathic physicians. That is, a 100X-potency remedy is considered stronger than a 1X potency, yet the 100X potency is much less likely to have even a single active drug molecule compared with the "weaker" 1X remedy. This process of progressive dilution and succussion, which is said to increase a remedy's strength of action, is referred to as potentization. Once a homeopathic dilution has been created using the aforementioned process, the liquid itself may be used to make a variety of forms of remedy applications. The potentized homeopathic dilution may be administered to the patient directly, as a liquid under the tongue, or more commonly, the liquid is added to a bottle of milk-sugar tablets, which become coated with the active homeopathic agent. Homeopathic tablets are the most common form of remedy prescribed by homeopathic physicians. Rather than swallowing the pills, patients place the homeopathic remedies in pill form under the tongue to dissolve in order to get the maximal effect. Alternately, a homeopathic dilution may be mixed with a neutral base to create homeopathic creams for direct application to the skin to treat various skin problems such as rashes, bites, injuries, or irritations.

Homeopathy: How Does It Work?

The fact that homeopathy continues to be used by practitioners throughout the United States and Europe attests to its effectiveness. A treatment based solely on the placebo effect could not enjoy such popularity for so many years if there were not something to it. Many different studies have been published examining the efficacy of homeopathy in treating various medical conditions. But most conventional medical practitioners remain unconvinced of its value because some studies on homeopathy's therapeutic benefits show a positive healing outcome whereas other studies do not. In other words, there is as yet no general agreement by orthodox medical practitioners as to whether homeopathy works at all. Whenever there have been strongly positive studies showing a beneficial effect of homeopathy, tremendous controversy has been the ultimate result. Assuming that one believes at least some of the studies showing a significant effect of high dilutions on people and cellular preparations, there must be some explanation for homeopathy's physiological and healing effects.

The reason homeopathy remains so controversial is that the dilutional principle underlying it appears to fly in the face of the established logic of Newtonian physics that forms the basis for conventional drug therapy. In order to understand this seeming contradiction, we need first to examine the kind of therapeutic logic behind drug dosing used by orthodox medical practitioners when prescribing conventional types of medications. In conventional medicine, drugs are prescribed for patients in very specific dosage amounts that have been shown to produce positive therapeutic effects in human beings with a minimum of side effects (in theory, of course). Many of today's newer drugs are often designed using computer-modeling techniques allowing pharmaceutical companies to create medicines that will interact in very specific ways with particular chemical receptors on the surfaces of cells throughout the body. For instance, in the treatment of hypertension, typical blood-pressure medications used are referred to as beta blockers, alpha blockers, calcium-channel blockers, and the like because they block the activation of certain types of cellular receptors that regulate functions related to blood-pressure control. These cellular receptors are like tiny switches designed to turn on or off certain aspects of normal cellular functioning. Receptor-specific drugs are created to bind to particular cellular-receptor switches with two main kinds of effects. Either the binding drug is highly active and enhances the cellular function normally turned on by the activated receptor, or else the receptor-bound drug blocks the receptor from being activated by naturally occuring chemicals.

Hypertension and other illnesses are amenable to treatment using receptor-specific medications. But let's use asthma to further describe receptor-specific medicine. The illness of asthma is frequently treated with a drug known as theophylline. Theophylline, a drug very similar to caffeine, binds to a specific cell receptor in the muscle cells surrounding the tiny bronchial airways of the lung. During a typical asthma attack, the muscles in the bronchial airways constrict, causing wheezing and shortness of breath due to a decrease in airflow through the lungs and a resulting drop in vital blood-oxygen levels. The administration of theophylline by pill or intravenous solution relieves asthma attacks because theophylline binds to a cell receptor on the bronchial muscles of the lungs, causing the constricted muscles to relax and allow better airflow through the lungs. Typically, theophylline dosages are calculated for each patient based upon age and body weight, in order to provide an optimal therapeutic drug level while avoiding toxic side effects caused by too much of the drug. Theophylline is a perfect example of a drug with a typical linear dose response. That is, the higher the amount of drug given, the more cell receptors are activated by the drug molecules binding to those receptors. If too little of the drug is given, too few receptors are activated and very little bronchodilation occurs, resulting in minimal improvement in the asthma attack.

When conventionally trained physicians try to apply this same drug logic to understanding the therapeutic use of homeopathy, much confusion and controversy arises. If, in homeopathy, the more dilute solutions of medicines are more potent, it would seem to go against the common sense of linear dose responses. That is, physicians are taught that the higher the drug dosage a patient is given, the more receptor molecules in the body become bound to the drug and the stronger the drug's effect. Conversely, the weaker a dose of medicine given, the fewer the number of drug-bound receptors and the smaller the drug's physiological effect upon the patient. Flying in the face of this linear, analytical reasoning is the thinking of homeopathic practitioners, who consider high-potency homeopathic remedies to be more powerful in their effects than low-potency remedies. The paradox here is that the higher-potency homeopathic remedies are less likely to contain *any* drug molecules of the original medicinal substance than low-potency remedies, yet they are considered more powerful in their effects upon the patient. How can such a contradictory phenomenon make any sense? If, as homeopaths contend, high dilutions really potentize the effects of a medicinal remedy, then where does the healing effect come from? There are a number of different theories as to how a homeopathic solution or remedy containing almost no drug molecules could possibly trigger a healing response within the body.

The key to understanding homeopathy is that it is primarily a nonphysical or energetic healing modality. The action of a homeopathic remedy does not occur through the same physiological pathways activated by the administrations of conventional medications. One way of conceptualizing homeopathy's mode of action is to consider human beings from the multidimensional, vibrational-medicine perspective again. We know that human beings are more than just a physical body. As suggested in Chapter One, we are unique energy systems with many complex energy-control systems helping to regulate and maintain the health of the physical body. One view of human beings as energy systems considers the physical body as a system with a predominant or resonant frequency of energy. Everything in the world—be it a plant, an animal, or even a rock—oscillates and vibrates to some degree. The key frequency of vibration of an object or animal is its resonant frequency. Vibrating or oscillating systems will maximally absorb energy if the energy is delivered in the resonant frequency of the system. For instance, if you have two well-tuned guitars in a room and you pluck the E string of one guitar, the sound frequency of the plucked string will make the E string of the second guitar vibrate as well, because the sonic energy is in the resonant frequency of the E string. Another famous example of the powerful effect of resonant-frequency transfer is the Memorex tape commercial where Ella Fitzgerald's voice shatters a wineglass by hitting the resonant note of the glass. In the case of the wineglass, the amount of energy resonantly absorbed is so great that it literally shakes the glass to pieces. The wineglass example might be considered a form of "*de*structive resonance." Conversely, homeopathy may be considered a healing modality that works according to the principle of "*con*structive resonance," where a resonant-energy transfer heals instead of destroys.

Although this will be a gross oversimplification, consider that the human body has its own inherent resonant frequency. Let us say a healthy body vibrates at 300 cycles per second (hypothetically speaking). The frequency of energy at which the body is vibrating or resonating is also a reflection of its current state of health or illness. In other words, when a person is sick, that person's resonant frequency might change from 300 to 350 or even 550 cycles per second, depending upon the nature of the illness. When an individual is sick, the body is doing everything in its power to throw off the illness and reestablish a state of equilibrium and balance. So while many people consider high fevers to be a "bad" symptom caused by invading germs, the higher body temperature actually stimulates the body's immune cells to eat up and destroy bacteria and other microbial invaders with greater efficiency. In other words, the fever experienced by a sick individual is actually a "good" symptom in that it is a strategic move by the body to activate a healing response to the illness. From an energetic standpoint, when a person is experiencing certain types of fevers or sweats, these symptoms may be an indirect reflection of the frequency of vibration at which the body is resonating, what might be called the "disease frequency."

For the purpose of this example, let us say a person exhibiting fevers and sweats is vibrating at a frequency of 350 cycles per second. In theory, if a person could get more energy in the frequency at which he or she is resonating—i.e., the disease frequency of 350 cycles per second—it might provide the activation energy needed to purge the toxicity of the illness completely from the body. By attempting to match the keynote prescribing symptoms attributed to a specific homeopathic remedy with the symptoms of an acutely ill patient, the homeopathic practitioner may actually be matching the vibrational frequency of a homeopathic remedy with the vibrational frequency of the patient. According to this "resonance theory" of homeopathy, a vibrating system will maximally absorb energy in its resonant frequency. Therefore, only a perfect match between the frequency of the sick patient and the frequency of the homeopathic remedy would provide a resonant-energy transfer to the patient and thus induce a healing response. In the case of the patient with fevers and sweats, if the homeopathic practitioner can empirically select a remedy whose resonant frequency is 350 cycles per second, the same as the sick patient's, the frequency match between patient and remedy will trigger a healing of the illness.

Only the correct selection of a particular homeopathic remedy will produce any effect upon the patient at all. Because homeopathic remedies are so extremely dilute, there are no side effects experienced from taking an "incorrect" remedy; there is merely no effect at all. If the correct remedy is selected and the illness is healed, it may be because the homeopathic remedy is providing a form of subtle energy to the body in the exact frequency of vital energy needed to bring about a healing of the specific illness. The homeopathic preparation process of progressive dilution and potentization is able to "extract" the higher vibrational qualities or vital life-energy patterns from the herb's plant substance directly into the water. As the dilutions (and potencies) of the remedies become higher and higher, with fewer and fewer physical molecules of plant actually left in solution, homeopathic remedies become less physical in nature and more ethereal and subtly energetic in their actions. In fact, homeopathic remedies may work primarily by energetically healing illness at the level of the etheric body, which then goes on to rebalance the physical body.

Another theory of how homeopathy works also has to do with the remedies providing a kind of subtle vibrational energy, but the mechanism is conceptually different. According to the resonance theory of homeopathy, merely providing the frequency of energy needed by the body is all that's required to heal using homeopathy. But what if the "something" being exchanged between the remedy and the patient is not strictly energy to revitalize the system but is actually a form of coded energetic biological information or "bioinformation" that may provide instructions to the body to help in the healing process? The "bioinformational theory" of homeopathy views the body not just as a biological-energy network but also as a complex information-processing system, a kind of biocomputer that uses coded information to regulate its many component subsystems. Most people think about information processing in the human body in terms of electrical impulses rapidly moving throughout the nerve cells of the brain. People also picture the impulses traveling along nerve pathways connecting the brain and bodily systems. Nerve cells transmit a form of electrical bioinformation that communicates messages between the brain and body. These bioelectrical nerve signals help us to think, to move about, to express ourselves creatively, and yet simultaneously regulate the organs, muscles, and bodily systems that help keep us alive. Our bloodstream also carries other types of coded bioinformational signals in the form of hormones, peptides, and other biochemicals that carry chemical messages back and forth between the brain and the organs and glands of the body.

Over the past twenty to thirty years, scientists have discovered that the biological information used to regulate the activity of the body's various systems exists in the form of physical codes, such as the biomolecules of hormones and neurotransmitters, but also in the form of energetic codes. These energetic codes of information may trigger the same cellular reactions that the molecular codes of information activate. As an example, the cells of our bones normally communicate with one another by both chemical and electrical signals. In recent years, orthopedic surgeons have begun to use electromagnetic treatment devices known as "bone stimulators" to accelerate the repair of especially difficult bone fractures that can't heal normally. The orthopedic bone stimulator energetically stimulates new bone growth and bone repair by transmitting healing messages carried in the code of pulsed electromagnetic (EM) fields tuned to a very specific frequency. For bone healing to work, researchers discovered, the frequency of the EM fields used to radiate the fracture site had to be so precise that a slight variation in frequency would change the message carried to the bones from "lay down new strengthening calcium matrix" to "reabsorb calcium from the bones to weaken the bone structure." In other words, any slight deviation in the coded energy message would not only fail to heal the broken bone but would actually weaken and aggravate the situation. In the case of bone stimulators, the body accepts a coded electromagnetic signal, in place of a normal biochemical or electrical signal, by using pure energetic bioinformation to instruct the bone tissue to activate the healing process. This is a wonderful example of how an information-carrying signal can exist not only as a chemical or hormonal signal but also as a vibrational message coded in the form of an electromagnetic field tuned to a specific frequency.

Now, how does this phenomenon of coded energetic information apply to understanding homeopathy? In the example of the bone stimulator and its ability to send energetic informational signals to accelerate healing in bone tissue, the healing bioinformational message is interpreted by the body just the same as if it were receiving the normal chemical and hormonal signals. It is a case illustrating how the body can input and use different kinds of bioinformational languages to trigger the same cellular healing reactions. Whether that signal is transmitted in chemical, electrical, or electromagnetic forms of coding seems to make no difference to the body. The end result appears to be the same. In the case of homeopathy, potentized homeopathic remedies may be carrying a kind of subtle-energetic bioinformational message that may stimulate certain aspects of the physical (and spiritual) body's energetic healing systems. An interesting example from homeopathic research may help to support and further illustrate this concept.

The drug cimetidine, more commonly known as Tagamet, is a medication used to treat stomach-acid problems. The mechanism behind the action of Tagamet is that it works as a kind of specialized antihistamine. When taken orally, Tagamet binds to histamine receptors in the stomach designed to normally trigger the release of stomach acid. By blocking the stomach's histamine receptors, Tagamet decreases the production of acid, thus relieving heartburn and a variety of other gastric disorders. Tagamet also binds to histamine receptors on special cells in the bloodstream known as basophils. Basophils are rich in histamine granules, and they release their load of histamine into the circulation when an allergic person comes in contact with ragweed pollen, dust, grass, or other typical allergy-producing substances. Allergic people frequently use antihistamines to relieve allergy symptoms. Antihistamines block the cellular release of histamine, the biochemical that produces most of the allergic symptoms of watery, itchy eyes, sneezing, and the like. One of the ways allergic reactions can be simulated outside the body is by drawing an allergic person's blood, extracting the basophils into a special preparation, and observing those basophils under the microscope as various solutions of allergens are added to the slide. If the individual's cells are allergic to a certain substance, the irritant will trigger the release of histamine granules. If, however, the basophils were pretreated with a histamine blocker such as Tagamet prior to the allergen's being added, no histamine granules would be released by the offending allergen.

The basophil-stimulation test was used to study the effects of homeopathic dilutions in a rather controversial study published in the prestigious scientific journal *Nature* in the late eighties. Preparations of harvested basophils were observed under the microscope while conventional-strength preparations and homeopathic dilutions of allergens were added to the basophils. When high molecular concentrations of allergens were added to the basophils, they easily released their load of histamine granules. But when progressively higher homeopathic dilutions of the same allergens were next added to slide preparations of basophils, the results were just as effective in causing histamine release. Researchers in several different international labs found that even ultra-high homeopathic allergen dilutions, containing fewer than one or two molecules of allergen, were able to trigger histamine release similar to the high-concentration allergen solutions. One of the implications of this experiment is that progressively higher dilutions of the allergen, nearly devoid of any allergen molecules, must still carry the "memory" of the allergen in the water. The fact that ultrahigh dilutions could produce observeable cellular changes seems to be a validation of the homeopathic effect. This memory effect of water suggests that some change in the very structure of the water molecules produced by the progressive dilution process seems to give the "treated" water an ability to influence cells, even in the absence of any original allergen molecules in the solution.

While the results of this controversial experiment continue to be debated, there was one interesting finding that never made it into the final published article. In a variation on the high-dilution experiment, one of the researchers tried pretreating the basophil preparations with the histamine-blocking drug Tagamet prior to allergen exposure. High concentrations of Tagamet, as well as homeopathic dilutions of Tagamet, were used. The results demonstrated that homeopathically prepared Tagamet blocked the release of histamine granules as effectively as high doses of the actual drug did. Even though the high homeopathic dilutions were unlikely to contain more than a few molecules of Tagamet, a similar reaction occurred. This unpublished finding suggests that the bioinformational message carried by the physical drug molecule could also be transmitted by a homeopathic dilution of the same pharmaceutical substance. In other words, the same message could be transmitted to cells of the body using two different forms of symbolic bioinformational language.

We know that concentrated Tagamet binds to histamine receptors on the surfaces of basophils, thus blocking the release of histamine in response to allergen exposure. But when using homeopathic dilutions of Tagamet to pretreat basophils, there are no physical Tagamet molecules present to bind to those cellular receptors. If this is a real phenomenon, how do the high dilutions of Tagamet achieve this feat? It may be that some other form of coded message—a nonmolecular message, perhaps a vibrational-energy pattern of the Tagamet imprinted into the water—affects the basophils. The homeopathic Tagamet may be sending the same instructional message to the basophils' histamine receptors. And even in an energetically coded form, the cells still recognize the Tagamet's "informational message" and react accordingly. There is a distinct similarity between the homeopathic Tagamet example and the case of the orthopedic bone stimulator used to accelerate the healing of bone fractures. In both cases the cells of the body are recognizing molecular and energetic bioinformational messages as if they were interchangeable. The key to the homeopathic effect may be the body's ability to recognize bioinformational messages in the form of subtle-energy signals carried by potentized homeopathic dilutions. The homeopathic messages seem to carry specific types of stimulating instructions to the body that trigger various energetic and molecular aspects of the healing response.

Both the resonance theory and the bioinformational theory of homeopathy may be applicable to explaining how homeopathy works. However, it is not always critical to understand exactly how something works for it to be a valuable therapeutic modality. After all, aspirin had been used by physicians for more than a hundred years before scientists began to establish its true mechanism of action. The two models presented here to explain homeopathy are primarily directed toward the energetic and metabolic systems of the physical body. They do not take into account the vibrational-energy components of the etheric body and the higher spiritual bodies. Nobody knows for sure how homeopathy affects these higher vibrational bodies, but it is suspected that most homeopathic remedies may have some kind of energy-rebalancing effect upon the etheric body as well as the physical body. Some sources have suggested that homeopathic remedies vibrationally duplicate the illness in the body in order to actually push it out of the body. To understand how this really works, we will have to extend our etheric-measuring technologies to much more sensitive levels of detection than are currently available. Some intuitive homeopaths believe that homeopathic remedies can work not only on the physical and the etheric bodies but on the higher spiritual bodies (such as the astral and mental bodies) as well. It's hoped that future research will help to firmly establish the true healing mechanisms behind this very important and very old vibrational-healing system.

The Many Methods of Homeopathic Prescribing

There are a number of different approaches used to prescribe homeopathic remedies for acute and chronic medical conditions. The most common form of homeopathic prescribing is known as "acute prescribing" or "homeopathic first aid." The term "acute prescribing" refers to the use of homeopathic remedies for treating acute medical conditions that have suddenly arisen. For instance, the sudden onset of a cold, a cough, diarrhea, a stomach upset, or a bruised muscle is frequently treated with common homeopathic remedies of a low potency, i.e., 12X to 30X or 12C to 30C potencies. Because homeopathic practitioners consider the low-potency remedies to be weaker in strength, they usually need to be repeated every two to four hours for an acute condition. The timing may vary, though, depending upon the patient's condition and response to a homeopathic remedy. One of the best homeopathic remedies for acute injuries is Arnica. Any injury to the body leaving a physical bruise or giving a sensation of feeling bruised in the muscles or joints will quickly respond to Arnica. Arnica may be given in a tablet form, which is dissolved under the tongue, or it can be massaged into the sore muscle in the form of an Arnica cream. Either form of Arnica can relieve the bruised feeling and can frequently accelerate the disappearance of new black-and-blue marks. In some cases, blood trapped under the skin in the form of a hematoma or blood blister will also disappear in a matter of a few days or less with the application of Arnica cream.

In general, homeopathic remedies are prescribed by matching the symptoms of the patient with the symptom indications listed for a particular remedy.

The Various Approaches to Homeopathic Prescribing

TYPES OF HOMEOPATHIC PRESCRIBING	TYPES OF REMEDIES USED	NATURE OF THE PRESCRIBING PROCESS	POTENCIES OF HOMEOPATHIC REMEDIES USED
Homeopathic first aid (classical homeopathy)	**Single remedies given for acute symptoms**	**Prescribing is guided by symptoms and "modalities" (aggravating and relieving factors)**	**Low-potency remedies given with frequent repeat dosing**
Complex Homeopathy	**Combination remedies given (contain multiple homeopathic remedies) For acute symptoms usually (and sometimes chronic symptoms)**	**Prescribing is guided by main symptom (e.g., back pain, allergy symptoms)**	**Combination remedies usually contain low potencies (especially over-the-counter remedies) Physician-prescribed complex remedies may contain high and low potencies**
Electroacupuncture-based homeopathy (EAV/EDI)	**Single or combination remedies used (alone, in combination, or in sequential order) For acute and chronic symptoms**	**Prescribing is guided by patient's electronic-resonance reaction to remedy placed in circuit with patient via acupoint connection on hands or feet (using electrodermal testing system)**	**Low or high potencies used**
Constitutional Homeopathy (classical homeopathy)	**Single remedies given for chronic symptoms**	**Prescribing is guided by lengthy homeopathic interview, likes and dislikes, background medical and symptom history, emotional constitution of patient**	**High potencies frequently used**

Homeopaths tend to use two main guides when prescribing remedies. The first is known as the homeopathic Materia Medica. The Materia Medica lists all the different symptom indications for each commonly used (as well as some rarer) homeopathic remedy. For each remedy, the Materia Medica lists different symptoms according to organ system. The other guide to prescribing is known as Kent's Repertory, named after a famous early homeopathic physician named James Tyler Kent. Instead of being organized according to remedies, the Repertory is a guide that lists symptoms of specific organ systems according to different categories. For instance, a symptom such as pain in the left foot might have five to ten different homeopathic remedies listed. The Repertory shows which remedy is most strongly associated with that particular symptom by printing its name in boldface type or italics to distinguish it from remedies that may be less likely to be of help for that particular symptom. Homeopathic practitioners will often consult Kent's Repertory first, noting each of the patient's symptoms while looking for any remedy that keeps coming up for each symptom. This practice of consulting Kent's Repertory to determine the correct remedy for a particular patient's illness is known in homeopathic circles as "repertorizing the patient's symptoms."

One of the ways of homing in on the correct remedy for an illness with different symptoms is based upon questioning patients about what things make their symptoms better or worse. The things that can improve or worsen the medical condition are known in homeopathic parlance as "modalities." Common modalities include such things as temperature preferences (does the condition feel better with hot or cold applications or when the patient is in a warm or cold room?) and food likes and dislikes (does the condition feel better with warm or cold liquids or after the patient has eaten bland or spicy foods?) as well as which side of the body may be affected (there are different remedies for conditions affecting the left versus the right side of the body). With regard to sidedness of symptoms, my wife once experienced a case of Bell's palsy, a condition that paralyzes the muscles on one side of the face, giving the stricken person an appearance resembling that of a stroke victim. Upon consulting Kent's Repertory, we found that there were different homeopathic remedies used to treat left-sided versus right-sided Bell's palsy. With the selection of the correct remedy, my wife quickly recovered without any residual symptoms. Each homeopathic remedy has particular modalities associated with it that help the homeopathic practitioner or the layperson using a homeopathic first-aid kit to select the most appropriate remedy for the situation.

Another style of homeopathic prescribing is known as chronic or "constitutional prescribing." This approach to homeopathy attempts to select the remedy for a particular patient that best fits his or her entire life history. Constitutional prescribing is a form of "old school" or classical homeopathy that involves extensive history-taking on the part of the homeopathic practitioner. Special attention is paid to unusual or "keynote" symptoms that stand out in the patient's history. One of the interesting things about homeopathic history-taking is that practitioners love to hear about strange, rare, and peculiar symptoms that most conventionally trained MDs don't quite know what to make of. Let's say, for example, you tend to wake up every night at four o'clock A.M. with feelings of dread and a craving for salty foods. These symptoms might actually have great significance in helping the homeopath to home in on a remedy that would be appropriate for you. Things like a driving need to be by the seashore or an aversion to sweets or various other strange preferences or aversions frequently have deeper meaning to a homeopath trying to repertorize a patient's symptoms.

Here's another interesting aspect of homeopathic history-taking: Mental and emotional symptoms are often given greater weight or meaning by homeopaths than are physical symptoms. Conversely, constitutional homeopathic remedies, when correctly prescribed, can sometimes produce major healing effects in patients that involve not only a disappearance of chronic physical symptoms but also a profoundly positive shift in mental and emotional states. Constitutional prescribing makes use of patients' mental and emotional patterns, their taste preferences and aversions, their unique climate preferences, whether or not they have left- or right-sided bodily symptoms, as well as their specific physical and mental/emotional complaints in order to select the most appropriate constitutional remedy. One unique difference between acute and chronic homeopathic prescribing is in the potencies of remedies usually given to patients for each style of prescribing. Acute homeopathic prescribing tends to use lower-potency remedies. Constitutional remedies for chronic health challenges are frequently given in very high potencies such as 1M ("M" equals one thousand dilutions) or higher. While low-potency remedies are closer to molecular concentrations and are thought to have greater effects upon the physical body and physical symptoms, homeopaths consider the high-potency remedies to have stronger effects upon the higher spiritual bodies and a greater impact in correcting long-standing dysfunctional mental and emotional symptoms.

The approaches to acute and chronic or constitutional prescribing are still based on what is referred to as classical homeopathy, in which remedies are selected primarily on the basis of history-taking and symptom analysis using the homeopathic repertory book. In more recent years, a number of other methods of homeopathic prescribing have arisen. One approach is based upon a technique that is variously referred to as "applied kinesiology" or "energy kinesiology" or simply "muscle-testing." This approach relies upon a kind of reflex testing of the body's muscle groups, a technique originally developed by Dr. George Goodheart, a chiropractic physician from Detroit. Since its development, many people have adapted muscle-testing in different ways, renaming it with their own particular nomenclature, but the technique is pretty much the same, with only minor variations. In standard kinesiology, the practitioner typically has a patient extend an arm, usually at a right angle to the body. The practitioner then asks the patient to resist while exerting a fixed amount of force against the muscle to try to force the patient's arm back down against the side of the body. A baseline test of the arm and of a particular muscle group's strength is initially performed to get a sense of the patient's strength. The strengths of different muscle groups are said to reflect the energetic and health status of specific organ systems because of acupuncture-meridian links shared by certain organs and muscles of the body. If the practitioner is unable to force down the muscle during a kinesiologic evaluation, the muscle is said to test "strong." The interesting part comes when different substances are introduced into the patient's energy field. When a noxious substance is introduced into the vicinity of the patient's energy field, as in having patients hold an unlit cigarette in their lips, retesting the muscle will now show it to be "weak." If the cigarette is removed from the patient's lips and the patient is immediately retested, muscle strength immediately increases, returning again to the normal baseline level.

In a way, kinesiology is supposed to be similar to a lie-detector test. The body literally reveals what it likes and doesn't like according to whether a substance tests strong or weak when held by the patient during muscle-testing. There are many variations on this technique. Also, different muscle groups must be tested to evaluate different organ functions. Certain advocates of the kinesiology approach attribute the mechanism behind kinesiologic muscle-testing to be the acupuncture-meridian system and its specific connections to particular organ systems. By placing a substance that is noxious to the body in close proximity to its surrounding auric field, the negative energy impact is said to affect the flow of ch'i energy to the muscle group being tested. The result is a temporarily weakened muscle. Conversely, if a patient's muscle group is somewhat weak during initial testing, placing a substance good for the patient's health may result in a strengthening of the tested muscle for as long as the substance remains close to the patient's body. Researchers who have attempted to validate kinesiology as a diagnostic technique have found the consciousness of the muscle *tester* to be a key factor in getting accurate information. There is some suggestion that the kinesiologic reflexes of the body may be a kind of unconscious "dowsing reaction," similar to the unconscious muscle movements behind dowsing and water-witching. Clearly, kinesiology tends to be a very subjective technique.

That having been said, it also appears, under appropriately controlled conditions, that kinesiologic muscle-testing is frequently able to provide accurate information about the health and energy status of the body's various physiological systems. Reliable data about the body's response to different medications, foods, and even allergic substances can also be derived via kinesiologic muscle-testing. Some practitioners tend to mix old-style classical homeopathic prescribing with newer muscle-testing techniques. I have known classically trained homeopaths who will take a detailed homeopathic history from a patient and then test their assessment of a selected remedy's appropriateness (as well as the correct potency of the remedy to give) based on kinesiologic testing of the patient. The patient will often be muscle-tested three times while holding one of three unmarked vials of homeopathic remedies, only one of which is the correctly selected remedy for the patient. The remedy is held by the patient (usually with the handheld remedy pressed against the left upper-abdominal region, the site of the spleen—a major immune-system organ) while the free arm's hand muscle is tested for a kinesiologic response. If the incorrect remedy is held, the muscle will test weak. However, when the appropriate remedy for the patient is held while muscle-testing, there will be an immediate strengthening of the muscle. The strengthening response indicates a positive kinesiologic reaction and a confirmation that the body "likes" or is energetically rebalanced by a particular remedy. This procedural example illustrates how classical homeopathy is being updated with newer and more modern testing methods that can directly (or indirectly) evaluate the body's energy response to homeopathic remedies.

Another technique for homeopathic prescribing that is becoming increasingly popular comes under the heading of "instrument-assisted homeopathic prescribing." In this approach, a patient's health status is evaluated using special computerized systems that can measure and quantify electrical-energy changes within the acupuncture merdians and acupuncture points of the body. The prototype for this method of testing is a machine sometimes referred to as the Voll machine or EAV (Electroacupuncture According to Voll). The Voll machine is an electrical-measurement device first developed by a German physician, Dr. Reinhold Voll, in the late 1950s and early 1960s. Voll initially developed his system as a way of electrically assessing the health status of the organs of the body through their link with the acupoints and meridians. Voll found that acupoints had particular electrical parameters easily measured using a handheld electrical probe that could be pressed against various key acupoints on a patient's hands and feet. According to classical acupuncture theory, each of the major organ systems of the body such as the heart, lungs, liver, and kidneys is connected to its own acupuncture meridian, which supplies each organ with vital ch'i energy. Each of the major meridians ends at key acupoints in each of the fingers and toes. By electrically measuring the energy status of these key control acupoints, Voll could determine ch'i energy imbalances in each organ. Voll recognized the detected imbalances as precursors to actual physical illness. Therefore, he used the energy-imbalance readings to diagnose illness before physical organ deterioration began. In retrospect, it appears that the readings from Voll's technology may have actually reflected energy disturbances at the level of the etheric body. From a diagnostic perspective, such organ-system-specific sensitivity to subtle-energy disturbances also allowed for testing the body's immediate energy response to different treatments of a homeopathic nature.

During a typical EAV session, a patient sits holding a metallic tube connected by a wire to the Voll system. At the same time, the practitioner completes the electrical circuit by touching one of the patient's acupoints with a handheld electrical probe also connected to the Voll machine. The electrical-energy characteristics of the acupoint are measured and displayed on the Voll system's electrical meter. Energy readings above or below a certain "average" or physiologically "normal" level are thought to reflect inflammation or deterioration or at least an energetic tendency toward illness of a particular organ.

As mentioned, there is a unique testing aspect of EAV-type electrodiagnostic systems. The effect of different medicinal substances introduced into the electrical loop between the machine and the patient can be immediately measured via the body's energy response. Various homeopathic remedies, nutritional supplements, flower essences, or other types of medicinal substances can be placed on a special metallic table in circuit with the patient while the patient's acupoints are tested. Through their electrical readouts, EAV systems show whether or not a patient responds favorably to a vibrational medicine placed in circuit with the acupuncture meridians. A kind of electrical-acupoint resonance reaction occurs when a healing remedy appropriate to the patient's illness is tested. An EAV practitioner might, for example, do acupoint-testing on a patient and find an energy imbalance such as low energy readings in a particular organ-linked acupuncture meridian of the body. If an appropriate rebalancing remedy is placed into the test circuit while the imbalanced acupoint is retested, the meridian disturbance present only a few minutes previously will immediately return to normal energy levels as revealed by the EAV meter. If a patient's kidney meridian tests abnormally low in energy, placing a compatible kidney-rebalancing remedy in the electrical circuit with the patient will temporarily correct the kidney-energy imbalance as revealed by a shift toward normal readings on the EAV meter.

EAV systems offer a more sophisticated form of energetic testing of patients' responses to various homeopathic remedies and other vibrational medicines without resorting to doing muscle-testing for feedback. Of course, for those health practitioners who do not own EAV-type systems, kinesiologic muscle-testing is a cheaper, simpler approach to measuring the body's energy response to various homeopathic remedies. Since the development of the early Voll machines, researchers have come up with many computerized EAV-type systems, including the Vega, the Computron, the Avatar, and the Bodyscan 2010 models. When used to test homeopathic remedies, these devices essentially function as a kind of lie-detector test that asks the body whether it responds favorably (or unfavorably) to a particular vibrational medicine. The questions are posed electrically and subliminally, without involving the conscious mind of the patient. Just as these devices reveal which substances positively stimulate the body, they can also reveal which chemical, biological, and environmental substances a patient may be allergic or hypersensitive to. The more sophisticated computerized EAV systems actually have a memory bank programmed with the coded energy patterns of hundreds and sometimes thousands of different homeopathic remedies, vibrational medicines, and nutritional, chemical, and biological agents. In other words, in these systems, because the memory bank contains the imprints of so many different remedies, there is no need to place the actual remedy upon an electrically connected metallic table to test a patient's response to it. But the computerized systems also have the ability to hook up an electrical sample table to the EAV-type device to test any medicine or substance in case a practitioner wishes to test a patient's response to a new remedy or material not already programmed into the device's memory bank.

An aspect concerning the medical/legal status of EAV-type devices should be mentioned. Many of these devices are registered with the Food and Drug Administration as biofeedback devices because they have the ability to measure different electrical rhythms from the human body. However, they are not approved as devices that may be used to diagnose illness or to test a patient's response to homeopathic remedies. A number of years ago, a small number of health-care practitioners, including some dentists and medical doctors, had their devices confiscated in raids by the FDA because, it was claimed, they were "illegally using" these devices to diagnose and treat illness. As such, many homeopathic practitioners who prescribe remedies using these devices often tell patients they are not curing illness, they are merely assisting patients in rebuilding their health. While it may simply be a matter of semantics, such is the state of our medical/legal system. Because of the hassles caused by federal agencies, most EAV practitioners tend not to advertise their work with these devices. It is often difficult to locate EAV-trained, bioenergetically oriented practitioners except through word of mouth. One exception to this rule is in the states of Arizona and Nevada, where there are homeopathic medical boards commissioned to oversee the quality of practice of homeopathic medicine. A more liberal use of EAV-type devices will be found in both states. EAV-type diagnosis is currently being taught in certain medical schools in Russia and various parts of Europe. But in the United States, this type of technology has tended to remain largely "underground." The EAV-type systems are futuristic technologies with tremendous diagnostic and therapeutic potential. Their effectiveness needs to be scientifically investigated for the benefit they can provide if used by appropriately trained health-care practitioners.

Homeopathy Today: The Resurgence of Homeopathic Medicine

Twenty to thirty years ago, one had to seek out a homeopathic physician to obtain homeopathic remedies for health problems. Today one can go to the local drugstore to obtain them. Within the last few years, homeopathic remedies have been turning up on drugstore shelves and in television ads for health-care products. Part of this shift toward homeopathic products is a reflection of the booming sales of homeopathic medicines in Europe. Profits associated with alternative therapies have not gone unnoticed by pharmaceutical companies, some of whom have actually purchased homeopathic drug companies in recent years. This phenomenon is somewhat akin to the way the oil and gas companies have pursued research into solar power alternatives. No successful enterprise can overlook future trends, be they alternate fuel choices or alternate health-care options.

As homeopathic remedies have begun to find their way into the mainstream, the approach to marketing homeopathy has changed the nature of the way it is used by most health-care consumers. In classical homeopathy, a homeopathic practitioner seeks to find the single remedy that will perfectly match all the patient's symptoms in order to heal a particular illness. In today's mass-marketing of homeopathic remedies, and even in many homeopathic medical practices, there has been a shift away from treating illnesses with "single remedies." The trend is toward what are known as "combination remedies." These combination remedies are actually a mixture of several different homeopathic remedies blended together to form a single medicinal preparation. Combination homeopathics often come in the form of a liquid tincture in a dropper bottle, but they may also be found as a body salve or cream. In the case of homeopathic remedies commonly found in drugstores or on the shelves of health food stores, remedies are labeled as being possibly useful for treating a particular medical problem. What homeopathic drug manufacturers and certain health-food chains have done is to create a line of homeopathics that each combine into a single medication a variety of different homeopathic remedies. This is more or less the "shotgun" approach to homeopathy. Let's apply "shotgun" homeopathy to a symptom many people experience—joint pain.

While a conventional anti-inflammatory drug such as ibuprofen will usually relieve typical joint pain for many different people, only the correct homeopathic remedy for a particular patient's energetic needs will be effective in reducing that person's joint pain. Classical homeopathy is so individualized to each patient's total symptom complex that, for instance, not every person with knee pain or hip pain will benefit from taking the same homeopathic remedy. The very fact that six different people with six different manifestations of joint pains might require six different homeopathic remedies to relieve their symptoms makes it very hard to do studies comparing the therapeutic benefits of any single homeopathic medicine with that of a conventional pharmaceutical drug. However, by combining a number of different homeopathic remedies, each of which has joint pains as a common hallmark symptom, there is an increased probability at least one of the medicine's homeopathic ingredients will provide the correct remedy to vibrationally match and relieve a variety of different patients' joint pains. While many of the ingredients in the combination homeopathic remedy might not fit the patient's total symptom complex, if at least one of the ingredients matches the patient's energy pattern, some symptom relief will be achieved.

This shotgun approach to homeopathy is not the same as the single "magic bullet" approach of classical homeopathy. With the single-remedy approach, you need a great deal of accuracy in homeopathic prescribing to "hit the mark" precisely in matching the patient with the correct remedy. However, when firing a shotgun at a bull's-eye target, less attention to accuracy is needed to hit the target. The wide dispersal pattern of buckshot spray from the shotgun is likely to hit at least part of the target. By combining many similar-acting homeopathic agents into a single medication, it is likely that at least one or more of the ingredients will hit the "energetic target" in fulfilling many different patients' energetic and homeopathic healing needs. While a patient's symptom relief might be attributed to the action of a particular single ingredient of a combination homeopathic remedy, there is also some evidence to suggest that combining various homeopathic remedies with similar therapeutic effects together into a "complex" medicine may actually provide additional benefits beyond single-remedy therapy. A number of homeopathic-medicine researchers and drug developers have discovered that by combining certain remedies into a special blend, they actually create a healing mixture possessing unique, synergistic therapeutic properties. The use of combination homeopathic drugs as opposed to single remedies is frequently referred to as "complex homeopathy" versus "single-remedy therapy."

What Kinds of Health Problems May Benefit from Homeopathy?

Homeopathic medicines may be of benefit for a wide variety of medical problems. However, some attention should be given to the types of medical conditions homeopathy will not heal. Acute severe trauma is not immediately healed by homeopathy, although homeopathic remedies may aid in the overall healing process. A bone fracture, for example, should be X-rayed and set in a cast to make sure proper bone alignment occurs as the fracture is healed by the body. Occasionally, pins or plates to reinforce the proper positioning of bone fragments may be needed via an operation by an orthopedic surgeon. But the use of acute homeopathic remedies for bruising, like Arnica and Symphytum (made from the herb comfrey), may help to accelerate the healing process in bone and soft tissue, thus lessening the time in a cast. Deep wounds extending into muscles, nerves, and tendons should be cleaned and treated surgically, using sutures to close the wounds.

In general, acute surgical problems such as appendicitis or a ruptured aortic aneurysm are best left to a competent surgeon to manage. Potentially life-threatening illnesses such as heart attacks should be managed in a coronary intensive-care unit. However, homeopathy may be useful in dealing with some of the physical and psychological aftereffects of a heart attack that sometimes arise once a cardiac patient has been discharged from the hospital. Severe illnesses like cancer are rarely healed by homeopathic remedies used alone and should be dealt with using a multidisciplinary approach. But a variety of natural healing methods and dietary changes can sometimes help to augment the effectiveness of conventional chemotherapy, surgery, and radiation treatments. Also, preliminary investigations have suggested that taking a 1M homeopathic potency of the chemotherapy drugs being administered for cancer can help to lessen drug side effects of highly toxic chemotherapy agents. In addition, some homeopathic physicians have found that homeopathic remedies can sometimes help patients with terminal cancer to make an easier and less painful transition, as in a hospice setting. Under the care of a skilled homeopath, homeopathic remedies can address many illnesses, especially those for which conventional medicines have failed. Homeopathic remedies can successfully treat psychological as well as physical problems, including anxiety, depression, and in certain cases even mania and schizophrenia, although the treatment of such severe psychological disorders is best left to a homeopathically trained psychiatrist.

Most people can use a homeopathic first-aid kit to deal with a variety of common acute problems, ranging from simple injuries to diarrhea, indigestion, nausea, and vomiting; from early colds and coughs to hay-fever symptoms; from insect bites and stings to sunburn. The following is a list of frequently used homeopathic remedies often found in a homeopathic first-aid kit, along with the specific indications for each remedy. Often when there are two or more remedies indicated for a particular acute medical problem, the modalities or modifying factors (such as whether a condition is improved by heat or cold applications) can help to pinpoint the most appropriate treatment for the problem.

Homeopathic Remedies for First Aid and Common Medical Problems

Aconitum napellus (Monkshood, commonly abbreviated as Aconite): Useful in common colds, croup, earache, eye injuries, fevers of sudden onset (especially after exposure to cold, dry winds), and bladder irritation. It may also be of aid in conditions of acute fright, such as after an accident or for pre-operative anxiety before surgery or dental extractions. For colds, Aconite may be of help in the first stages if there is lots of sneezing, a drippy nose with hot, watery nasal drainage, and eye iritation where the eyes are actually reddened and watery. Croupy, barking, dry coughs that occur suddenly late at night and are associated with anxiety may also benefit from Aconite. Modalities: Problem feels better in the open air, lying on the affected side, as well as when listening to music. Problem tends to feel worse at night, in a warm room, with exposure to cold and dry winds, when lying down for long periods, or in the case of eye injuries.

Aesculus hippocastanum (Horse chestnut): May be helpful in treating hemorrhoids and varicose veins. Symptoms include a sense of fullness in various parts of the body (such as in the rectal area and in the legs). The moods of patients who may benefit from this remedy can sometimes be those of irritability and outward depression. Modalities: Symptoms are often worse with walking, motion, standing for long periods, and after eating. Symptoms are frequently better being in cool, open air.

Allium Cepa (Red onion): Used primarily for colds associated with copious amounts of burning, watery nasal drainage (sometimes dripping from only one nostril), stuffy nose, soreness and redness of the upper lip, dull headaches in the forehead region, early signs of hoarseness, and a burning sore throat. Modalities: Condition tends to be better in open air or in a cool room. Symptoms are worsened in the evening and in a warm room. Allium cepa can sometimes help allergy sufferers who present with some of these same symptoms.

Antimonium crudum (Antimony sulfide): Useful in chickenpox, heartburn, gas, and indigestion. Patients who may benefit may exhibit symptoms of fatigue or exhaustion and may sometimes have a white-coated tongue. Emotionally they may have irritability, be upset or sulky for no apparent reason, and often they can't bear to be touched or even looked at. Modalities: Symptoms are aggravated when patient bathes in cold water or spends time in an excessively warm room, and during the evening and nighttime hours. Symptoms often improve in a warm bath, with warm applications, or in open air.

Apis mellifica (Honey bee): This remedy is often helpful in insect bites and bee stings, but may also be of aid in treating heatstroke and urinary problems causing painful, scanty urination. Patients who may benefit from Apis often present with puffiness and swelling (often at the site of the insect bite) as well as localized or generalized redness to the skin. They may also have stinging pains, itching, a sense of constriction, or a general feeling of exhaustion. Emotionally, patients for whom Apis is helpful may be listless, apathetic, sad, tearful, or whining, or they may merely demonstrate poor concentration (as in the case of heatstroke). Modalities: Symptoms feel worse in late afternoon, in an overly warm room, and with application of any heat, pressure, or even the slightest touch.

Arnica montana (Leopard's bane): Arnica is a first-aid homeopathic remedy for any kind of traumatic injuries that produce bruising, soreness, swelling, and bleeding into tissues. Arnica may also be helpful for conditions in which muscles feel sore and bruised (as in cases of overexertion). Arnica is often used to treat falls, blows to the body, and muscle and joint sprains, and it is sometimes given after labor and delivery because of the excessive muscular exertion associated with expelling the newborn during childbirth. Arnica has also been used to treat nosebleeds as well as acute emotional (as opposed to physical) trauma. Emotionally, a typical Arnica patient will deny anything is wrong, yet may appear to be in a state of shock or obvious mental strain. The patient can also be nervous and will often want to be left alone. Modalities: Patients feel worse with the slightest touch, with motion, and in damp and cold conditions. They feel better lying down or with the head held low. For acute injuries, Arnica may be taken as an oral remedy in the form of tablets dissolved under the tongue or applied as a cream to the injured area of the body.

Arsenicum album (Arsenic trioxide): While this remedy is indeed made from the poison arsenic, it's in a potentized homeopathic form, which rarely contains any original arsenic molecules. It has therapeutic value in treating nocturnal asthma, diarrhea (following eating or drinking), influenza, burning sore throats, vomiting associated with nausea, and a burning pain in the stomach. Patients most likely to improve with Arsenicum tend to be restless, to feel chilly, to feel extremely weak, and often thirst for small sips of liquids. Emotionally, Arsenicum patients feel anxious, restless, and irritable and may sometimes demonstrate a fear of being alone as well as a fear of death. Modalities: Symptoms are often worse at night, when the patient is alone, and when drinking cold drinks or eating spoiled food and excess alcohol consumption. Symptoms may improve with heat or warmth or with elevation of the head.

Belladonna (Deadly nightshade): Belladonna may be useful in the treatment of crampy abdominal pain, boils and abscesses, high fevers of sudden onset, earaches (especially if they are right-sided), styes and pinkeye, throbbing headaches, sore throats, and (right-sided) toothaches. Emotionally, the Belladonna patient may complain of nightmares and tend to be excited, violent, or even delirious (as in the case of high fevers). During a typical Belladonna fever, patients may have a reddened face, hot and dry skin, with cold hands and feet. Belladonna headaches are often accompanied by head congestion, worsened by lying down, and may sometimes be accompanied by dilated pupils. Modalities: Symptoms are aggravated by noise, touch, bright lights, the patient's being jarred, exposed to drafts, and lying down. Symptoms may improve with warmth or when the patient is standing.

Bryonia alba (Wild hops): Bryonia may be of aid in stomachaches (when the stomach is extremely sensitive to the touch or there is also vomiting of bile); constipation (with dry, hard stools); colds that travel from the head into the chest; dry, painful coughing that makes the stomach feel sore; influenza (accompained by aching, stiffness, stitching pains, and a sense of faintness on arising); joint pains (when the joints are swollen, reddened, and hot with shiny overlying skin); fevers (accompanied by great thirst for cold liquids); and headaches (often described as splitting in nature, worse with stooping, coughing, and during the evening hours). Modalities: Symptoms are worsened by exertion, light touch, hot weather or a warm room, minimal motion, noise, and light. Symptoms are often improved with rest, cold air, cool applications, firm pressure, or the patient's lying on the affected side.

Cantharis (Spanish fly): Can be helpful in burns including sunburn and scalding of the skin (that feels better with cold applications), stings (that are swollen, inflamed, and associated with burning pain), bladder irritation (scalding, painful, sometimes bloody urination), reflux esophagitis (associated with burning in the stomach and esophagus), and marked bleeding from any orifice. Emotionally, Cantharis patients are frequently anxious, restless, crying, sometimes in a frenzied rage or at times, in a state of increased sexual desire. Modalities: Symptoms are worse from touch, after the patient has drunk coffee or cold water, and with urination. Symptoms are improved by gentle massage and cold compresses applied to the affected area of the body.

Carbo vegetalis (Vegetable charcoal): This remedy is of great help in cases of indigestion (with abdominal distention and crampy pains), heartburn (with sour belching), gas, hemorrhoids, and spasmodic, rattling coughs (sometimes associated with gagging and vomiting up mucus, as well as painless hoarseness). Carbo vegetalis patients are often sluggish, with weak, icy-cold hands and feet, and may complain of burning pains. Hemorrhoids treated by Carbo vegetalis tend to bleed easily with oozing dark blood. Emotionally, the Carbo vegetalis patient seems mentally slow or lazy or may exhibit a fear of the dark. Modalities: Symptoms are frequently aggravated in warm and humid weather; at evening; after the patient has eaten rich, fatty foods, or drunk coffee or wine; and after lying down. Symptoms may be relieved by burping, by being fanned, in cool conditions, and after drinking milk.

Ferrum phosphoricum (Iron phosphate): May be of aid in simple colds, in the early stages of fevers (of gradual onset, associated with paleness, restlessness, and rapid, weak pulse), with headaches (throbbing in nature, head often sore to touch, sometimes from excess sun), nosebleeds (with profuse, bright red blood that clots easily), earaches (with throbbing pains and sense of ear's being plugged), hemorrhages (from any orifice), coughs (especially short, tickling coughs or hard, dry coughs associated with chest soreness), and insomnia (accompanied by restlessness and anxiety). Emotionally, the Ferrum phos patient may be nervous, sensitive, or forgetful, many have difficulty concentrating, complain of restless sleep, and may often want to be left alone. Modalities: Symptoms are usually worse at night (especially from 4:00 to 6:00 A.M.), in open air, in cold conditions, and may be aggravated by jarring or any kind of motion, and may occur more on the right side of the body (right-sided chest discomfort, nosebleeds from the right nostril). Symptoms often improve with application of cold compresses.

Gelsemium (Yellow jasmine): This remedy may be useful for summer colds (accompanied by sneezing and watery, burning nasal drainage), influenza (characterized by extreme tiredness, achy and bruised-feeling muscles, chills up and down the back, lack of thirst, sleeplessness, and heaviness of the arms, legs, eyes, and head), heat prostration (with tight, bandlike headache and heaviness of the eyelids), sore throat (with swollen red tonsils, ear pains on swallowing, and difficulty swallowing), dizziness (associated with light-headedness, blurry vision, and balance problems), and anxiety (associated with trembling, weakness, or diarrhea, especially anxiety preceding dental work or surgery). Emotionally, Gelsemium patients may seem apathetic, dull, listless, or may just wish to be left alone. They may demonstrate ill effects from fear, fright, or news of an exciting nature. Modalities: Symptoms are aggravated by emotion, too much thinking, fixating on one's ailments, smoking cigarettes, and damp weather. Symptoms may be improved by open air, bending forward, or just moving around.

Hypericum (St. John's wort): An excellent remedy for nerve injuries, especially those involving the fingers and toes, or in injuries to the tailbone. May help with any kind of nerve problems associated with numbness, tingling, or shooting and burning pains. It is also useful for treatment in any type of wounds, especially in any postoperative pain. Also helpful in falls, bruises, any blows to the body. May be of great help after dental extractions (to prevent painful dry-socket syndrome) and after root-canal work. This remedy has also been used for treating pain from splinters, bee stings, jagged cuts, bites, and burns. Emotionally, the Hypericum patient may exhibit sadness, melancholy, aftereffects of shock, and in some cases, a fear of heights. (With regard to its use in treating painful conditions associated with sadness, it is interesting to note that the herb from which Hypericum is made homeopathically, St. John's wort, is gaining increasing acceptance as an effective herbal therapy for depression.) Modalities: Symptoms may be aggravated by touch, cold and damp conditions, or in a closed room. Symptoms may improve when bending the head backward.

Ignatia (St. Ignatius bean): This is a wonderful homeopathic remedy for treating the emotional aftereffects of grief and loss, characterized by worry, sadness, disappointment, moodiness, emotional lability, rapidly changeable or hysterical emotional and physical states, broken sleep with frequent waking, sharp headaches on stooping over, and a general oversensitivity to any stimuli. Emotionally, Ignatia patients may exhibit uncontrollable sobbing, frequent sighing, anxiety, and may appear high-strung and overwrought. They may demonstrate both mental and physical exhaustion. Modalities: The Ignatia headache is often made worse by cigarette or cigar smoke. Other symptoms may be worse during the morning hours, after drinking coffee, with warm applications, or in the open air. Symptoms frequently improve after eating or changing position.

Ipecac (Ipecac root): This homeopathic remedy has application for treating persistent nausea (unrelieved by vomiting), morning sickness, vomiting (of slimy, white mucus) and gastroenteritis. (This remedy offers a perfect example of the homeopathic principle of treating with a substance that reproduces a patient's symptoms in a normal person. Here homeopathically potentized Ipecac is used for treating conditions associated with nausea. Ipecac is a medicine routinely given to individuals to induce nausea and vomiting, especially in cases of poisoning. When Ipecac is given in homeopathic form, it relieves those very same symptoms.) Homeopathic Ipecac may also be useful in treating nosebleeds (with bright red bleeding), profuse dental bleeding, hoarseness (following a cold), violent recurrent coughing, asthma attacks (of sudden onset, associated with wheezing, gasping, chest constriction, and an inability to breathe lying down), abdominal pain (of a cutting nature, especially in the vicinity of the navel), as well as diarrhea (associated with green, frothy stools and crampy abdominal pains). Emotionally, the Ipecac patient is often irritable. Modalities: Symptoms may be aggravated in a periodic fashion, after lying down, and after motion or exercise.

Ledum (Marsh tea): This is a good remedy for falls, blows (especially in black eyes caused by a blow to the eye), bruises (with prolonged skin discoloration), insect bites and stings, sprains and strains (especially of the foot and ankle), but especially puncture wounds. There may be associated swelling of the face or arms or legs. Ledum may also be helpful in treating gout-related pains in the big toe, arthritic pains in swollen ankles, rheumatism associated with painful soles and bruised-feeling heels, as well as shooting pains in the small joints of the hands and feet. Modalities: Symptoms may be worse at night or with the heat of the bed. Symptoms frequently improve with cold applications or in cold water (as with foot pains).

Magnesia phosphorica (Phosphate of magnesia): This remedy may be of help in treating abdominal pains (especially crampy, spasmodic, sharp, shooting pains that feel better when walking around), muscle spasms (such as in leg cramps and writer's cramp), menstrual cramps (that improve with the beginning of the menstrual flow), and colic in infants. From an emotional and mental perspective, patients likely to benefit from Mag phos will often have an inability to think clearly, complain constantly about their pain, or suffer from insomnia due to feelings of recurrent indigestion. Modalities: Symptoms are frequently worse on the right side of the body or during the night, and may be aggravated by touch, by exposure to cold air or by bathing in cold water. Symptoms often improve with gentle pressure and with warm applications (such as a heating pad to the soothe the crampy abdominal pains or leg cramps).

Mercurius vivus (Quicksilver): This remedy has application in treating earaches, colds and sinus infections (associated with chilliness, sneezing, profuse watery or greenish nasal drainage, sore throats (associated with ear pains, swollen neck glands, profuse salivation, and burning in the throat, frequently worse on the right side), toothaches (especially with pulsating, tearing, or shooting pains radiating into the face or ears), and gum disease (where the gums may be swollen and the teeth feel loose). Mercurius vivus is also useful in treating any kinds of boils, diarrhea (accompanied by slimy or bloody stools), vaginal infections (with foul-smelling discharges), as well as bladder irritation (associated with urinary urgency and painful urination, often burning at the beginning of urination). Emotionally and mentally, the Mercurius patient may exhibit irritability, distrust, a weariness of life in general, and a sluggishness of mind characterized by a weakened memory and a slowness in responding to questions. Modalities: Symptoms are often worse at night, in a warm bed, in damp weather, and during perspiration. Symptoms frequently improve with rest.

Nux vomica (Poison-nut): Nux is useful in treating a variety of digestive disturbances, ranging from heartburn; abdominal bloating (after eating); sour belching and acid reflux; alternating constipation and diarrhea; the aftereffects of rich foods, coffee, or liquor (a good hangover remedy, especially hangover headaches accompanying stomach upsets following a night of dietary overindulgence). Nux is also helpful in treating hemorrhoids (especially those associated with itching and relieved by cool bathing), early colds (associated with sneezing attacks and daytime runny noses), asthma attacks (often following stomach upsets and accompanied by frequent belching and a feeling of oppressive, shallow breathing). Nux vomica may also relieve coughs that are dry, tight, and hacking in nature, associated with soreness of the throat and chest or coughing spells that end in retching. Emotionally, the Nux patient may be irritable and sullen and does not like to be touched. Modalities: Symptoms are usually aggravated by spicy foods, stimulants, during the morning (the so-called morning after), and after eating in general. Symptoms usually improve after a nap, during the evening hours, and during wet, damp weather.

Rhus toxicodendron (Poison ivy): This is an excellent remedy for sciatica (with tearing pains down the thighs), back spasms, sprains and strains, joint pains (that feel better with motion and limbering up), wry neck, as well as falls, blows, and bruises. Rhus tox is also helpful in treating itchy skin conditions (such as hives, poison ivy, chickenpox, shingles, and any skin conditions associated with itching, skin redness, and a physical and mental restlessness). This remedy may also bring relief in influenza (accompanied by aching bones and restlessness), dry coughs (a teasing, dry, nocturnal cough usually associated with chills), and cold sores following the flu or colds. Emotionally, the Rhus tox patient may be sad, listless, extremely restless (with constant changing of body position), and may demonstrate great fearfulness at night. Modalities: Symptoms are worse in cool, damp, or rainy conditions, during sleep, and on first moving about. Symptoms may be improved by warm applications, change of position, continued motion, rubbing the affected area, walking, and with stretching of the limbs.

Ruta graveolens (Rue bitterwort): This is another good homeopathic remedy for sciatica (associated with back pain radiating down the hips and thighs, often worse at night while lying down), backaches (with deep pain or a bruised sensation in the lower back, relieved by pressure or lying on the back), strained and sprained muscles and tendons, tennis elbow, and for general problems that may affect the muscles, tendons, joints, and bones. Ruta may even cause ganglion cysts to recede and dissolve. It may also be helpful for eye injuries as well as for eyestrain (where the eyes are red and feel hot and burning, with pressure deep in the eyeball or over the eyebrow). Headaches caused by eyestrain (with a sense of weight in the forehead) are also amenable to treatment with Ruta. Ruta may also relieve the pain of dry socket following tooth extraction. Emotionally, Ruta patients may be quarrelsome, irritable, restless, anxious, or even sad. Modalities: Symptoms may be exacerbated by lying down, from overexertion and straining, while sitting, at rest, and during cold, wet weather. Symptoms may improve when lying on the back, with motion, and with warm applications.

Sulphur (Elemental sulphur): Sulphur is an important homeopathic remedy for rashes and skin problems where there are itchy skin eruptions (such as in chickenpox), dryness, and painful burning (often made worse by itching or bathing). Sulphur patients will frequently have offensive-smelling sweat. Eczema accompanied by dry, scaly, flaky, itching skin and burning discomfort may respond well to Sulphur. Pinkeye and styes where the eyelids burn and itch (especially in chronically inflamed eyelids) may also improve with Sulphur. Sulphur may also be helpful in gastroenteritis and digestive problems (where individuals tend to drink, especially alcohol and cold drinks, much more than they eat; and when they do eat, they crave sweets, fatty foods, and meats). Also hemorrhoids (accompanied by burning and itching), diarrhea (especially in the morning, often alternating with constipation), constipation (associated with dry, hard, large, painful stools or pain from rectal fissures), and pregnancy-associated nausea may respond to Sulphur. Emotionally, Sulphur patients may be irritable, depressed, selfish, and may seem busy all the time. Mentally, they may seem forgetful or have occasional difficulty thinking. Modalities: Symptoms are made worse by bathing, a warm bed, with sweating or dampness, and in the mornings around 11 A.M. Symptoms improve in the open air, in warm and dry weather, and while standing or lying on the right side.

While this list of common homeopathic remedies is fairly extensive, it is by no means exhaustive. Most first-aid kits of homeopathic remedies will contain the aforementioned in a low potency (12X to 30X or 30C potencies). For the acute medical conditions listed, the individual should dissolve three or four of the tablets under the tongue every three to six hours until relief is achieved.

Homeopathy is one of the most common forms of vibrational medicine available. It is of great benefit to keep a homeopathic first-aid kit on hand and experiment with treating friends' and family members' illnesses, colds and coughs, and aches and sprains on a limited basis to convince yourself of the efficacy of these remedies. Once you have had a healing experience with homeopathy, whether it is relief from the pain of a stubbed toe after taking Arnica or improvement of abdominal discomfort after ingesting Carbo vegetalis, you will be convinced of the vibrational healing power of homeopathic remedies. Also, homeopathy may relieve illnesses for which modern medicine has yet to come up with a solution. While there is still nothing in conventional medicine for the common cold (except for selective symptom relief), homeopathy offers a number of different remedies for colds, although each person's unique symptom complex must be individualized with the specific remedy to obtain the maximum benefit. Homeopathic first-aid kits are frequently available from health-food stores. They usually come with booklets describing the therapeutic indications and modalities that will help you to find which remedy fits the acute medical condition you wish to treat. If you are unable to find them locally, a list of homeopathic companies offering such kits is available along with other homeopathic resources in Appendix 1.

A Trip to a Homeopathic Physician

While homeopathic remedies can be self-prescribed for acute illnesses by following the information in a homeopathic first-aid kit, more complex and chronic problems may require a visit to a homeopathic physician. The issues of which potency of a remedy to choose, how often to repeat the dosage, etc., are often a matter of practitioner training and experience. While homeopathic physicians may prescribe low-potency remedies (as are commonly available at health-food stores), many homeopaths prescribe high-potency remedies, usually available only through a homeopathic pharmacy or by direct-ordering from a homeopathic drug company. One of the problems in choosing remedies and potencies is the phenomenon of the homeopathic "aggravation." When people take homeopathic remedies, they will sometimes experience a brief worsening of their symptoms prior to complete resolving of their illness. Such an exacerbation of symptoms is referred to by homeopaths as an aggravation. It is said that certain potencies (usually higher potencies) are more likely to produce aggravations than others. This is why it is sometimes better to have a trained homeopathic practitioner select the remedy, the potency, and the dosage schedule for each medication.

Your experience at a homeopathic physician's office can be quite variable, depending upon the individual's training and diagnostic methods. Classically trained homeopaths rely primarily on extensive history-taking from patients. A first session may typically last up to an hour or an hour and a half. The kinds of questions asked will depend upon the complexity of the health challenge. For an acute condition, it is likely the history-taking will be briefer. But the interview will be long enough to establish the "keynote" symptoms of the patient's condition as well as which modalities appear to help or worsen them. When classical homeopaths are focused on analyzing chronic medical conditions, so-called chronic or constitutional homeopathic prescribing, they tend to ask many questions about the patient's specific problem as well as general physical, emotional, and mental makeup. The questions asked will focus on specific symptoms, on which side of the body the symptoms tend to occur, preferences in foods and whether certain foods or beverages tend to aggravate conditions. Other questions focus on temperature and climate preferences, mental and emotional tendencies, and any history of early psychologically traumatic experiences. The homeopathic physician is especially interested in the "strange," rare, and peculiar symptoms that will help single out the constitutional remedy best fitting the patient's total physical, mental, and emotional picture. In this sense, homeopathy was probably one of the first forms of medicine to be truly holistic in nature, in that it has always focused on the mental, emotional, and physical totality of the patient in selecting the most appropriate treatment.

There are other homeopathically oriented practitioners who use a more eclectic approach to homeopathic prescribing. As discussed earlier, there are a number of health-care practitioners, especially chiropractic physicians, who use kinesiology or muscle-testing to select homeopathic remedies for patients. The selection of remedies to muscle-test is often based partly on a brief patient history in order to establish the nature of the symptoms. After the symptoms are analyzed, one or more symptom-matching remedies are kinesiologically evaluated. That is, while the patient holds the remedy, the practitioner tests a particular muscle group to see if the patient tests strong or weak. In essence, the patient's body is giving unconscious feedback to the practitioner by revealing whether a homeopathic remedy is energetically appropriate for the patient at the time of testing.

After a patient takes a remedy, symptoms will often change. As new symptoms arise, a patient's energetic frequency will shift, and so a remedy that tested positive a week ago may no longer show a positive muscle-test response a week later. This is one of the unique aspects of homeopathy. As the patient's energy state shifts, the homeopathic practitioner may then prescribe a new remedy based upon new symptoms. In some ways, homeopathy may energetically rebalance the physical and subtle bodies by neutralizing or rebalancing and peeling away various levels of energetic disturbance as one might peel away the many layers of an onion to get to its center. As we grow older, our bodies encounter various childhood and adult illnesses that we process and get over to varying degrees. However, each bacterial, viral, emotional, or physical trauma to the body is processed differently. And sometimes the illnesses are not processed very well. Our physical and subtle bodies may develop layers of energetic armoring based upon old, partially healed traumas. Sometimes childhood illnesses may seem to resolve, but an energetic as well as a physical taint of the illness may persist at some level of the physical or subtle body. It is now, for example, well established that shingles, the painful skin eruption that often occurs in adulthood, may be linked to the chickenpox virus a person encountered and held on to from childhood. Similarly, certain bacterial and viral illnesses, as well as chemical poisonings, may be energetically carried in the auric field and higher spiritual bodies as vibrational patterns that can reenter the physical body years later. Such vibrational patterns are referred to in homeopathy as "miasms." A miasm is not so much an illness as an energetic tendency toward illness. Miasms may or may not manifest in times of physical, emotional, and energetic stress to a person. They may affect a person many years after the original illness taint was acquired, or they may actually be passed on to later generations as "inherited miasms." Let's apply this profound concept to venereal diseases such as syphilis and gonorrhea. While easily treatable with modern antibiotic therapy, venereal diseases are thought by homeopaths to create miasms that may affect the patient years later, or the patient's children. Diagnosis of the presence of a miasm or of its contribution to a patient's current medical problems is challenging. Classical homeopaths may focus on parental illnesses in order to gain an understanding of potential inherited miasms, but such diagnosis involves a great deal of experience and clinical judgment.

This brings us to the modern approach to homeopathy that patients are encountering more and more, the field of instrument-aided homeopathic prescribing. As discussed earlier, instrument-aided homeopathic prescribing is derived from the EAV-type testing methods established by Dr. Reinhold Voll using his Voll machine. Since the original Voll machine first came into use, there have been many new computerized systems created by homeopathic medical researchers. Typically, a patient who goes to see a practitioner using EAV or EDI (electro-dermal information-gathering) systems will have an initial interview with the practitioner to establish the medical problems which the patient is seeking help with. Unlike conventional physicians, many EDI-based practitioners will obtain medical and psychological histories but never actually physically examine the patient (except to look at the eyes, tongue, and nails for additional clues to the origins of the problems). The exception to this is when a conventionally trained MD or DO uses an EDI system. Patients of such practitioners may undergo the more traditional history and physical, followed by some standard medical testing including bloodwork, hair analysis, EKGs, stress-testing, etc. In some cases, conventionally trained MDs and DOs will have an EDI-trained technician on their staff who will perform the EDI patient evaluation and then report any findings and recommendations to the prescribing physician.

During an EDI evaluation, the patient sits in a chair (usually a nonmetallic one situated over a plastic insulating mat) and holds a brass metallic tube covered with damp gauze in one hand. The brass tube is connected via a wire to the EDI instrument. The wet gauze around the brass tube assures a good electrical connection between the patient, the metallic tube's electrode, and the EDI instrument. Thus when a patient's acupoints are tested by the practitioner's penlike electrical probe (also connected to the EDI instrument), the completed circuit ensures that an accurate electrical acupoint reading will be obtained. A patient sits in the chair, usually with shoes and socks taken off beforehand, so the acupoints on the hands and feet are easily accessible to the EDI practitioner. Patients generally undergo inital EDI testing in a two-step procedure. First, the practitioner places the acupoint probe at key acupoints on the fingers and toes to get an electrical measurement of the health status of all of the key acupuncture meridians for a baseline reading of the body's major organs. Any meridians showing significant deviations from the established normal parameters for acupunture meridians may be considered to have actual or potential organ problems developing. Remember, the meridian system, linked not only to the physical organ but also to the corresponding part of the etheric body, will sometimes reveal only an energetic disturbance in an organ system that has not yet manifested as physical illness. This is one of the reasons a trained health-care practitioner needs to evaluate the readings from these devices, since the readings do not always indicate active illness in the body. Sometimes the meridian disturbance is an indicator of a weak link in the meridian chain for a particular patient. The weak link indicates where the patient is most likely to develop problems if the organ energy imbalance is not appropriately corrected.

Following the intial evaluation to determine which meridians show energetic imbalances, the EDI practitioner will then move on to the second phase, that of specific acupoint analysis. In the second phase, the practitioner seeks to pinpoint the nature of the energetic disturbance in the meridians that tested weak (such as in degenerative disorders) or overly energized (as in the case of inflammatory diseases). Because of the way testing is done, discovering the source of the meridian and organ problem will also identify vibrational and homeopathic remedies that will correct the imbalance. In the original Voll machine, the practitioner actually placed a vial of a specific homeopathic remedy in circuit with the patient while testing the imbalanced acupoint, looking to see if a particular remedy would rebalance the electrical reading for the acupoint of interest. When an EAV-acupoint reading that was originally too high or too low reads normal after the homeopathic remedy is placed in circuit with the patient, the practitioner knows that the appropriate remedy has been selected.

Let's say a patient shows a disturbance in the liver meridian. The practitioner places different key homeopathic remedies associated with liver problems in circuit with the patient and retests each time, looking for a remedy that will correct the acupoint-energy imbalance. After identifying the remedy, the practitioner may also be able to tell what caused the organ and meridian disturbance. For instance, the patient with the liver imbalance might show the liver acupoint rebalanced when a 30X potency of hepatitis-B virus is placed in circuit. Homeopathic remedies made from disease-producing agents are known as nosodes. In the hypothetical case here, an acupoint energy correction by the hepatitis-B nosode would not only identify hepatitis-B as the probable source of the patient's liver-meridian/organ imbalance but would also indicate that giving the patient this homeopathic remedy would correct the energy disturbance as well. EDI practitioners will test patients against classical homeopathic remedies, such as the common remedies previously listed, as well as against various homeopathic nosodes made from many viruses, bacteria, or chemical and environmental toxins. If another patient with a liver-meridian imbalance showed acupoint-energy correction with placement of a 200C potency of formaldehyde into the EDI circuit, this would indicate that the second patient's liver imbalance was due to chemical exposure as opposed to contact with a virus. Similarly, taking the homeopathic formulation of formaldehyde (which contains only the potentized energetic signature of formaldehyde) would rebalance the second patient's liver and also allow the patient to push any remaining traces of formaldehyde out of the body. The giving of a homeopathic remedy made from a specific toxic agent in order to detoxify from that same toxic chemical is known as isopathy. Some homeopaths believe that isopathy works because the homeopathic dose of the toxic agent raises the vibratory rate of the physical toxin in the body to very high levels, pushing it out of the physical body entirely.

Because there are so many potential pathogenic bacteria, viruses, fungi, chemicals, and environmental toxins that can be an underlying cause of illness, it is often impractical to place hundreds of tiny vials of homeopathically potentized microbes and toxins in circuit with patients as was done in the original Voll machines. So today's newer EDI systems have hundreds if not thousands of different remedies, viruses, toxins, chemicals, and other causative agents vibrationally encoded and stored in the magnetic memory of their computer hard drives instead. These computerized EDI systems allow a practitioner to test multiple remedies simultaneously and then to narrow down the testing field to one or several homeopathic remedies in a much shorter period of time. Many of the homeopathic remedies stored in the computer systems of EDI devices are actually complex homeopathic remedies consisting of many different remedies combined in a single dropper bottle of homeopathic liquid. The frequent use of combination homeopathics (complex homeopathy) by EDI-based homeopathic practitioners points out a big difference between classical homeopathy and the new computerized homeopathic prescribing. Yet another big difference between classical and EDI-based homeopathy is that computerized EDI systems allow the practitioner to energetically interrogate the body directly, asking it, in a subtle bio-electrical language, exactly what it requires to rebalance itself, while the classical homeopath prescribes primarily by analyzing a patient's reported symptoms.

One of the advantages to EDI-based homeopathy over classical methods is that computerized electroacupuncture diagnostic systems provide a practitioner with the ability to test quickly for hundreds of environmental toxins and chemicals that can not only produce illness but can also block the effectiveness of classical homeopathic remedies. The world we live in today is much different from the 1880s world when classical homeopathy predominated. The presence of environmental pollutants, radiation, and electromagnetic fields in our homes and work environments, the creation of homes and offices with potentially toxic building materials, and numerous other factors have left many people with traces of toxic substances in their bodies that can block the effectiveness of classical single-dose homeopathic remedies. In Germany, where many of the EDI systems have been developed, a number of homeopathic practitioners have developed special complex nutritional, herbal, and homeopathic remedies designed to remove the blockages from the toxically impaired enzymes and metabolic pathways in the body so patients can better respond to homeopathic and other vibrational therapies. While homeopaths have long known that certain substances like camphor, caffeine (especially in coffee), and strong perfumes may inhibit the effectiveness of homeopathy, a number of prescription drugs such as prednisone and illicit drugs like marijuana and LSD may also inhibit homeopathic remedies.

Recently, European researchers have discovered that geopathic stress (abnormal earth energies) may also block the effectiveness of homeopathic medicines. A number of EDI systems developed in Europe can actually test patients for chronic exposure to geopathic stress. If you live in a house situated in a geopathic-stress zone, long-term exposure to geopathic energy can block the effectiveness of homeopathic remedies, flower essences, and even acupuncture treatments. The identification of geopathic stress as a disease-causing factor is somewhat new to health-care practitioners in the United States, but such a possibility should be suspected when a person fails to respond to (appropriately prescribed) homeopathic remedies or vibrational therapies. While electroacupuncture diagnosis may identify geopathic stress as a causative factor in a person's illness (or failure to respond to homeopathy), EDI systems will not tell you exactly where the geopathic-stress zone is located. The only way to evaluate your house or office for geopathic stress is to hire a professional dowser who specializes in detecting (and neutralizing) this abnormal earth energy. (For help in finding a dowser, see Appendix 6.)

Needless to say, a homeopathic prescription from an EDI-based practitioner will often be quite different from that obtained by classical homeopathy, although some EDI practitioners still use the old style of homeopathic prescribing along with the newer computerized methods. As mentioned, EDI-oriented homeopaths will often give patients a dropper bottle containing a homeopathic combination remedy to which the patient had a positive EAV-type acupoint response, while the classical homeopaths tend to give the traditional single remedy in varying potencies. Following the initial homeopathic prescription, whether it is a single remedy or a complex combination remedy, patients will go back to the practitioner for a reevaluation. During a follow-up session, patients are asked whether their symptoms improved, became worse, or disappeared entirely. The classical homeopath may prescribe another single remedy if a patient's symptoms begin to shift. Sometimes there may be a series of single remedies given over a period of weeks to months before a patient's condition undergoes a major healing. In contrast to the classical homeopath, the EDI-based homeopath will retest the patient's acupoints and meridians at their follow-up visit, looking for a change in the energy status of the organs and meridians that were out of balance during the first EDI evaluation. Depending upon whether the meridian energy (as well as the patient's symptoms) have shifted toward a healthier state or not, the EDI practitioner will retest the patient for different homeopathic remedies in an effort to correct the energy imbalance. Patients' symptoms may improve after only one or several sessions. But in the case of long-standing medical conditions, positive health changes tend to occur gradually over time.

Homeopaths will tend to vary as to how they work with patients using modalities besides homeopathy. Some homeopaths are also trained in herbal and naturopathic medicine, iridology (the evaluation of past, present, and potential future illness based upon evaluating the iris, the colored portion of the eye), acupuncture, or other healing traditions they might tend to mix with their use of homeopathy. There are still many classically trained homeopaths who are essentially purists. They do not mix or match more than one homeopathic remedy at a time, let alone combine it with other healing systems. With regard to using single versus complex homeopathic remedies, there is still considerable disagreement and debate between the classical and the EDI-based homeopaths. Yet there is no right or wrong answer, merely different professional opinions and different patient experiences between the two groups. What is probably most important in selecting a homeopathic practitioner is the individual practitioner's level of training in using homeopathy as well as the word of mouth from other satisfied patients.

Whether you decide to go to a homeopathic physician or merely try self-prescribing from a homeopathic first-aid kit, homeopathy is an important vibrational-healing approach that can be of great help in a variety of medical and psychological conditions. In Appendix 1 you will find a list of resources about homeopathy, locating a practitioner, and obtaining homeopathic remedies for personal use.

Five

~~~

ACUPUNCTURE AND CHINESE MEDICINE: AN ANCIENT APPROACH TO VIBRATIONAL HEALING

ACUPUNCTURE IS THE art of needling the body to produce healing. Medical historians have found records documenting the existence of traditional acupuncture treatments in China dating back five thousand years. It was not until the early 1800s, though, that traditional acupuncture treatments made an appearance in Western medicine. To the allopathic mind-set of American physicians, needle healing was a mystery not understood and certainly not universally embraced. Despite resistance, acupuncture healing eventually took hold of America during the 1970s. Nowadays, even your next-door neighbor has probably heard of and maybe even tried acupuncture therapy. Yet in spite of increasing acceptance, Western doctors still debate and speculate as to how and why acupuncture works at all.

What Is Acupuncture and How Does It Work?

Acupuncture treatments usually involve the careful insertion of extremely fine needles into locations on the skin known as acupuncture points. Ancient Chi-
~~~

nese acupuncture maps reveal a unique system of specialized points running along channels that feed energy to the organs of the body. Acupuncturists refer to these channels or lines as meridians. Traditional Chinese medicine teaches that the placement of acupuncture needles into specific acupoints affects the movement of ch'i, a unique form of life energy, which flows through the meridian channels to nourish and support the various organs of the body. Although this "energy flow" explanation for acupuncture's effects has long been acceptable to practitioners of traditional Chinese medicine, it does not sit well with Western physicians. Western doctors and researchers who have studied acupuncture have tried to explain the efficacy of acupuncture therapy in terms of nerve-related, biochemical-related, and even psychologically related mechanisms. They have tried to "fit" acupuncture into the mechanistic medical model of the body that still occupies a major position in the practice of twentieth-century Western medicine.

When reports of acupuncture's effects in relieving pain and its use in anesthesia during operations made front-page news in the *New York Times* during the early 1970s, doctors in the United States were either incredulous or inclined to believe that acupuncture's benefits were due to the placebo effect. If a new pain-relieving drug is tested, it usually is compared with a placebo pill or injection in order to distinguish how much of the new drug's effects are merely due to a patient's belief in the healing power of the medicine. Western doctors knew that placebos, as simple sugar pills or saline injections, could produce therapeutic improvements in more than 30 percent of patients who were given them in place of real drugs. Therefore, it was easy to see how many physicians would believe in the power of the placebo before changing their medical thinking to accept the reality of meridians and life energies flowing through the body. Having never been educated in meridians and acupoints in their medical anatomy courses, the American doctors dismissed the Chinese doctors' claims of manipulating ch'i flow in the body. It was much easier to assume that the Chinese population had been brought up to believe in the power of acupuncture as a healing modality and that this powerful belief was responsible for any of acupuncture's therapeutic effects. But one of the problems with attributing all of acupuncture's therapeutic benefits to the placebo effect was the fact that veterinarians in China had used acupuncture successfully to treat animals. After all, it is difficult to believe that horses, cows, or other farm animals could "believe in" the power of the needle as a healing tool!

Many Western scientists sought to fit acupuncture's effects into the realm of established human physiology. In the early 1970s, two researchers, P. Wall and R. Melzack, put forth the idea that acupuncture pain relief might work by stimulating nerves in the skin to affect the flow of pain impulses to the brain. The researchers hypothesized that the insertion of acupuncture needles into areas of skin rich in nerves might send a series of signals to the spinal cord, where there was a gate that, if closed, could turn off the flow of pain-message-carrying nerve impulses traveling to the brain. The theory was known as the "Gate Control theory" and was initially thought to be more plausible than the concept of meridians and acupoints. As it turned out, the Gate Control theory did not explain acupuncture's many other therapeutic effects. However, researchers did discover that you could indeed shut off the flow of pain impulses to the brain. All you had to do was influence pain-carrying nerves traveling through the spinal cord by using the modality of electrical stimulation. Based on this theory, neurosurgeon Dr. Norman Shealy developed the Dorsal Column Stimulator, an implantable electrode system that was surgically attached to the spinal cord. It was quite effective in managing certain types of chronic pain. Later, the neurosurgical procedure became unnecessary, as one could achieve similar pain relief through electrically stimulating nerves in the skin to achieve the desired effect. Thus, Transcutaneous Electrical Nerve Stimulators or TENS devices were born and have been used for treating chronic pain ever since. Unfortunately, though, why acupuncture worked still remained a mystery to Western science.

During that period in the late 1970s, research on acupuncture's mechanism of action began to move away from nerves and to focus upon hormones and neurochemicals. Neurological researchers had recently discovered the powerful naturally occurring chemicals in the body known as endorphins. Endorphins behaved like the narcotic drugs opium and morphine. By binding to opiate receptors throughout the nervous system, endorphins were found to produce their pain-relieving effects. Pretty soon acupuncture researchers sought to connect acupuncture pain relief with the release of endorphins. Dr. Bruce Pomeranz, working at the University of Toronto, found he could block the effects of traditional acupuncture analgesia (pain relief) in rats by pretreating them with a drug called naloxone. Naloxone was a medicine that was known to block the effect of endorphins in the body. Sure enough, acupuncture pain relief could not be achieved in the naloxone-treated rats. This finding suggested that endorphin release caused by traditional acupuncture needle stimulation was indeed an important mechanism behind acupuncture's pain-relieving effects. Since this initial research seemed promising, Pomeranz and others began to study the effects of electroacupuncture (electrically stimulating acupoints) on pain relief in rats. As Pomeranz proceeded, he soon found that naloxone did not always work as predicted in blocking pain relief, especially when he varied the oscillating electroacupuncture current from low- to high-frequency pulsing. It seemed other neurochemicals, such as serotonin, must be involved in high-frequency acupuncture's pain-relieving effects, thus complicating the picture of acupuncture analgesia. Another big problem with trying to attribute acupuncture's effects only to endorphins or serotonin changes in the nervous system was the fact that acupuncture has been used in China for many years to treat many nonpainful conditions such as infectious diseases and chronic degenerative disorders of different organ systems. For instance, even the "barefoot doctors of China" who treated Mao Tse-tung's Red Army during the early 1930s used acupuncture effectively to treat illnesses like malaria. Time and again, the numerous examples of successful treatment via acupuncture indicated that this ancient "needling therapy" definitely was more complicated than Western science ever suspected.

While researchers in the West were busy investigating acupuncture's neurochemical basis, Soviet scientists were taking a very different approach. In the early 1970s, Russian researchers noted that acupoints had unique electrical characteristics that distinguished them from surrounding skin. There was a ten- to twenty-fold drop in electrical skin resistance over acupoints that could easily be measured using conventional electrical recording equipment. This drop in electrical resistance meant that acupoints also conducted weak electrical currents more easily. From this finding, Russian scientists developed the first handheld electrical acupoint locators that could reliably locate acupuncture points in the skin based upon lowered electrical skin resistance. The acupoint locators were a boon to acupuncture research. In fact, soon after their creation, another (more electrical) theory about why acupuncture works was proposed. Some researchers went so far as to suggest that the healing effects of acupoint stimulation by classical acupuncture needle insertion were really due to a phenomenon known as the "current of injury." Scientists have long been aware of an electrical phenomenon that occurs whenever tissues of the body are traumatized or undergo microscopic damage. For example, when skin cells are pierced, they begin to leak electrically charged ions into surrounding tissues. So when the skin is pierced by a metallic acupuncture needle, a weak electrical batterylike effect is created. We call this weak electrical current the current of injury. The current of injury is known to stimulate a healing response from nearby cells. It was theorized that electrical stimulation of local tissues by the current of injury, created by acupuncture needles piercing the skin, was the true source of acupuncture's healing effects. While this explanation is potentially valid for certain types of acupuncture treatment, it does not explain how stimulating acupoints with low-level, nonpiercing lasers (such as those found in a laser pointer or a bar-code scanner) could achieve the same therapeutic effects as needle stimulation of acupoints. Again, it seemed that the mystery of how acupuncture worked was not totally resolved.

The fact that acupuncture points possess unusual electrical characteristics that distinguish them from surrounding skin suggests there should also be something anatomically unique about acupoints. When European and American scientists sought to study acupoints by dissecting cadavers or by observing thin slices of acupoint tissues under the microscope, they noted that acupoint regions were rich in nerves and blood vessels. Unfortunately, no one in the West seemed to be able to microscopically identify anything remotely resembling a meridian connecting the acupoints. However, anatomical acupoint research in the Far East eventually seemed to provide the missing link in identifying the existence of true anatomical meridians. Much earlier, in the 1960s, a Korean researcher by the name of Dr. Kim Bong Han had claimed to have developed special tissue-staining techniques that allowed him to identify meridians. Under the microscope, these meridians appeared as tiny microtubular structures traveling beneath the skin and throughout the organs of the body. In an attempt to verify meridian pathways, Dr. Bong Han injected radioactive tracers into the acupoints of rabbits. Following acupoint injection, he found that the tracers always remained highly localized and concentrated in this unique microtubular meridian system. Remarkably, the meridians followed the same pathways shown in ancient veterinary acupuncture diagrams of rabbit meridians.

Many Western scientists found the Korean claims difficult to believe, and thus ignored Bong Han's findings until the early 1980s, when two French researchers, Drs. Claude Darras and Pierre De Vernejoul, tried the radioactive-tracer experiment again, but this time on human beings. They injected radioactive technetium (a common radioactive isotope used in bone scans and thyroid scans) into the acupoints of patients and used nuclear scanning equipment to noninvasively follow the flow of technetium away from the injection site. In order to make sure the doctors were really measuring meridians and not blood vessels or lymphatic channels, some patients received technetium injections into adjacent non-acupoint skin regions as well as into nearby blood vessels and lymphatics. When non-acupoints, blood vessels, or lymphatic channels were injected, the radioactive tracer tended to diffuse outward from the injection site into a typical small circular pattern. However, when true acupoints were injected, the radioactive technetium followed the exact pathways as the acupuncture meridians described and illustrated in several hundred-year-old acupuncture charts of the human body! Even more amazing was the fact that when acupuncture needles were inserted into distant acupoints along the same tracer-labeled meridians and then twirled, a change was produced in the rate of flow of technetium through the meridians. In other words, needle stimulation of an acupoint on the meridian changed the flow rate of "something" (including the technetium tracer) along the meridian channel. This research provided the first believeable evidence supporting the ancient Chinese claim that acupuncture-needle stimulation affected the flow of ch'i through the body's meridians. It now seems acupuncture became a little less mystical as a result of Darras's and De Vernejoul's work.

In another recent technological verification of acupuncture meridians, Dr. Zang-Hee Cho, a physics professor at the University of California at Irvine, found an amazing correlation between needle-acupoint stimulation in the body and the activation of specific areas in the brain. Using a special brain imager known as functional MRI (fMRI), Cho studied the brain activity of volunteers who received classical needle stimulation of acupoints of the feet. These acupoints occurred along the bladder meridian, a channel that acupuncturists say travels from the outside of the foot, up the legs and spine, over the head, all the way to the eye region. To Cho's amazement, needling of the acupoints on the foot that were said to stimulate the eyes caused the visual cortex, the brain region associated with vision, to light up in all the volunteers tested. Sticking an acupuncture needle into the acupoint on the foot had the same effect on the brain as shining a light into a person's eyes! As a control, needle stimulation of nearby non-acupoint skin regions in the foot failed to produce the same response. This was yet another verification of how needling an acupoint in one part of the body produced a distant organ effect, similar to what was predicted by classical Chinese meridian theory. Based on Cho's findings, it was suggested that there was indeed some previously unknown physiological pathway connecting the feet to the eyes, just as the ancient Chinese had shown in their acupuncture-meridian maps of the human body.

While the American and French research does not necessarily prove the existence of ch'i, it goes a long way toward validating the existence of acupuncture meridians as discrete anatomical pathways running through the body. If researchers could accept the possibility that meridians were genuine structures in the body, the Chinese theories of acupuncture therapy might be more correct than previously believed. According to traditional Chinese acupuncture theory, illness is ultimately caused by a disturbed energy state (an imbalance between the opposing forces known as yin and yang) or by an abnormal flow of ch'i moving through specific organ-linked meridians in the body. Since the meridians are thought to supply life energy to each of the organ systems, an excessive or diminished flow of ch'i to any organ would result in abnormal organ function and ultimately lead to physical disease. Also, a diminished flow of ch'i energy could result in an increased susceptibility to external disease-causing agents such as harsh environmental conditions, bacteria and viruses, toxic substances, and the like. Needle stimulation of the acupoints seems to act like a regulatory key that selectively turns up or down the amount of ch'i energy flowing though acupoint "valves." The valves, in turn, act like power-relay stations along the body's meridian channels or life-energy distribution network that runs throughout the body.

In classical acupuncture, either of two different needle-stimulation techniques, known as tonification and sedation (or dispersion), may be selected by the practitioner. The techniques tend to increase or decrease, respectively, the flow of ch'i energy along selected organ-linked meridians. Which of the two techniques of needle stimulation will be used by the traditional Chinese medicine (TCM) practitioner is determined only after careful study of a patient's case and an analysis of what type of energy-balancing is required at the current time in order to restore health and balance to the patient. Tonification is achieved by gently rotating or twirling the acupuncture needles between the practitioner's fingers. While tonification will tend to increase the flow of ch'i through the meridian, the technique of sedation tends to release excess ch'i energy from an acupoint, almost like venting steam from an overpressured steam pipe.

This explanation is really a gross oversimplification of acupuncture. But it does provide a clearer idea of how classical Chinese medicine views acupuncture's mechanism of action compared with other neurological, bioelectrical, and biochemical theories put forth to explain the physiological effects of acupuncture. Considering the fact that modern nuclear technetium scans have actually verified the existence of meridian pathways in the body, it is highly likely that there really is something to the Chinese model of acupuncture after all.

The Many Flavors of Acupuncture: A Look at the Different Acupuncture Approaches

There are a wide variety of therapeutic approaches to healing with acupuncture. While the act of inserting a fine needle into an acupuncture point may be considered an acupuncture treatment, there are many different techniques and schools of acupuncture training that determine how and why one goes about selecting specific acupoints to treat for different health conditions. Probably the simplest form of acupuncture is sometimes referred to as "symptomatic acupuncture" or "formula-based acupuncture." Practitioners trained in this approach will consult a manual for a specific acupoint formula. For example, a doctor who specializes in treating chronic pain conditions will look up specific points to needle or stimulate for arm, leg, or low-back pain. Acupuncture pain manuals have pictorial diagrams and acupoint descriptions indicating which points should be treated with acupuncture needles for a specific area of pain in the body. Such an approach is strictly symptom-driven and does not require any deep analysis of a patient's medical or emotional history in order to achieve results. While this type of treatment can be successful for a variety of painful conditions—including sciatica, tendinitis, bursitis, and even painful arthritis—it is also limited in what it can achieve. Doctors who perform acupuncture for weight loss or for quitting smoking also tend to follow this type of formula-based approach.

The treatments for weight loss and smoking cessation often involve placing small acupuncture needles in specific acupoints in the ears. Ear-acupoint treatment is in itself a specific subcategory of acupuncture, known as auricular therapy or "ear acupuncture." According to this theory, the ear is mapped out with acupuncture points that correspond to all the organs and structures of the entire body. In other words, there is an acupoint map of the entire body located upon the ear. If you look at an acupuncture chart of the ear, the body map of acupoints actually looks as if it were a drawing of a baby lying upside down in fetal position. On an ear body map, the region of the earlobe corresponds to the acupoints for treating the head, while the outer curved part of the ear corresponds to the acupoints for treating the spine and back. There are auricular-medicine specialists who do nothing but treat acupuncture points on the ear in order to achieve therapeutic results. To an auricular acupuncturist, the ear is like a holographic map of the body, providing an amazing gateway to both diagnosing and treating various medical conditions. An auricular acupuncturist will tend to select points to treat based upon which areas of the body are affected by pain or illness. In selecting treatment points, auricular therapists also look for imbalanced ear acupoints that are tender to the touch or more electrically reactive than the surrounding skin. A common method of finding abnormal or pathological acupoints on the ear is through electrical testing of the skin of the ear. Many auricular specialists use penlike handheld electrical acupoint locators to locate acupoints by touching the skin of the ear. These point locators register the presence of an acupoint by a change in an audible musical tone or by the lighting of a special light-emitting diode. A clearly distinguishable change in the emitted sound occurs when the detector registers a drop in electrical skin resistance (such as when the point locator's metal probe comes in contact with an ear acupoint). When an electronic point locator detects an ear acupoint with very high electrical reactivity, it reveals to the acupuncturist that there is an associated imbalance in the organ or body region corresponding to that "reactive" zone. In other words, diseases in the body often show up as electrically imbalanced acupoints on the ear.

Another form of acupuncture relies on holographic body maps located on the hand rather than on the ear. Korean hand-acupuncture maps show acupoints on the hand that correspond to all of the different parts of the body. The hand body map has definite similarities to the body map found on the ear.

The fields of both hand and ear acupuncture are somewhat specialized and not the garden variety of acupuncture therapy most patients would encounter when seeking out an acupuncture practitioner. In contrast, most acupuncturists tend to perform "whole-body acupuncture" that uses acupoints throughout the entire body, including those specialized points on the ears and hands. For example, classically trained acupuncturists tend to focus on treating acupoints throughout the entire body that are part of a general treatment program for a specific health problem. But at the same time they may also add the corresponding ear acupoints for treating that specific problem in order to enhance the acupuncture treatment's efficacy. The use of body acupoints combined with ear acupoints often increases the effectiveness of an acupuncture treatment above and beyond what might be expected from doing strictly body acupuncture. However, the selection of acupoints is entirely dependent upon the school of training of the acupuncturist. As we have mentioned, there are those acupuncture practitioners who tend to consult "point formulas" that list recipes of specific acupoints to needle for healing specific types of disorders. In most conventional medical settings, acupuncture treatments are usually focused on obtaining relief from acute and chronic pain that has failed to respond to more conventional medical or surgical treatments.

One of the more popular forms of whole-body acupuncture is based on the TCM approach to illness. It is known as "five-element-based acupuncture" and is derived from the five-element model of physiology. This model is based upon viewing the human body as a microcosm that functions according to the same principles that rule the macrocosm, the world around us. The five elements from which all things are said to be composed (in TCM terms) are earth, water, wood, fire, and air. According to this Chinese viewpoint of the universe, everything in nature is thought to relate to one of these five elements. Ancient Chinese medical practitioners and philosophers who studied the relationship between the different elements found that each element related to another in very specific ways. The relationships worked out among the five elements were then extrapolated to human beings and their internal workings as a way of understanding the intricate relationships between the body's organ systems in various states of health and illness. The ancient Chinese medical practitioner classified different organs according to which of the five elements they represented or corresponded to. A working scheme was then created to show the patterns of how energy (ch'i) tended to flow from organ to organ in balanced as well as unbalanced interrelationships. Each of the five elements was said to be associated with specific taste or flavor preferences, a season of the year, specific emotions, colors, sounds, and other unique qualities. Using this ancient model today, five-element acupuncturists analyze patients not only by their symptoms but also according to their body types, their lifestyles, their taste preferences, color likes and dislikes, as well as other seemingly unrelated qualities. This attention to taste preferences is done to determine in which of the five elemental groups each patient belongs. Some have referred to this system as Chinese elemental "biopsychotypes." Having established a particular patient's elemental type, the five-element practitioner then attempts to determine the exact nature of the dysfunctional relationship between the patient's organ systems and where the greatest energetic weakness or imbalance lies.

Essentially, the five-element analysis uses many unique aspects of the patient history, as well as observations of the patient's face, voice, specific tastes in food, seasonal preferences, as well as their emotional temperament. By understanding the areas of greatest weakness in the body's organ systems, five-element analysis helps to determine the nature of ch'i imbalance that is resulting in the patient's outward health problems. Thus, a person who has an

The Chinese Law of the Five Elements and Their Correspondences

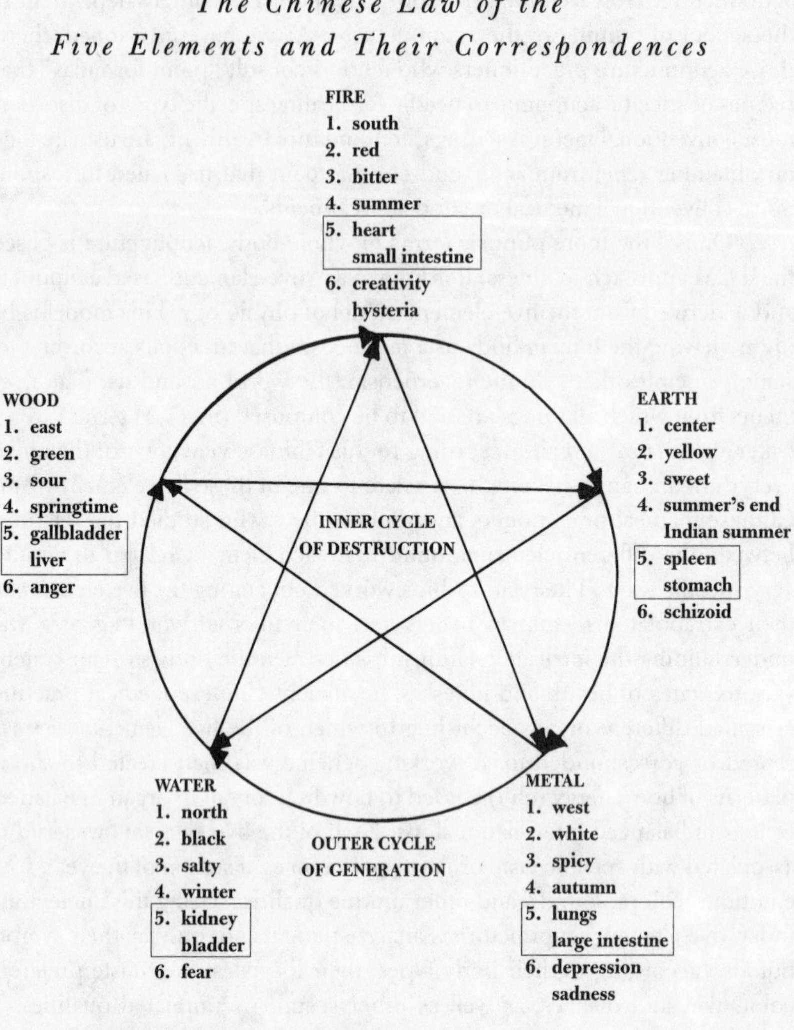

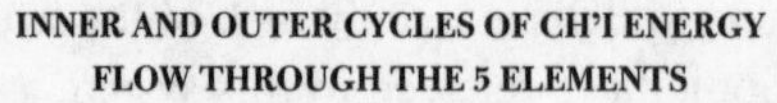

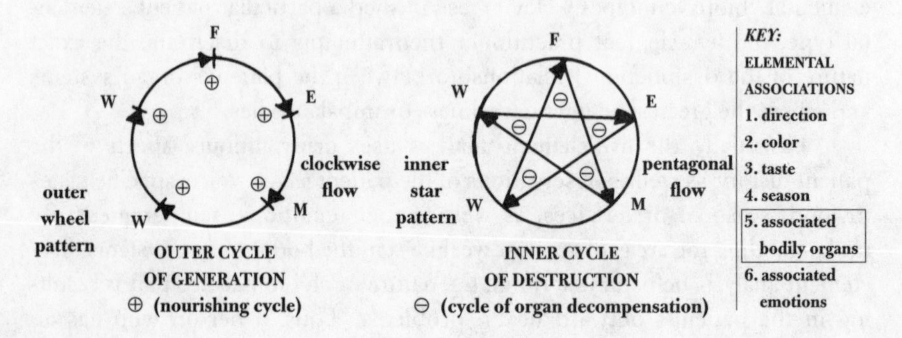

imbalance in the flow of ch'i through the kidney meridian might exhibit water-element-related symptoms such as salt cravings, a preference for or intense dislike of the color blue, excessive fear, dislike of cold weather, sexual potency problems, or even pains along the distribution of the kidney meridian itself.

Another aspect of Chinese medicine that may be used in conjunction with five-element-based acupuncture is pulse diagnosis. Traditional Chinese medicine teaches that there are different kinds of pulses that can be appreciated by specially palpating the wrist. Different pulses indicate different things about the energy and health status of the various meridians and the five-element-linked organ systems. A good practitioner trained in pulse diagnosis can evaluate not only the adequacy of ch'i flowing to the different organ systems but also the effectiveness of an individual acupuncture treatment. It seems a shift in meridian energy following an acupuncture treatment causes an immediate but subtle shift in the qualities of the pulse (at least to the trained pulse diagnostician). In Chinese pulse diagnosis, training allows the experienced practitioner to detect just such subtle changes. In addition to analyzing the patient's symptoms, element-related preferences, and pulse, practitioners trained in traditional Chinese medicine also study the patient's tongue. According to Chinese teachings, the tongue is mapped out in zones, again corresponding to different organs of the body. The tongue, like the ear and the hand, serves as a kind of holographic micromap of the body. While the ear and hand have maps of the entire body, the body map on the tongue tends to give information primarily about the health of the internal organs.

Following this five-element-oriented analysis, the acupuncture practitioner will study which acupoints to stimulate in order to help a patient's ch'i energy move in the appropriate direction for achieving equilibrium and health. In addition to inserting acupuncture needles in the acupoints that fit the patient's specific elemental- and organ-dysfunction pattern, traditional acupuncturists may also stimulate the acupoints with heat, using a technique known as moxibustion. Moxibustion involves placing a small amount of mugwort plant, referred to as moxa, on the free end or "handle" of the metallic acupuncture needle, then lighting the moxa so it will burn slowly. Thus a gentle heat is sent down the acupuncture needle into the acupoint, creating an additional stimulating effect. Practitioners of five-element acupuncture will often recommend to patients the ingestion of particular Chinese herbs. Prescribing Chinese herbs in conjunction with acupuncture is also based on five-element analysis, as well as an assessment of the patient's energy status relative to the internal balance of yin and yang. According to traditional Chinese medicine, everything in nature maintains a certain balance of opposing forces, classified as either yin or yang. Yin is considered the feminine energy and is associated with water, quietness, the moon, and nighttime, while yang, the masculine energy, is associated with fire (as opposed to water), noise (versus quiet), the sun, and daytime. Each of the organs of the body also has certain yin and yang qualities, but one energetic quality usually predominates over the other. The Chinese believe that when there is an imbalance between the forces of yin and yang in the body, the individual becomes more susceptible to diseases caused by external pathogens as well illnesses due to internally generated disharmony. Specific Chinese herbs are prescribed according to which of the five elements the patient best fits, which organ system is weakened by disease, and also by analyzing the patient's yin/yang status. The practice of five-element acupuncture follows very specific rules of acupoint selection and treatment for guiding the practitioner in his or her therapeutic approach. This complex form of acupuncture is taught in schools that focus primarily on the TCM approach. The most famous of the five-element-oriented schools is the College of Chinese Acupuncture in England run by Professor J. R. Worsley.

Another approach to acupuncture comes from France and might be classified as a kind of specialized energetically oriented acupuncture. One might wonder how an ancient Chinese medical practice like acupuncture found its way to France. In fact, some of the earliest Western accounts of the practice of acupuncture in China came from French Jesuit priests who wrote about this unique form of healing through needling the body. The French have embraced acupuncture in a form that is based upon both mathematics and complex "formulas," sometimes utilizing calculations of meridian relationships using specially coded formulas derived from the trigram symbols of the I Ching. The I Ching is an ancient Chinese divination tool that is still used to this day to gain intuitive and synchronistic insights from the higher self. Even though the French use these special formulas to calculate which acupoints to treat for a particular patient, they still employ electrical stimulation of acupuncture needles as well as moxibustion in order to enhance the efficacy of the acupuncture treatment and to get the ch'i flowing. Needless to say, the calculations and thinking behind the French kind of energetic acupuncture often appear extremely esoteric to the layperson, perhaps even more so than traditional Chinese medicine. However, practitioners trained in the French form of acupuncture, just like their traditional acupuncture counterparts, are often successful in treating medical conditions that do not respond to conventional medical treatment. Consider the acupuncture approach to treating strokes as an example. When a person has a stroke, the typical course of allopathic medical therapy prescribed to enhance recovery is an intensive physical-therapy program geared to reeducating the areas of the brain unaffected by oxygen starvation and nerve-cell damage. There are still no available medications known to restore health and normal function to damaged brain cells. While clot-busting drugs administered early in the course of a stroke might limit damage to brain cells, they cannot reverse the damage. The response to physical therapy is variable and depends upon the patient's age and the extent of the damage. However, when acupuncture treatments oriented toward stroke recovery are added to a physical-therapy program, much more dramatic and rapid recovery is often seen. Physical and cognitive functioning in severe stroke cases that might otherwise have never recovered often improve greatly, sometimes at a dramatic rate, when acupuncture treatments are added to the physical- and occupational-therapy regimen.

A number of important innovations have enhanced the practice of acupuncture within the last two decades. The first innovation concerns the use of pulsed electrical currents to stimulate acupuncture points. Patients undergoing surgery in China will often have acupuncture anesthesia instead of a general anesthetic. After the patient receives acupuncture by an anesthesiologist, the needled acupoints are electrically stimulated to prevent patients from experiencing pain while under the knife. Patients have been known to undergo abdominal surgery while awake and free of pain during the entire surgical procedure. The use of presurgical acupuncture anesthesia in China is quite different from the way surgical patients are chemically anesthetized for surgery in the United States and Europe. While electroacupuncture anesthesia has not yet made its way into American hospitals, many acupuncture practitioners commonly use electrostimulation of ear and body acupoints for treating common medical problems.

A number of devices are now being marketed for both professional and layperson use in non-needle electrostimulation of acupoints for pain relief. Some of the more expensive instruments available to health professionals, like the Electroacuscope, offer a variety of different electrical-stimulation modes. A physical therapist or a sports medicine physician equipped with such a device uses a handheld electrical probe to stimulate specific acupoints directly in order to relieve certain types of muscular and skeletal pain. Other devices, such as the Healthpoint stimulator developed by Dr. Julian Kenyon in England, offers non-needle electrical acupuncture therapy for health-care practitioners and for patients who wish to administer self-treatment. The Healthpoint stimulator has been shown to be effective in relieving many different types of painful conditions, as well as insomnia and even addiction disorders. At the lower end of the cost spectrum are the piezo-acustimulators like the Acu-Spark, which rely on electrical currents generated by crystals under pressure (the piezoelectric effect). When a quartz crystal is squeezed, it generates an electrical charge. Many people are familiar with the cigarette lighters that generate a small electrical arc to ignite a butane-gas jet (using an internal piezoelectric mechanism). This same principle is adapted to small handheld acupoint stimulators whose electrical probes are placed in contact with anacupoint while a button is squeezed. Squeezing the button produces a brief electrical discharge (caused by the piezoelectric effect), which sends a small electrical current into the acupoint, resulting in a localized electrostimulation to the acupoint. While the piezo-type acupoint stimulators are somewhat effective, the Healthpoint stimulator tends to be more efficient in relieving a wider variety of painful conditions. This is because—unlike the piezo-acustimulators, which produce a fixed-voltage stimulus—Healthpoint stimulators can be tuned to a wider range of electrical frequencies, waveforms, and intensity settings that may be more effective in relieving certain types of pain.

Another technological innovation to be adapted for acupuncture therapy provides yet another form of needleless acupuncture. For the needle-reticent patient, acupoint stimulation with sound in the form of finely focused ultrasonic waves may be something to try. Sound acupuncture, sometimes referred to as "sonopuncture," has been used in England and the United States with varying degrees of effectiveness for relieving pain caused by arthritis, menstrual cramps, and even lower-back pain.

Continuing down the needle-free acupuncture path leads us to another from of energetic acupoint stimulation that employs the use of tiny magnets, taped to key acupoints of the body, to produce a stimulation effect. This practice has been called "magnetopuncture," although the use of the term "puncture" is something of a misnomer. No needles are usually inserted, and thus there is no real puncturing of the skin. Sometimes acupuncture needles may be used in conjunction with tiny magnets. For instance, a number of acupuncturists will also sometimes tape very small but powerful magnets over the acupoints that have been needled following a more traditional acupuncture treatment. Acupuncturists who use this combined approach feel that the magnetic stimulation of the acupoints seems to continue the acupuncture treatment for a significant period of time, long after the needles have been removed.

Dr. Peter Mandel of Germany has developed still another form of needle-free acupuncture. He calls his procedure "colorpuncture." Mandel created a special handheld light device that has different-colored transparent probes capable of projecting various colors of light directly into acupoints. This specialized form of color therapy stimulates the acupoints in highly specific ways. Mandel has actually determined that different colors of light appear to "tonify" (stimulate) or "sedate" various organ-associated acupoints. A small number of complementary-medicine practitioners in both Europe and the United States are experimenting with Mandel's system of colorpuncture with varying degrees of therapeutic success.

One of the more common forms of light used to stimulate acupoints is laser light, in what has sometimes been called "laserpuncture." "Soft" laser beams do not cut or burn tissue but still emit an extremely bright coherent light beam that can be focused directly into acupoints. The lasers used for laserpuncture vary from fairly potent professional laser systems to battery-operated, handheld, solid-state models (that employ a laser LED or light-emitting diode). Much research has been carried out in Russia on the use of soft-laser stimulation of acupoints for healing various medical problems. Russian researchers have claimed to successfully treat pain disorders, metabolic problems, and even hypertension with the use of laserpuncture. Some of the early work on laserpuncture carried out in the former Soviet Union also involved measuring the energy field around patients using sophisticated whole-body Kirlian scanners. The Kirlian electrical photography enabled researchers to graphically record energy emanations from the body that appeared to correlate with different patterns of illness. Electronography, another offshoot of Kirlian scanning developed in Romania by Dr. Ioan Dumitrescu, has also been able to record glowing acupoints along meridians connected to organs affected by existing or impending disease. Interestingly, much of Dr. Mandel's research with colorpuncture in Germany is based upon his diagnostic use of Kirlian photography of his patients' hands and feet. Mandel uses Kirlian photographs of the fingers and toes to monitor the health and energy status of acupuncture-meridian energy in a fashion similar to the Voll machine and electrodermal (EDI) technology. Kirlian techniques merely show energy changes in the acupoints by their different visual patterns of "spark discharge," in contrast to reading electrical acupoint changes on a voltage meter (of a Voll machine or EDI device). In addition to being useful for diagnosing energetic imbalances in the body, Kirlian imaging has also been used by Mandel to follow the subtle-energetic effects of colorpuncture, homeopathic medicines, and nutritional therapies on his patients. Currently, there are only a few practitioners in the United States who use Kirlian imaging to gain additional energetic information from patients to guide diagnosis and therapy with homeopathy and acupuncture.

Besides using needles or different forms of energy to stimulate acupoints, simple finger pressure or acupressure massage can also be a useful treatment modality. Applying pressure to specific acupoints on the body can be extremely helpful in temporarily relieving certain types of pain. Acupressure massage might be considered a cousin to hand and foot reflexology, since reflexology therapists use different massage techniques to stimulate points on the hands and feet that correspond to different regions of the body. Whether patients have their acupoints stimulated by acupressure, acupuncture needles, electrical probes, or laser beams, any acupoint stimulation can have therapeutic benefits if performed by a skilled practitioner.

What Medical Problems Can Be Helped by Acupuncture?

Before being treated for any disease, it is important that individuals receive a proper diagnosis of their medical condition. Conventional medicine offers excellent therapy for many illnesses and should be tried prior to pursuing acupuncture as a primary treatment modality. Of course, many of the people who seek out acupuncture have already been to a number of different medical specialists. And a correct diagnosis may have been made. But in such cases, the conventional medical treatments were unsuccessful in alleviating the condition to the individual's satisfaction. Whereas acupuncture is not 100 percent effective all the time, certain medical conditions do appear to respond favorably.

The most common use of acupuncture today is for the treatment of various pain syndromes. One typical pain condition frequently amenable to acupuncture therapy, in conjunction with physical therapy, is lower-back pain. A number of insurance companies are now actually referring patients with recurring low-back-pain syndromes to acupuncturists early in the course of their pain management. Insurance companies are, of course, financially invested in getting workers back on the job and off the disability payroll as soon as possible. While most insurance companies do not pay for acupuncture treatments as a primary treatment modality, more and more insurers, including some HMOs, are beginning to offer this therapy to patients because of the increasing popularity of alternative medicine in the United States.

Other types of pain syndromes, whether they are caused by tendinitis, arthritis, or neurological problems, have also been shown to be treatable by acupuncture. There is currently research funded by NIH's National Center for Complementary and Alternative Medicine (NCCAM) that is attempting to evaluate the effectiveness of acupuncture as a therapy for painful osteoarthritis of the knees. In China, arthritis is commonly treated with herbal medicine and acupuncture. Medical research in China suggests 70 percent or more of patients with osteoarthritis respond favorably to acupuncture treatment. The Western experience places the figure closer to a 50 percent response rate. Again, results seem to depend upon the practitioner's skill in acupuncture. However, the combining of both acupuncture with Chinese herbal medicine seems to be helpful in relieving pain for many arthritis sufferers. Pain relief from acupuncture therapy can sometimes last for up to six months. Some patients report decreased joint pain for up to two years after a series of acupuncture treatments. On the other hand, rheumatoid arthritis in the acute inflammatory stage does not usually respond well to acupuncture. There is even some evidence that acupuncture treatments during the inflammatory stage of rheumatoid arthritis might actually worsen joint pain. However, when the acute inflammation dies down, acupuncture may then become effective in relieving pain caused by joint erosion from the primary rheumatoid process. Simple painful strains and sprains of muscles and tendons also respond favorably to acupuncture, especially if it is used early in the course of treatment. Early acupuncture treatment can sometimes avoid a long drawn-out course of recurring pain due to a strain or minor tearing of ligaments and muscles. Chronic pain due to disease of the musculoskeletal system, whether it has been present for weeks or months to years, appears to respond equally well to acupuncture.

It may be worth giving acupuncture treatments a try for any kind of chronic joint, tendon, or muscle pain that is not in the acute, inflammatory stage.

Acupuncture has been used to treat a variety of headaches, especially migraines. Migraine headaches seem to respond to acupuncture as well as, if not better than, other forms of headache (such as tension headaches.) In general, the published work on acupuncture's success in treating headaches suggests that from 65 to 95 percent of headache sufferers may achieve some degree of relief from acupuncture, regardless of the cause of the headache. Another neurological condition associated with head pain is trigeminal neuralgia or tic douloureux, a form of severe, one-sided facial pain that can be quite excruciating. While the Chinese claim to improve the pain from this condition in up to 70 percent of cases, the Western experience in treating trigeminal neuralgia with acupuncture places the figure much lower. Acupuncture, though, can certainly help the condition when standard drug therapies (or even homeopathy) have failed to provide relief. Post-herpetic neuralgia, a form of pain that persists after an episode of shingles, can sometimes be amenable to acupuncture treatment as well. In China, most patients under treatment for shingles are treated daily with acupuncture. The acupuncture-treated patients suffer only rarely from the more chronic pain of post-herpetic neuralgia. It is possible that the acupuncture treatment of shingles might actually prevent the persistent pain that sometimes remains long after the skin lesions of shingles have healed. In any case, reports suggest that up to 40 precent of patients with shingles may achieve some degree of pain relief with acupuncture treatments. Some acupuncturists have found that combining ear acupuncture with body acupuncture is particularly helpful, especially when ear acupuncture is done with both needle and soft-laser stimulation of ear acupoints. In general, most types of pain considered to be neuralgic or nerve-related may be helped to varying degrees by acupuncture. There is no way to predict in advance which neuralgia patients will respond to treatment. Sometimes five to ten treatments are necessary before any significant change in this chronic form of neuralgic pain is seen.

Select neurological disorders associated with varying degrees of paralysis can also be improved by acupuncture. In the United States, the mainstay of therapy for paralyzing strokes is intensive physical therapy, speech therapy, and occupational therapy. Yet in China, almost all stroke patients are treated using acupuncture. The Chinese feel that physical therapy is of less benefit to stroke victims than acupuncture. Some of the published literature on acupuncture for strokes seems to back this up. Clinical trials of acupuncture for strokes by the Chinese claim to have achieved some degree of improvement in neurological function in up to 80 percent of stroke patients. Although the Western experience in treating strokes in this fashion is much less extensive, anecdotes by individual acupuncture practitioners seem to support the Chinese claims. Research by the Chinese using scalp and body acupuncture for stroke treatment suggests that acupuncture actually increases blood flow throughout the brain. This enhanced blood flow effect appears to aid in stroke recovery. If acupuncture therapy is given for a stroke, it should ideally be given within six months of the stroke. Patients may benefit from acupuncture up to two years after the stroke first occurred, though. In general, acupuncture treatment for strokes beyond the two-year mark seems to offer little benefit. Since Western medicine has very little to offer stroke victims besides intensive physical therapy (or the initial administration of intravenous blood-clot-busting drugs to prevent the neurological consequences of a full-blown stroke), acupuncture should be considered in stroke rehabilitation during the early phase of recovery.

Another form of paralysis that may be helped by acupuncture is traumatic paralysis due to accidental destruction of nerve pathways in the nervous system. Such damage might occur in a motor-vehicle accident or some other traumatic injury. While a number of traumatic injuries to the nervous system may improve on their own over time, acupuncture therapy may significantly accelerate recovery from this type of neurological damage. In China, acupuncturists claim that up to 50 to 60 percent of such trauma-induced injuries will demonstrate significant recovery of movement and motor functioning with intensive acupuncture treatment. Typically, patients may receive acupuncture treatments almost daily for periods of up to six months. One might also consider acupuncture for spinal-cord injuries and closed head injuries. Of course, one must find a skillful acupuncturist who is willing and able to perform intensive acupuncture therapy on a long-term basis. But considering that Western medicine has little to offer certain patients other than physical therapy and occupational therapy, acupuncture may improve a patient's final outcome. Bell's palsy is another common form of paralysis involving the muscles of the face. The exact cause is unknown, but a viral inflammation of the nerves of the face has been suspected to play a role. The Chinese claim that acupuncture with moxibustion (the burning of mugwort on the free end of the acupuncture needle) can result in complete recovery in up to 75 percent of patients. Please keep in mind, though, that many cases of Bell's palsy resolve on their own. With regard to other neurological problems, some acupuncturists claim that Parkinson's disease can be effectively treated with acupuncture. However, I have personally found no formal studies that support these claims. Other neurological disorders that may improve symptomatically with acupuncture treatment include cerebral palsy and ALS (amyotrophic lateral sclerosis or Lou Gehrig's disease), involuntary tremors, as well as urinary and bowel incontinence from a variety of causes. Since acupuncture is a fairly benign therapy, individuals who have not responded to conventional medical approaches to managing such neurological disorders might consider acupuncture as an alternative treatment.

Acupuncture can sometimes be helpful in treating hearing loss. Most cases of hearing loss are neurogenic in nature, meaning they are caused by nerve damage to the auditory nerve or any of the nerves that involve the processing of sound information from the ears. Such damage may occur before birth or at any time during a person's life and can affect hearing in one or both ears. Hearing aids can increase the loudness of sounds, but they don't allow users to distinguish between relevant sounds and background noise. Therefore, hearing aids do not necessarily make discrimination of words any easier. Interestingly, acupuncture treatments for hearing loss can improve an individual's ability to discriminate between different sounds, words, and background noise even if the therapy fails to increase a person's overall sensitivity to hearing sounds. Several clinics doing acupuncture for hearing loss report that up to 90 percent of patients with some residual hearing showed some degree of improvement in speech discimination after acupuncture therapy. In the same patient group, over 70 percent had at least a ten-decibel improvement for hearing in four or more frequencies during audiogram testing (after ten acupuncture treatments). The number of acupuncture treatments needed to see an improvement in hearing and speech discrimination can vary with the patient's age and with the cause of the deafness. Most patients will need at least daily or weekly treatments. Patients who receive more frequent acupuncture treatments tend to get better results. But even once-a-week therapy can result in significant improvement by the tenth treatment session. Some patients may even notice a dramatic improvement in hearing in fewer than five treatment sessions. Patients with hearing loss who are more likely to see rapid improvement in their hearing after acupuncture are older people who have lost their hearing gradually. While some teenagers with hearing loss can benefit after only a few sessions, children with hearing loss tend to improve more slowly and can sometimes require up to forty treatments to see optimal improvement.

Acupuncture can also be effective in treating a condition known as tinnitus in which the hearing of persistent unwanted background noise occurs. Since certain drugs—including quinine, aspirin, and alcohol—can produce tinnitus, an acupuncture practitioner might ask the tinnitus patient to eliminate suspected aggravating agents before starting acupuncture treatments. Acupuncture seems to have a high rate of effectiveness for treating tinnitus. In fact, the condition may totally disappear after about ten sessions. Relief from the unwanted noises can sometimes last for a number of years. Another neurological condition associated with tinnitus is Menière's disease, a disorder of the inner ear. Patients with Menière's disease complain of tinnitus, hearing loss, dizziness, and loss of balance, as well as nausea and vomiting. After patients with Menière's disease have failed to obtain relief from antibiotics (to clear up infections), antihistamines, and anti-dizziness, a trial of acupuncture can frequently provide lasting relief. At least six to ten acupuncture treatments are required. However, a repeat series of sessions after months or years may be needed, as Menière's disease tends to be a chronic, recurring disorder.

Asthma exacerbations and allergies such as hay fever can sometimes be significantly improved by acupuncture. Asthma attacks are caused by a muscular contraction of the airways of the lung and may be stimulated by exposure to inhaled dust, pollen, food allergies, and upper-respiratory infections. In asthma, the initial acupuncture treatments will often relieve the wheezing and shortness of breath. Then a series of six to ten treatments may be necessary for long-lasting relief. Acupuncture has been shown to dilate the constricted bronchial passageways that are usually tightly narrowed during acute asthma attacks. The exact mechanism responsible for this effect has yet to be entirely explained in conventional medical terms. One Chinese trial of acupuncture for asthma treatment found that up to 70 percent of asthmatics had a positive response to a series of ten asthma treatments given once a year. The asthma patients in China who were treated with acupuncture tended to have fewer asthma attacks and less intense attacks over the period of a year. Chronic bronchitis is another form of wheezing caused by muscular contraction and inflammation of bronchial airways and lung tissue. Often smokers who have already acquired some degree of lung damage experience chronic bronchitis. Chinese acupuncture researchers claim that up to 50 percent of patients with wheezing and chronic bronchitis experience an improvement in their condition with acupuncture therapy. Acupuncture treatments for the chronic bronchitis sufferer must be repeated regularly, though, in order to maintain the improvement in breathing. In the acupuncture treatment of hay fever, symptoms such as nasal congestion, runny nose, and itching are often relieved after the first treatment session. At least six treatments are routinely needed to give longer-lasting relief. Hay-fever patients usually will need to return each year during allergy season. Acupuncture can also relieve the nasal congestion associated with sinus infections. But acute sinusitis is still best treated with antibiotic therapy.

Acupuncture can usually lower blood pressure acutely, but there is no good research to demonstrate its effectiveness in long-term control of hypertension. Diet and exercise, along with the judicious use of blood-pressure-lowering agents, may be needed for long-term control of high blood pressure. However, acupuncture treatments may help to decrease the amount of medicine required to maintain a normal blood pressure. Another cardiovascular condition that may be amenable to acupuncture treatments is angina, a condition associated with exertion-related chest pain usually caused by decreased blood flow to the heart muscle. The Chinese have carried out studies on the effects of acupuncture upon the heart that seem to demonstrate a marked improvement in the functional ability and efficiency of heart muscle after acupuncture therapy. Clinical work by acupuncturists in treating angina suggests that up to 80 percent of angina patients may benefit from acupuncture. Usually, patients with angina are given a series of acupuncture treatments that are then repeated every four to six months in order to maintain the therapeutic effect. Angina patients who are receiving acupuncture should still continue to take their regular medication and sublingual nitroglycerin if needed while remaining under a cardiologist's care. In many cases, acupuncture treatments decrease the frequency of chest-pain attacks. And those episodes of angina that do occur tend to be of much less severity. Heart patients will frequently benefit from a low-fat, low-sugar diet, as well as from taking vitamin supplements such as vitamin E (400 to 1,000 International Units [IU] daily), vitamin C (2,000 milligrams daily), folic acid (1 milligram daily), a B100 complex, niacinamide (1,500 milligrams daily) and a good multivitamin with trace elements. All these nutritional factors help to retard the atherosclerotic process that contributes to the clogging of the arteries supplying blood to the heart. There is even evidence that acupuncture treatments may actually lower cholesterol. But one should not rely solely on acupuncture to maintain lowered cholesterol. Acupuncture may also be helpful in treating the anxiety, depression, and insomnia sometimes affecting heart patients. Acupuncture can sometimes be effective in treating abnormal heart rhythms as well. While long-standing arrhythmias such as chronic atrial fibrillation (a chaotically beating heart rhythm) respond to acupuncture perhaps less than 2 percent of the time, new-onset arrhythmias may improve with acupuncture therapy in up to 70 percent of cases.

Peripheral vascular disease (PVD), a narrowing of the arteries delivering blood supply to the legs, is also amenable to acupuncture therapy. Frequently occurring in smokers, peripheral vascular disease can also affect individuals with chronically high cholesterol levels. Individuals with this type of vascular problem often experience leg pains whenever they walk any significant distance. The pain is usually relieved by resting. The medical term for the symptom of periodic leg pain is intermittent claudication. Since medications to improve blood flow to the legs do not always help, many PVD patients end up having vascular surgery to remove severely blocked arteries, which are replaced with synthetic blood-vessel grafts. If a PVD patient chooses acupuncture therapy, significant relief of leg pain may be experienced after only a few acupuncture treatments. Circulation to the legs will often begin to improve, and the patient may enjoy months to years without recurrence of claudication symptoms. Patients will require several acupuncture treatments every month or two to remain in a state of remission.

If the peripheral vascular disease resulted from smoking, acupuncture treatment for smoking addiction can also be helpful for the claudication patient. In addition, a good multivitamin, a B complex, vitamin E (at least 400 to 1,200 IU daily), and vitamin C (at least 3,000 to 10,000 milligrams a day in divided doses) should be started at the beginning of acupuncture treatments and continued daily thereafter. Exercise to increase collateral circulation (new blood-vessel growth) should be attempted gradually as the leg-pain symptoms are relieved.

Raynaud's syndrome is another vascular problem that affects certain individuals by decreasing blood flow to the fingers and sometimes the toes as well. Raynaud's patients experience cold hands and feet, pale coloration, and even severe pain in the hands or feet. A Raynaud's patient's fingers will often turn a bluish color when exposed to even the slightest cool conditions (due to constriction of the small blood vessels in the fingers). Although this syndrome may sometimes be due to significant arterial disease or connective-tissue disorders, it can occur in young people, especially young women, without any known cause. The common term for Raynaud's syndrome that affects young people and women is idiopathic Raynaud's syndrome. Patients with Raynaud's symptoms frequently wear gloves or take vasodilating medications (to keep their hands warm and their blood vessels open). A number of acupuncturists have found that ideopathic Raynaud's syndrome appears to respond well to acupuncture treatments. In those idiopathic Raynaud's patients who respond (usually after a series of six acupuncture treatments), relief from painful symptoms can sometimes last for up to a year. The series of treatments may need to be repeated yearly in order to keep the condition under control. Raynaud's syndrome due to more severe vascular disease or connective-tissue disorders sometimes responds as well, but in such cases acupuncture treatments will often need to be continued for a much longer period of time.

Other foot and leg problems, such as conditions causing chronic edema or swelling of the legs, can sometimes be effectively treated by acupuncture. Remember, attention to the cause of the swelling is important. It must first be determined that there are no serious heart, kidney, or liver problems causing the fluid retention before turning to acupuncture as a primary therapy. While doctors often prescribe diuretic medications to relieve swelling and fluid accumulation, such medicines may be associated with a number of unwanted side effects. It must be kept in mind, however, that while a series of six acupuncture treatments directed toward the symptomatic treatment of edema might relieve ankle swelling for many months, acupuncture treatments aimed at decreasing swelling do not necessarily correct the disease process originally responsible for the fluid accumulation. In order for acupuncture therapy to heal such a primary disease process, a five-element analysis of the patient's condition would be required to determine the best course of treatments aimed at correcting the ch'i imbalance within the various organ systems (that led to the fluid buildup in the first place).

Acupuncture has also been used to treat a variety of skin problems. In general, skin diseases should first be examined by a dermatologist and a trial of standard medical therapy attempted before turning to acupuncture. Chronic itching has been known to respond to acupuncture treatments, albeit quite slowly. Patients with chronic itching may need to undergo up to ten or more acupuncture treatments before obtaining significant relief of their symptoms.

As part of their treatment, any potential offending allergens should be removed from the patient's home environment. As a rule, acupuncture needles should never be inserted directly into skin lesions, but instead inserted in such a way as to surround the lesion. As mentioned previously, shingles, a painful skin disorder caused by a herpes-virus-type nerve inflammation, is treated primarily with acupuncture in China, and there are few reported cases of post-herpetic neuralgia, a potential complication of shingles. Eczema may also be amenable to acupuncture therapy. Acupuncture treatments can reduce allergic reactions, improve body physiology, and be helpful in relieving the anxiety, depression, and insomnia that can sometimes afflict eczema sufferers. At least six to ten acupuncture treatment sessions are usually required to see improvement in eczema. Patients may be gradually weaned off medications under medical supervision (if no recurrence of symptoms or skin lesions occurs). An attempt to identify food allergies or other potential allergens should also be made. In addition, nutritional supplementation with vitamins such as vitamin A (20,000 IU daily), vitamin C (2,000 milligrams daily), and pantothenic acid (200 milligrams daily) can offer additional help to patients with eczema. Acupuncture can support psoriasis sufferers as well. But usually many treatments are required. Those who treat psoriasis with acupuncture suggest that patients will generally need a series of ten treatments three or four times a year to achieve significant improvement in their condition. Since psoriasis sufferers often experience an exacerbation of their skin lesions with emotional stress, acupuncture treatments for anxiety, depression, insomnia, and itching, along with psychotherapy and stress-reduction exercises, can offer additional benefit.

Alopecia or hair loss is just one more skin-related disorder that is sometimes helped by acupuncture. Before beginning acupuncture treatment for hair loss, a dermatological evaluation should be made to rule out scalp infections, parasitic infestation, or other easily treatable conditions. There are now a number of topical drugs being offered for hair loss as well. The response to these drugs is quite variable. One specific form of hair loss, known as alopecia areata, characterized by asymmetric patches of baldness on the head, sometimes responds well to acupuncture treatments, especially when combined with soft-laser stimulation of acupuncture points. Since there is little to offer medically for this specific cause of hair loss, acupuncture may be worth a try. Acupuncture may accelerate the regrowth of new hair in bald patches on the scalp and can frequently prevent new bald spots from occurring. With regard to the more common forms of hair loss in men and some women, acupuncture has prevented further hair loss and has promoted new hair growth. A course of six treatments, done daily or at least weekly, followed by treatments every two months, may be helpful in restoring a more normal hair distribution. Unfortunately, even with this regimen, the success rate for acupuncture treatment of typical male-pattern baldness is perhaps only 30 percent at best.

Acupuncture has also had some success in dealing with various gastrointestinal problems, ranging from peptic ulcers to constipation and diarrhea. While Zantac, Tagamet, and Prilosec can give relief from abdominal discomfort, acupuncture sometimes provides additional benefits in achieving a remission from recurrent stomach and duodenal ulcers. Research by Chinese physiologists has demonstrated that acupuncture treatments for stomach problems actually decrease stomach acidity. In certain cases, acupuncture or electrostimulation of acupuncture points for peptic ulcers can provide relief when conventional drug therapy proves ineffective in managing symptoms. Of course, recurrent symptoms should be evaluated by a physician or a gastroenterologist using upper-GI X rays or gastroscopic evaluations in order to rule out the rarer forms of malignant ulcers that can sometimes occur in patients with chronic symptoms. Symptomatic stomach pain, abdominal cramps, nausea, and even the flatulence of irritible bowel syndrome have been known to respond to acupuncture. While the Chinese claim that acupuncture treatments cause people to pass sizable gallstones through their intestinal tracts and out through their feces, there is little Western experience with this approach. In general, acupuncture should not be used to treat inflamed gallbladders, severe bacterial infections, parasitic infestations, or tumors of the gastrointestinal tract. On the other hand, acupuncture can frequently be effective in relieving recurrent diarrhea when the aforementioned medical conditions have been ruled out as potential causes. With regard to bowel infections such as diverticulitis, the Chinese report that acupuncture may accelerate recovery time from diverticulitis with fewer complications. However, in spite of these reports, oral or intravenous antibiotics should still be given for infectious bowel problems along with acupuncture as adjunctive therapy. There is increasing evidence that acupuncture stimulates the body's natural immune defenses in infectious diseases by enhancing the production of antibodies against various pathogenic microbes. Acupuncture can sometimes be of benefit in treating diarrhea and abdominal pain caused by inflammatory bowel disorders such as Crohn's disease and ulcerative colitis when conventional medical treatments have failed. Even constipation of a chronic nature often responds to at least six sessions of acupuncture treatments. Along with acupuncture treatments, an increased water intake and an augmentation of the diet with unsweetened grape or prune juice, raw vegetables, and increased dietary fiber can offer additional benefits in relieving chronic constipation.

With regard to the treatment of diseases that have an infectious basis, I have personally seen some rather remarkable health changes in a patient with liver and kidney failure caused by a chronic hepatitis-B infection. The patient had acquired a viral infection years earlier and had developed chronic, progressive hepatitis as well as severely weakened and diseased kidneys due to the persistence of the hepatitis-B virus in his system. Along with chronic fatigue and other symptoms, he experienced recurrent swelling of the legs. His liver- and kidney-function tests remained severely abnormal in spite of high-dose steroids and immunosuppressive drugs prescribed by a kidney specialist who told the patient he would eventually require dialysis. Following a series of five-element-based acupuncture treatments, the patient's tests started normalizing (to the disbelief of the patient's kidney specialist, who claimed that the improvement was nothing to get excited about and that it was most likely due to "lab error"). Within a year of acupuncture treatments, the patient's ankle swelling had disappeared and his kidney and liver functions returned to near normal. The patient's nephrologist was amazed and bewildered, as he exclaimed to the patient, "You shouldn't be getting better!" This case illustrates how a disease caused by a virus can still be reversible with acupuncture treatments (in spite of medical opinion to the contrary).

Interestingly enough, acupuncture has also been of value in treating mental-health problems. For instance, acupuncture is often helpful in relieving anxiety, depression, and even neurotic disorders. Many patients will report a feeling of well-being after only a single acupuncture session. This phenomenon is probably related to the fact that acupuncture elevates endorphin levels in the brain and spinal fluid. Acupoint-needling can also affect the balance of other brain neurochemicals like serotonin. Decreased levels of serotonin have been implicated as a contributing factor in certain types of depression. The therapeutic effects of acupuncture for relieving anxiety and depression can sometimes last for several years. There are reports of manic-depressive disorders that have been amenable to acupuncture therapy as well. Acupuncturists who have had success in treating manic-depressives suggest at least ten acupuncture treatments (on an almost daily basis at first), along with a high-protein diet and high doses of nutrients, including niacinamide, vitamin C, vitamin E, calcium, zinc, and a multivitamin containing minerals and trace elements. Patient response to acupuncture for manic-depressive illness seems to be quite variable. Some manic-depressives may require only a few more acupuncture sessions (beyond the initial ten treatments) to maintain their new emotional stability, while others need to have acupuncture treatments repeated weekly or every two weeks for several months. Acupuncture can also be effective in treating situational and endogenous (internally generated) depression. When acupuncture is used for treating depression and anxiety, patients will often note a dramatic decrease in anxiety attacks, along with a general feeling of well-being. The addition of electrostimulation to acupuncture needles placed in acupoints for the treatment of depression will usually increase the effectiveness of the treatment. Many patients may feel some partial relief from the depression after only a single acupuncture treatment. At least six treatments are needed to obtain long-lasting relief. While depression and anxiety may respond to acupuncture, psychotherapy is frequently beneficial as an adjunct. For treating anxiety, megavitamin therapy has been reported to be of additional benefit when used in conjunction with acupuncture. The Chinese also claim to have successfully treated schizophrenia with acupuncture. Acupuncture treatment for such severe mental disorders is still in its infancy in the West. However, a six-month study conducted at a Texas mental-health facility on the effects of ear acupuncture for conditions that included paranoid schizophrenia and borderline personality disorders did show a drop in the length of hospital stays from twenty-eight to eight days when researchers compared hospital records of patients from the previous year with the average length of stay for the acupuncture-treated group.

The category of psychological problems that seem to be most amenable to acupuncture therapy are the addiction disorders. Forms of addictions responsive to acupuncture include not only narcotic, cocaine, amphetamine, and street-drug addictions but also drug dependence on cortisone, prescription tranquilizers, nicotine, and alcohol, and even food addiction. Acupuncture treatments for weight problems caused by food addictions are often quite effective. As a result, a number of acupuncture-type approaches to weight loss have appeared on the health-care scene. Yet not all of these acupuncture-oriented approaches are of equal therapeutic value, mainly because their acupuncture component is only partially executed. I am referring here to the practice of placing staples in the ear to promote weight loss. The metal staples stimulate only one or two of the ten or twenty acupoints that need to be stimulated for an optimal response. Haphazard ear-stapling is only minimally effective, and the stapling process can pose a significant risk of infection to the ear. For those people who undergo complete acupuncture therapy for weight loss, six to ten treatments are usually necessary, with treatments given once or twice a week. A typical pattern of weight loss expected with acupuncture therapy is two to four pounds per week. The effects of acupuncture for weight loss can last up to six months. Retreatment for several acupuncture sessions will be necessary if the excessive hunger returns.

In alcoholism, acupuncture helps with both the addiction as well as with relieving symptoms of alcohol withdrawal. The effects of acupuncture treatment for alcohol addiction vary considerably. Some patients notice benefits after only a single treatment session, while others seem to require an additional series of acupuncture treatments after several months of abstinence. A 1989 study published in the prestigious British medical journal *Lancet* noted that when acupuncture was added to the treatment program of chronic alcoholics, it significantly increased the number of patients completing the program and decreased the number of relapses and readmissions to alcohol-detoxification centers. Acupuncture programs in New York that have been established to treat pregnant mothers addicted to crack cocaine show promise in getting patients free of their cocaine habits. One favorable side effect of the treatments is the increased birth weights of children whose mothers participated in the acupuncture and counseling programs. Some acupuncturists claim success in treating nicotine addiction as well as heroin and opium dependence, with success rates reaching 90 percent or better. Results were especially positive when electroacupuncture and ear acupuncture were used along with body acupuncture. Curently, there are nearly three hundred acupuncture-based treatment programs for substance abuse in the United States.

A Visit to an Acupuncture Practitioner

Depending upon the acupuncturist's training and particular orientation to acupuncture therapy, sessions with an acupuncturists can vary widely. In general, most acupuncturists will ask the patients for a history of their complaint in terms of specific symptoms and how those symptoms affect them. For acupuncturists who specialize in pain management, there will be little history-taking other than a description of the nature and location of the patient's pain and perhaps some attention to any conventional medical testing or treatments that have already been tried. Since pain management is directed primarily toward pain relief, selection of acupuncture treatments may be based on specific acupoint recipes used for treating back pain, leg pain, etc. Acupuncturists trained in the French school may also spend some time analyzing the problem in terms of which meridians flow through the painful area of the body. They'll use special energetic calculations, too, before deciding upon a course of acupuncture treatments. If pain management is to be handled by a five-element acupuncturist, much more time will be spent on history-taking with regard to the problem and the patient's past medical and psychological history. Five-element analysis takes into account food preferences and dislikes, temperature and seasonal preferences, favorite colors, and a variety of matters that would not seem to be of immediate relevance to the medical problem for which relief is being sought.

Following history-taking, most acupuncturists will decide upon the optimal placement of acupuncture needles and will usually do a treatment during the first session. The needles used in acupuncture are exceedingly thin and unlike the hypodermic needles most people have seen used in a doctor's office. The treatment needles are usually about three inches in length with a special handle that may be metallic or plastic. Acupuncture needles with metal handles are often made of two different metals, one for the needle and a different metal for the handle. Some theorists have speculated that the needle's dual-metal configuration, combined with the patient's body heat, actually creates a weak electrical stimulation to the acupuncture point when the needle is inserted. But the plastic-handled needles seem to be equally effective, as long as the needles are placed into the appropriate acupoints. Some acupuncturists will use gold or silver needles. The choice of precious-metal needles depends upon the need for either stimulating or sedating (dispersing excess ch'i energy) at a particular acupoint. There is usually little pain associated with needle placement. A few individuals may experience a mild light-headedness or even near fainting after needle insertion. More often the type of person this happens to is not the frail, elderly female patient but the young, muscular, stoic male who professes no fear of needles. Rarely, in extreme reactions, the needles may need to be removed. A skilled acupuncturist is able to deal with the problem quickly and without incident or serious side effects. Needle-insertion techniques will vary from practitioner to practitioner. Some will insert the acupuncture needles directly, using a quick twirling motion, while others use a fine plastic tube to guide the needle. The acupuncturist merely taps lightly on the handle of the needle to insert it gently into the appropriate acupoints. In general, disposable needles are used, so the fear of disease transmission from recycling and reusing needles is unfounded.

With body acupuncture, the traditional steel needles are used. But in ear acupuncture, the type of needles used can vary. Some acupuncturists use the traditional acupuncture needles, while others insert a tiny needlelike tack with a circular base. This needle, only about a millimeter long, is inserted into the appropriate ear acupoint. Many times auricular therapists will leave some of these ear needles in place for an extended period of time and place a small piece of flesh-colored adhesive tape over the needle to keep it from falling out (and to disguise its presence). A minority of auricular therapists tape tiny gold- or silver-coated metallic balls, only a millimeter in size, over the key ear acupoints in need of stimulation. Interestingly enough, patients on certain types of medications, such as steroids, may have a diminished response to body acupuncture but are more likely to have a positive response to ear-acupuncture treatments. Whether this reflects different physiological or energetic mechanisms is still unclear.

Many auricular acupuncturists use electronic acupoint locators to find key acupoints on the ear. Point locators are able to zero in on acupoints because they have a lowered electrical skin resistance compared to non-acupoint skin areas. A point locator often looks like a fat ballpoint pen with a smooth metallic tip. The tip is used to touch the skin and to register the skin's electrical characteristics. Most acupoint finders also have a second dual function of performing as an electrical acupoint stimulator. This dual function-point finder delivers a weak, pulsed electrical current to selected acupoints, eliminating the need for needle insertion. The current used to stimulate an acupoint is extremely weak, as most devices run off 9-volt batteries. There are a number of good point locator/stimulators available to patients for home use in needleless electrostimulation of both body and ear acupoints. One of the best is the HealthPoint system developed by Dr. Julian Kenyon of England, a leading practitioner and researcher of complementary and vibrational-medical therapies. A number of physical-therapy departments around the United States are experimenting with using acupoint electrostimulators for pain relief, with promising initial results.

After needle insertion into either ear or body acupoints, many practitioners will provide some additional stimulation to the points. Most will connect key acupoint needles to wires from small electrical stimulator boxes, similar to TENS devices, which deliver a weak, pulsed electrical current to the selected acupoints. More traditional acupuncturists will use heat to stimulate the needle by burning a small ball of moxa or mugwort herb wrapped around the acupuncture needle's handle. In other words, no burning herb actually comes in contact with the skin. Only the heat of the burning moxa is transmitted through the needle into the acupoint. Thus, thermal instead of electrical energy is used to provide additional stimulation to the acupoints. In a typical acupuncture session, the needles are left in place for ten to twenty minutes and then removed by the acupuncturist. Following needle removal, many acupuncturists may rub a small amount of commercially available Tiger Balm ointment into the point, bringing a warm sensation to the skin. Some acupuncturists feel that this continues the stimulation of the acupoints. A few acupuncturists will tape very small magnets, only a few millimeters in diameter, directly over the acupoints into which needles have been inserted. The magnetic stimulation seems to continue the acupuncture treatment at an energetic level for some time after the needles have been removed. While electrical, magnetic, and thermal stimulations of acupoints are among the more common forms of energetic acupuncture aids, an increasing number of acupuncturists are now working with soft-laser stimulation. The soft-laser devices are frequently solid-state lasers (based on light-emitting diodes or LED chips) similar to the penlike laser pointers in common use. Other practitioners use the more conventional helium-neon laser devices similar to the lasers used in grocery-store bar code scanners. Because the lasers are of low energy output, there is no cutting, burning, or even any sensation of heat when the laser beam is focused upon acupoints. Laser stimulation of both ear and body acupoints has been used. However, traditional needling of acupuncture points tends to be much more common.

In addition to receiving acupuncture treatments, patients who see acupuncturists trained in five-element-based acupuncture and traditional Chinese medicine may also be prescribed specific Chinese herbs to help rebalance their energy systems. Herbs are selected according to how they may help to restore the appropriate yin/yang balance within the body, as well for their effects upon specific organ systems of the body. Some acupuncturists, as mentioned, may also prescribe homeopathic remedies in conjunction with acupuncture therapy. If one goes to a physician trained in acupuncture, it is conceivable that other, more conventional medical therapies might also be suggested for a patient's particular medical problem. In a variation of the old B12 shots given by doctors in days gone by, a number of MDs will actually inject specific acupoints with vitamin B12, with varying degrees of therapeutic success.

Whether one goes to an acupoint-recipe-based acupuncturist for pain management or to a five-element-based practitioner who will analyze the whole person, acupuncture in its many forms can be extremely helpful for a wide variety of disorders. Even the National Institutes of Health has begun to recognize acupuncture's value by finally declassifying acupuncture needles from their former status as experimental and investigational devices. One of acupuncture's greatest strengths is that it is a totally natural system of subtle-energy rebalancing, with literally centuries of experience. In addition, acupuncture is often capable of treating disorders for which modern medicine has no cure or even any symptomatic treatments available. Regardless of disagreements between clinicians on the mechanisms behind acupuncture's actions, this vibrational-medical therapy continues to enjoy greater acceptance and increasing validation of efficacy in the mainstream medical community.

Six

THE ART OF HEALING WITH FLOWER ESSENCES

FLOWER ESSENCES OR flower remedies are a unique and subtle form of vibrational medicine that has become increasily popular over the last sixty to seventy years. Although most people familiar with this healing approach think that flower-essence therapy originated during the beginning of the twentieth century, there is evidence suggesting that flower essences are an extremely ancient healing modality. While flower essences may be foreign to most American doctors, they have been successfully used throughout Europe (especially in England) by many health-care practitioners. This unique form of vibrational medicine can be quite powerful in rebalancing the many subtle-energy systems of the multidimensional human being.

Phytomedicine: The Many Forms of Plant-Based Therapies

Over the course of the last several decades, the interest in phytomedicine, the use of medicinal plants in healing, has literally exploded. Recently there has been a boom in the sales of plant-derived herbal remedies such as echinacea for immune enhancement, St. John's wort for depression, and saw palmetto for prostate problems. Even the respected *Journal of the AMA* has published an article on the use of saw palmetto for treating prostate enlargement, which noted that the herbal therapy was as effective as current drug therapy and was associated with fewer side effects than conventional medicines. It is ironic that our high-tech medicine is now shifting back to an interest in plant-based remedies that were, after all, the predominant form of medical therapy prior to the pharmaceutical revolution following World War II. Plants have unique healing properties just waiting to be discovered and used in a wide variety of ways. The ingestion of the crude plant substance in capsules or tinctures forms the basis for herbal therapy. Plants contain a wide variety of chemicals and medicinal properties in unique combinations that have healing effects as well as subtle-energetic effects upon the human body. Some of these subtle-energetic properties can be seen in the use of Chinese herbal medicines prescribed to restore harmony and balance to a patient's ch'i energy. Certain herbs seem to have an energetic affinity for particular organ systems of the body, working not only at a physiological level through various naturally occurring medicinal substances but also at a subtle-energy level as well. This may be one reason herbal medicines are sometimes superior to drugs in treating certain health problems.

Different parts of a healing plant—the leaf, the root, the stem, the flower, and in some cases the fruit—may all possess slightly different beneficial healing properties. As we learned earlier, when a healing-plant tincture is diluted and potentized, a homeopathic remedy of the plant can be created to provide additional healing benefits beyond the crude plant extracts normally used for basic herbal medicine. But if we take the same plant and extract additional healing substances using different methods, we can obtain a special medicinal substance called an essential oil. The art of compressing plants in special presses, subjecting them to distillation systems, and utilizing other extraction methods allows us to acquire a plant's essential oils in a concentrated form. Essential oils are extremely potent substances with unique medicinal properties and powerful subtle-energetic-rebalancing effects that work upon the physical as well as the lower subtle bodies, especially when used in aromatherapy and bath therapy. But in addition to the aforementioned plant-based therapeutic modalities, the flower of the plant can also be used to create special vibrational remedies capable of healing or rebalancing not only the physical body but also the chakras, the meridians, and the higher spiritual bodies as well. Of course, we are referring here to the use of flower essences as a subtle yet potent vibrational-healing tool.

What Are Flower Essences?

Flower essences (not to be confused with essential oils) are specially prepared liquid tinctures made from all sorts of flowers. Even the mundane dandelion has its special place alongside rare Amazonian orchids in essence therapy. Flower essences are not a physical medication in that they do not contain specific molecules of medicinal substances taken from ground-up flowers. Rather, flower essences are prepared by picking fresh flowers, still wet with dew, and placing them in a clear glass bowl of spring water during the early-morning hours. The bowl of flowers is left in bright sunlight for a period of several hours. During this time, a unique process takes place. The energy of sunlight appears to transfer a certain aspect of the flower—the very pattern of its life-force energy—directly into the water. Unlike homeopathy, which hinges upon succussion and dilution to imprint water with the vibrational pattern of different substances from nature, flower-essence preparation usually depends upon the energy of sunlight to imprint water with the healing vibrational properties of flowers. With certain select flowers, an alternate method of preparation is used, whereby the flowers, including leaves and twigs, are placed in a saucepan with two pints of spring water, which is then brought to a boil. Next, the liquid is decanted and filtered through three pieces of filter paper. In both the sunlight and boiling methods, brandy is added to the water in a one-to-one ratio. The sun-drenched (or boiled) liqueur is often referred to as the "mother tincture."

The development of flower essences as a healing modality is often credited to an English physician by the name of Dr. Edward Bach, who lived in the early 1900s. Bach was actually a bacteriologist. His field of study today would be comparable to that of an infectious-disease specialist. He studied diseases that were linked to infections caused by specific types of bacteria. Bach had worked for years trying to find treatments for patients that were less toxic than the therapies of his day had to offer. During his search, he came across the idea of giving liquid tinctures prepared from flowers. Bach was a radical for his time because he strongly felt that the mind and the emotions played a substantial role in most illnesses. Bach came to believe that if one could help rebalance a patients' emotions, then their illnesses, regardless of the cause, would most likely improve. It seemed to Dr. Bach that remedies borrowed from nature might hold the answer to the less toxic emotional-rebalancing therapy he sought. As it turned out, Bach was not only a trained physician, he was also an intuitive or "psychically sensitive" human being. One day, during a morning stroll through the English countryside, he happened upon what eventually came to be known as the Bach Flowers. As he stopped in front of a particular flower, he suddenly became overwhelmed with strong emotions that seemed to come from outside himself. Based on an intutitive hunch, Bach spread some of the morning dew that had collected upon the flower directly on his lips. As the dewdrops from the flowers touched his mouth, the strong emotion quickly vanished, returning him to his previous state of calm. It appeared that Bach could sense from different flowers the kind of emotional disturbances which each specimen was capable of neutralizing and rebalancing. He soon discovered that by ingesting a small amount of a flower's sun-drenched dewdrops, that flower's healing and rebalancing properties could be easily transmitted to an individual.

Based upon this discovery, Bach set out to discover which flowers had the ability to rebalance those key emotional-energy patterns he believed were behind many people's illnesses. In all, Bach found thirty-eight flowers whose liquid tinctures came to be known as the Bach Flower Remedies. The final flower remedy, perhaps the most popular flower essence currently in use, became known as Rescue Remedy. Essentially, Rescue Remedy was a combination flower essence made from five different flowers, which Bach found to calm and rebalance individuals who had just suffered from a sudden shock, a physical trauma, an accident, or some emotionally stressful situation. Although many new flower essences have been discovered since the time of Dr. Bach, the Bach Flower Remedies still remain the most widely prescribed of all flower essences in use today.

How Do Flower Essences Work in Healing Body, Mind, and Spirit?

Even though flower essences have been in use for many years, an exact understanding of how they work has been slow in coming. Just as homeopathy has been used as a healing modality for more than a hundred years without an exact scientific explanation, so too have flower essences been prescribed without a complete understanding of exactly how they work. Much of the research on flower essences has been carried out through various forms of psychic and intuitive exploration. Part of the reason for this is that most flower essences appear to work less on the physical body directly and more at the level of human subtle anatomy, such as the chakras, meridians, and the spiritual bodies. To date, very few device technologies have been developed that can accurately measure the effects of vibrational remedies upon the etheric body, the astral body, or the chakras. On the other hand, there are many clairvoyants and sensitives who can directly perceive those aspects of human subtle anatomy. However, researching the healing actions of flower essences is a case where the psychic observations of trained intuitives will still need to be validated using newer and more sensitive vibrational-measurement technologies. The ability to quantitatively as well as qualitatively measure and record changes in the higher-dimensional components of human anatomy will play an important part in the vibrational medicine of the future. That having been said, what has come through various intuitive sources regarding flower essences is both insightful and intriguing.

Through his research, Bach eventually came to the conclusion that flower essences worked on the level of the spiritual bodies. He also believed that our human subtle anatomy was magnetic in nature. Much of the modern research on life-force energy (emitted by healers' hands) suggests that the higher realm of human spiritual anatomy Bach spoke of is composed of an energy and structure that is indeed magnetic in nature. However, the magnetic nature of the spiritual bodies and the life force is different from the "iron filings attracting" kind of magnetism. The energy of the life force is a distant cousin of the magnetic energy associated with permanent magnets. Bach felt that our emotional expressions were reflections of subtle-energy patterns we carry in our "magnetic" spiritual bodies. When people became fixated upon a certain way of emotionally reacting to the world around them, it was because their magnetic emotional bodies were also fixed in similar emotional-energy patterns. Following this same line of reasoning, if the energy patterns of our astral, emotional, and mental bodies can indirectly influence our etheric and physical bodies, then emotional disturbances could eventually lead to physical illness through the subtle-energetic connection between the physical and the spiritual dimensions. Bach, as well as current flower-essence practitioners, always believed that flower essences possess the capacity to vibrationally rebalance disturbances in the subtle bodies and the chakras in ways that lead to greater psychological and physical health. This belief is based in part on clairvoyant observation of flower essences' actions upon human physical and subtle anatomy, as well as upon clinical observations of patients' responses to flower-essence therapy. One reason flower essences appear to affect human subtle anatomy so strongly is that they tend to contain more actual life-force energy from plants than do other vibrational medicines.

Flower essences contain a kind of liquid tincture of the plant's life-force energy in addition to the healing energetic patterns possessed by the plant. The flower is the crowning achievement of the plant, containing the highest concentration of its life-force energy. After all, the flower is the site at which most plants are pollinated and are able to reproduce. When a flower essence is produced using the sun method, the pranic energy of sunlight transfers or imprints a small amount of the life-force energy's pattern into the water. In some ways, this is analogous to making an old-fashioned sunprint by placing a leaf on photosensitive paper and exposing the paper to bright sunlight. By bathing the sun-exposed photo paper in developer solution, a photographic image of the plant leaf which lay upon the paper is quickly revealed. Similarly, the energy of sunlight is able to make a kind of etheric imprint of a plant's flower in water. The water does not actually contain medicinal molecules taken from the flower. Instead, it contains a subtle-energy pattern that can produce profound effects in human beings if appropriately prescribed for various medical and emotional conditions.

The use of flower essences in healing represents an approach based upon a spiritual understanding of human illness. Dr. Bach considered illness to be a reflection of some form of divine disharmony between the soul and the conscious personality. He put forth the concept that each soul incarnates into a physical body, bringing with it a "divine purpose" that would manifest during an individual's lifetime. According to Bach's understanding, the personality becomes disconnected from its true soul pupose, either from outside influences or from internal forces and emotional disharmony. He felt that all illnesses were learning experiences meant to help people recognize the error of their perceptions, their misguided patterns of thinking, and their negative emotional expressions. In Bach's worldview, disharmony of the mind and emotions was thought to be the breeding ground for all illnesses, regardless of their supposed bacterial, viral, genetic, or environmental origins. The typical emotional patterns of mental and emotional disharmony that Bach believed were precursors to illness included impatience, being overcritical, persistent grief, excessive fear, extreme terror, bitterness, lack of self-esteem, overenthusiasm, excessive restraint, indecision, doubt, ignorance, denial or repression, resentment, restlessness, apathy, indifference, weakness of will, and guilt (among others). Bach felt that his thirty-eight flower essences provided vibrational patterns needed to help neutralize or serve as an antidote to the negative subtle-energy aspects associated with each of these different negative emotional and mental states. Ingestion of the flower essence was followed by a flooding of the body with positive, healing, rebalancing subtle-energy vibrations. In other words, Bach believed that his flower remedies would not only neutralize negative emotional- and mental-energy patterns but also infuse positive vibrations associated with specific virtues into an individual, such as the virtues of love, peace, steadfastness, gentleness, strength, understanding, tolerance, wisdom, forgiveness, courage, or joy. After a period of taking the flower remedies, Bach noted profound emotional and mental changes occurring in his patients, along with improvements in their physical symptoms. With time, patients enjoyed improved overall physical health as they became rebalanced at the mental, emotional, and spiritual levels. With regard to their therapeutic actions, it is important to recognize that flower essences do not actually suppress negative emotions. They merely act as catalysts in assisting the spiritual bodies to release unwanted negative emotional- and mental-energy patterns.

The flowers Bach used to make his flower remedies were thought to be from plants of a "higher order." These higher-order flowers embodied certain soul qualities represented by specific energy wavelengths and frequencies of subtle-energy vibration contained within the plants. Each of the soul qualities "captured" by the flower remedies resonated with or was in tune with a particular frequency in the multidimensional human energy field. According to Bach, each human soul contained within itself thirty-eight different soul qualities embodied by the various Bach flowers. These qualities might be considered analogous to energy potentials, virtues, or even "divine sparks" of intelligence. Through his intuitive research, Bach came to believe that if conflicts arose between the intentions of a person's soul and the actions of the conscious personality, the frequency of the energy field would become distorted, out of harmony, or even slowed down. This vibratory distortion in a person's energy field was thought to create a negative effect upon the person's entire psyche. When correctly prescribed for a particular negative emotional pattern, the Bach flower remedy was thought to "resupply" the person with the emotional vibratory frequency that had become changed or distorted in the first place. In a sense, Bach flower-essence therapy might be considered a way of infusing the personalilty with a "missing" frequency-specific emotional-energy pattern. While the essences seemed to deliver emotion-specific frequencies of subtle energy, they also appeared to have definite energy affinities for particular soul qualities that were out of balance. On a higher level, each essence was thought to strengthen contact between the conscious personality and the vibratory qualities of the soul. Thus, a flower remedy would act as an energetic catalyst that helped to reestablish contact between the personality and the soul or the higher self.

While the Bach Flower Remedies may work in reconnecting the personality with the higher self, there are now many newer flower essences that supposedly work by providing a wider range of therapeutic actions upon the physical body as well as the spiritual bodies. In addition, a number of the newer flower essences are also said to work by rebalancing the emotional and astral bodies of people in order to create harmony in the face of various disharmonious energy patterns. Like the original Bach Flower Remedies, each of the newer flower essences is quite specific in its vibrational actions upon particular emotional- and mental-energy patterns. Some of the newer flower essences are said to have an affinity for specific aspects of human subtle anatomy (such as particular chakras or meridians), or they may help to create a stronger alignment between the physical body and the spiritual bodies. Among the newly produced flower essences are a group of essences that seem to work primarily at the level of the cellular and molecular structure of the physical body in order to bring about healing. This is in contrast to the original thirty-eight Bach Flower Remedies, which heal because of their ability to rebalance the disturbed astral, emotional, and mental patterns that might have originally led to the physical ailment. There is still not much hard research to document the physical cellular effects of flower essences upon living systems. However, the clinical observations of flower-essence practitioners around the world, combined with a great deal of intuitively derived information about flower essences, provides therapeutic guidelines for their usage. The scientific validation of flower essences seems to be coming about almost as slowly as the validation of homeopathy in conventional medical circles. Nevertheless, at least one double-blind study has been conducted showing that flower essences perform better than a placebo. So at least preliminary evidence seems to support a real therapeutic effect.

The Bach Flower Remedies: A Major Contribution to Flower-Essence Healing

As we have already mentioned, the most popular and widely used form of flower-essence therapy is the Bach Flower Remedies. Each of the thirty-eight remedies has a specific indication for a particular emotional pattern of disharmony. The following is a list of the different Bach Flower Remedies and their therapeutic indications (as suggested by Dr. Bach).

Agrimony: For individuals who put on a cheerful face while repressing or denying their feelings of pain and torment. Addictions to drugs, food, or even television can sometimes be a way of numbing their inner pain or repressing feelings too difficult to face. Those requiring Agrimony tend to avoid arguments and confrontations at all costs. As such, Agrimony may be quite helpful in counseling if one is trying to get in contact with buried and repressed feelings and painful memories. Agrimony is said to carry a vibration of joy.

Aspen: Useful for individuals who are experiencing fears of unknown origin, such as free-floating anxiety, uneasiness, apprehension, a sense of foreboding, as well as panic attacks. Aspen may help to transmit a vibration of fearlessness.

Beech: May be helpful for people who are overly critical and intolerant of others. For those who are constantly finding fault in things and in others, or who tend to be overly judgmental, arrogant, or extremely prejudiced. Beech carries the vibration of tolerance.

Centaury: For people who are unable to stand up for themselves or to refuse others' requests of them. For individuals who are always trying to please others or to fulfill other people's needs at their own expense, often behaving like human doormats. For those who are unable to tell others that they are feeling resentful. For types who don't like to make waves, or to displease others for fear of being disliked. Codependency issues are frequently present, as in those who are always playing the victim. Centaury has the vibration of self-determination.

Cerato: For people with an extreme lack of self-confidence in their ability to make good decisions, always looking to others for guidance and direction. For inability to trust one's judgment. Cerato carries the vibration of inner certainty.

Cherry Plum: For those who fear losing control. Also for those who take unneccesary risks or act rashly, without thought for possible harm they might bring to themselves. For situations of extreme emotional crisis or possibly even threat of suicide. Can also be helpful for those who lose control of their tempers and become abusive or may be out of control in their addictions to drugs, food, gambling, or even spending excessive amounts of money. It has proven to be of great benefit in treating individuals with obsessive-compulsive disorder. Cherry plum transmits the vibration of composure.

Chestnut Bud: For those people who seem unable to learn from their past mistakes, habitually repeating the same patterns over and over again, unable to see their lives in a truly objective light. Especially useful for those individuals who are constantly involved in abusive relationships. May also be very helpful in recovery from addiction disorders or releasing addictive behaviors. This essence carries the vibration of the capacity for learning.

Chicory: For the overly possessive and overly nurturing types, the overly smothering mother or spouse, who tend to have a smothering kind of love and attempt to gain absolute control over the lives of those they love. Love is given, but conditionally, with strings attached. It can be helpful for those who seem to act like martyrs but expect to be appreciated for all the sacrifices they have made for others. Often these types can be extremely possessive, self-centered, and manipulative, with a strong need for attention. They love to be told how much they are appreciated. Chicory has the vibration of selfless love.

Clematis: For those who are daydreamy and lack the ability to concentrate. For people who have lost interest in everyday life. For individuals who seem to be preoccupied with their fantasy lives because they are usually unhappy with their real lives. It may be of help to artists who lack the ability to express their artistic gifts in practical ways due to lack of focus. This essence can be extremely grounding, helping the individual to become more focused and to take an active role in everyday life. As such, Clematis may be of great benefit to children with learning disabilities. Clematis carries the vibration of creative idealism.

Crab Apple: This essence is a great cleanser and may be of use in treating those who are unhappy or disgusted with their body image. Also useful for those people who feel ashamed of their illness or infirmity. For people who focus on their flaws, often blowing things out of proportion, overly concerned with their appearance to the exclusion of everything else. It may be of great benefit to those with anorexia and bulimia and for those who have obsessions about their weight. Crab Apple is also helpful to victims of rape and incest who feel "unclean," helping to release the emotions of guilt, shame, and self-loathing. It may also be of use to people who are "germophobic," afraid of any little bacterium or microbe that might contaminate them or make them ill. Crab Apple carries the vibration of purity.

Elm: Elm flower essence can be of great help for people who experience an overwhelming sense of responsibility, for those with too much to do and not enough time or energy to do it, overcome with exhaustion. This is usually a temporary state, often experienced by people who are normally responsible individuals but who now find themselves feeling overburdened. Elm carries the vibration of right responsibility.

Gentian: This essence is indicated for those who feel despondent, often over repeated setbacks and delays in home or work projects. People who need this essence frequently feel tired, depressed, frustrated, and discouraged, sometimes thinking, "What's the use anyway?" This is another of several Bach remedies that can be helpful for children with learning disabilities who experience a daily sense of struggle with schoolwork. Gentian is associated with the vibration of faith.

Gorse: This flower remedy is for those who feel truly hopeless, overcome with a sense of extreme dispair. The depression is much deeper than that associated with Gentian. People who need Gorse often feel that there is nothing that can be done to make their situation any better. Gorse carries the vibration of hope.

Heather: This remedy is for people who are entirely preoccupied with themselves. They seem caught up in their own problems and may frequently be hypochondriacs over the slightest problem with their health. These people tend to talk mostly about themselves, while listening very little to what others have to say. There may be a sense of loneliness and a starving for attention. Being around such people is often a draining experience because they can be so needy. Heather carries the vibration of empathy.

Holly: Holly is indicated for people who experience hatred, envy, jealousy, suspicion, or thoughts of cruelty or revenge toward others, for instance in relationship breakups and divorces (which often stir up feelings of rage and negativity). Holly can assist in releasing these negative emotions. It may also be of great help for people who experience feelings of paranoia. Holly carries the vibration of divine love.

Honeysuckle: This flower essence can help those who long for the past, the so-called good old days. For people who dwell on past memories and nostalgia, especially older individuals whose friends and loved ones may have already passed on. It may also be applicable for individuals who have undergone a divorce, yet cling to past memories of things, keeping them from moving forward with their lives. It can help with separation anxiety and homesickness, especially in children who find it difficult to go to school or to summer camp. Honeysuckle carries the capacity for change.

Hornbeam: This essence is useful for people who have a sense of weariness and fatigue. It may be either mental or physical weariness. It is for people who are okay once they start moving but find it difficult to make that initial effort to get going, that "Monday-morning feeling." It can be helpful for those who procrastinate about things as well as for those who feel a need to revitalize their lives. Hornbeam is also indicated for those who feel that they are not strong enough physically (often leading to a need to constantly work out and build up their muscles). Hornbeam transmits the vibration of inner vitality.

Impatiens: As its name implies, this essence is for those who are impatient. They tend to be assertive, independent individuals who move much faster than those around them and are intolerant of people who do not possess the same motivation, drive, and speed in getting things done. These people may be restless and are easily irritated. Impatiens essence carries the vibration of patience.

Larch: Larch is for people who lack self-confidence and have chronic low self-esteem. Such people often feel inferior to those around them. They frequently have a sense that success can never be theirs, so why bother making the effort? Larch carries the vibration of self-confidence.

Mimulus: Mimulus essence is for fears of known things, such as fear of illness, fear of death, fear of intimacy, fear of being financially insecure, etc. It may be of great help in treating phobias such as fear of heights or of specific animals such as dogs or snakes. It can also be of assistance to those who are shy and timid (which can stem from a fear of rejection). Mimulus has the vibration of courage.

Mustard: This essence is for treating a slightly different type of depression from the ones mentioned previously. It is indicated for a sudden onset of depression from no known cause, associated with an overwhelming sense of doom and gloom. Such depression may descend suddenly and then dissipate just as quickly. It may also be helpful in treating the postpartum depression that sometimes affects new mothers. Mustard carries the vibration of cheerfulness.

Oak: Just like the tree, people who benefit from Oak flower remedy are usually strong, capable, and self-reliant individuals. However, they may be too strong, finding it difficult to take a break and rest, to ask for assistance, or to just let go of things. The classic workaholic type can benefit from Oak. These types will keep their nose to the grindstone, persevering in spite of any obstacles in their way. Whether they are suffering from coughs or colds, they keep right on working, ultimately being much too hard on their minds and bodies, sometimes to the detriment of their health. As with other flower remedies, Oak flower remedy helps to bring balance to the negative emotional state, but it does not change a person's basic personality type. Oak carries the vibration of endurance.

Pine: This essence is indicated for those who are perfectionists. They are never entirely satisfied with their own performance, usually being highly self-critical (although they are usually not critical of others; their perfectionism applies mainly to themselves). Such individuals often berate themselves with guilt for not having done better. They seem to maintain a standard for themselves that they can never quite live up to. At times they will even blame themselves for others' mistakes, often being quite apologetic. Pine carries the vibration of forgiveness.

Red Chestnut: This essence is indicated for those who worry excessively about their loved ones, often with much anxiety and overconcern in a way that is usually out of proportion to reality. There may be a constant fear for the welfare of others, that something bad will happen to them. It is indicated for that sense of stress, anxiety, and fear that some mothers experience in excessively worrying over a child's being run over by a car on the way to school or a wife's concerns that her husband's thirty-minute delay in getting home from work must be due to a catastrophic car accident. Red Chestnut carries the vibration of solicitude.

Rock Rose: This essence is of assistance in emergency situations, helping to release the fear, sheer terror, and panic that often accompanies such crises. It can also be helpful in calming people after they have had nightmares. The emotional state this essence treats is usually a temporary one, but Rock Rose can be of great benefit in getting through the emergency with less stress. Rock Rose is associated with the vibration of steadfastness.

Rock Water: Although one of the Bach remedies, technically speaking, Rock Water is not actually a flower essence, nor is it even made from a plant. This essence is made from the healing waters where Dr. Bach worked, and he found it to have unique healing properties. Rock water is indicated for individuals who are rigid and self-denying, sometimes to the point of martyrdom. They tend to be too hard on themselves, living a life that sometimes seems without joy. These individuals tend to adhere strictly to a specific lifestyle or personal or religious discipline. They frequently have an overly rigid set of personal ideals that they try to live up to. This essence is said to bring a sense of balance, conferring a greater capacity for gentleness and acceptance. Rock water is said to carry the vibration of adaptability.

Scleranthus: This flower remedy is for those who have a hard time making up their minds. They tend to be very indecisive, often waffling back and forth one way or another before making choices. This essence can also be helpful for individuals who are subject to mood swings, vacillating from optimism to pessimism or from joy to sadness or vice versa. It is a flower essence that helps to bring balance on a number of different levels simultaneously. And in its capacity to restore balance, it may actually be of assistance in treating motion sickness, a disorder of the inner ear, which regulates our sense of equilibrium. Scleranthus is said to carry the vibration of balance.

Star of Bethlehem: This particular essence is especially helpful in dealing with situations that involve grief and the loss of a loved one, through either separation or death. It is also useful in treating any type of physical or emotional trauma, whether it is recent or years old, especially abuse of any kind. Old, unresolved traumas often remain vibrationally stored within the body for years and later resurface as physical symptoms or even emotional ones, such as anxiety and depression. This flower essence seems to work even down to the cellular level in releasing the vibration and cellular memory of any traumatic experience that the body may be holding on to. (Such old traumas encoded in cellular memory can frequently block the healing process.) Star of Bethlehem may be especially helpful in individuals who have experienced rape, incest, physical or sexual abuse of any kind, and even post-traumatic stress disorder. Star of Bethlehem carries the vibration of restoration.

Sweet Chestnut: This is another of the Bach Flower Remedies useful in treating depression. Sweet Chestnut is indicated for treating depression that is usually much deeper and more severe than the indications for Gentian or Gorse essence. Such severely depressed individuals appear to be stuck in a great black hole, a void of darkness and despair that seems never-ending. They are so weak and despondent they haven't even the energy to try to end it all. This flower essence may be useful in dealing with the "long dark night of the soul," a kind of spiritual depression and sense of isolation where even God appears out of reach. Sweet Chestnut carries the vibration of release.

Vervain: This essence is helpful for strong-willed individuals who feel the need always to be right, no matter what. They tend to argue fanatically for their beliefs, as they try to convince others of the truths that to them seem so self-evident. Because these types tend to be high-energy, with their enthusiasm going toward the hyper side, they may become burned out over time. Vervain types are frequently teachers, prophets, or proselytizers. Vervain carries the vibration of restraint.

Vine: This flower remedy is indicated for those whose personalities tend to be ruthless, domineering, controlling, and inflexible. They truly believe "my way or the highway." Such types are often in a position of authority or power and will frequently crave even more power. They also tend to be quite rigid and dictatorial in their thinking. Many Vine types are actually born leaders who, when balanced, can be organized, take charge, and get things done. Vine essence helps them to recapture that inner balance and flexibility. Vine carries the vibration of the right use of authority.

Walnut: Walnut is an important flower remedy for transitions in life. These transitions include teething, puberty, marriage, divorce, empty-nest syndrome, menopause (male and female), and any major kind of potentially stressful life change. The main indication for Walnut essence is the need to adapt to new situations. It can provide greater ease and confidence in moving through these transitional periods. Walnut can help in breaking old ties to the past that may be keeping people from moving on with their lives. It can also be helpful in releasing addictive behaviors as well. It helps to provide the strength for people to follow their own true path and not be swayed by the opinions of others who might disagree with their course of action. In this capacity, it offers protection from outside influences, even to the extent that it is helpful for people with allergy symptoms (an immune reactivity to "outside influences" such as potential allergens). It can be also be helpful for individuals who work in the healing professions, as it may protect against the potentially draining subtle-energy effects of certain patients. Walnut carries the vibration of unaffectedness.

Water Violet: This essence is useful for individuals who tend to be proud and aloof, doing their work quietly, yet expecting to be left alone. Frequently these types of individuals are special teachers and leaders. Others look to them for their calmness and apparent self-assuredness. Sometimes there may be feelings of superiority, often deservedly so. They tend to be loners, finding it uncomfortable relating to others, especially when stress pushes them off balance. However, this isolation may actually prevent them from truly knowing themselves, because we all need human contact in order to grow. Water Violet carries the vibration of humility.

White Chestnut: This flower remedy is helpful for persistent, unwanted, obsessive thoughts. It can be helpful for those times when the mind is racing, replaying a sequence of thoughts over and over again. Such repetitive, obsessive thoughts can keep people from falling asleep or may awaken them in the middle of the night. Concentration is poor in this distracted state, and peace of mind is nowhere to be found. White Chestnut aids in releasing unwanted, obsessive thoughts and restoring calm and peace of mind. It is another of the Bach remedies that may be useful for children with learning disabilities because it can help improve concentration. White Chestnut carries the vibration of tranquillity.

Wild Oat: Wild Oat can greatly benefit people who feel dissatisfied that they have not found their goal in life. It can also be helpful for people suffering from a kind of "divine discontent," in that the work they are doing is not something that causes their "soul to sing." It is of aid for people who question the meaning and purpose of life and are looking for greater direction. Some of the telltale signs of a need for Wild Oat are boredom, dissatisfaction, and a lack of fulfillment from one's present working situation. Wild Oat either may make one less restless and more content with present circumstances or it may provide the courage to go out and discover one's true calling and soul path. Wild Oat carries the vibration of purposefulness.

Wild Rose: This flower essence can be of help to those people who feel that their lot in life can never change. They have a sense of apathy and resignation that nothing can be done to improve their situation. They seem to have accepted their circumstances as fixed and irreversible and feel that there is no hope of improvement in the foreseeable future. Individuals who may benefit from Wild Rose include those with various handicaps or developmental disabilities, individuals suffering from a potentially terminal illness, invalids dependent upon others for their basic needs, as well as people who perceive themselves as victims of abuse. The Wild Rose essence carries the vibration of inner motivation.

Willow: This flower remedy can be of great help for people who harbor bitterness and resentment for past hurts and perceived wrongs that have been perpetrated against them. They blame others for their current situation, seeing themselves as victims, without taking any personal responsibility for where they are in their life. They may blame their parents or others for events that may have taken place decades ago, never forgiving or releasing the past and accepting responsibility for moving on with life. Willow essence helps in releasing bitterness and resentment and brings greater awareness of one's personal responsibility and of the potentially valuable life lessons and growth opportunities inherent in all life experiences, both negative and positive. Willow carries the vibration of personal responsibility.

Rescue Remedy: This essence is the only combination flower remedy created by Dr. Bach. It is also one of the only Bach remedies to have immediately perceivable effects, usually felt within fifteen to twenty minutes or faster. It is a blend of Clematis, Cherry Plum, Impatiens, Rock Rose, and Star of Bethlehem. It is used primarily in situations involving acute stress, shock, or trauma, but its applications are actually much more widespread. Clematis is for dizziness or loss of consciousness. Cherry Plum is for loss of mental or physical control. Impatiens is to treat emotional tension and pain. Rock Rose helps to address fear and panic. Star of Bethlehem is of benefit in emotional or physical trauma, as well as grief or loss. Typical major and minor emergencies that may benefit from Rescue Remedy include accidents (such as after a fall or a car accident), any kind of shock and trauma to the body or mind, acute grief reactions, severe nausea with dizziness, anxiety and panic attacks, hysteria, after arguments, before a stressful interview or public-speaking engagement, and even before or after visits to the doctor or dentist. Rescue Remedy is also available as a cream, which contains Crab Apple added as a cleansing agent. The cream may be helpful when used for insect bites and bee stings, bruises, burns, cuts, and skin irritations (including nipple soreness associated with breast-feeding) and may be quite soothing when applied to tired eyes. Rescue Remedy is perhaps one of the best general-purpose flower essences available, and a bottle or two should always be kept on hand for emergencies.

There are a number of different techniques that have been used in selecting which of the Bach Flower Remedies, or any other flower essences, may be of greatest healing benefit for the individual. The classical method of remedy selection is derived from simply reading the descriptions and indications for each of the Bach flowers and deciding which sounds most like you at a particular time in your life. There are Bach flower questionnaires that ask questions relative to the issues and indications of each different flower. Often it is useful to answer such questions in terms of a relative-rank order (on a one-to-ten scale) denoting how strongly one resonates with the issues of a particular flower essence relative to one's current life circumstances. In general, it is best to try to limit the number of Bach flowers taken to no more than five to seven essences at a time. This can be done by selecting only those flower essences to which you have assigned the highest rating (and thus the most pressing emotional issues for you at the current time). You can always come back to other essences in the future. Also, you don't want to present too many vibrational-healing frequencies to the body at one time, or the (multidimensional) body may not be able to properly assimilate and effectively use the energies of the Bach remedies.

An alternate method for selecting remedies is to do kinesiologic testing to determine your response to individual flower remedies. This can be done in several ways. The traditional kinesiological method involves having the test subject hold a bottle of a selected Bach Flower Remedy in one hand while the tester attempts to gently force down the subject's (opposite) outstretched arm. If there is a strengthening of the arm muscle so that it cannot be forced down while holding on to a particular flower essence, this is taken as a positive response. Two people are required to do this type of flower-remedy kinesiologic testing. Alternately, you can try kinesiologic self-testing using the so-called bidigital O-ring test. In this single-person test, the thumb and pinky finger of one hand are pressed together to form a circle or O ring. The thumb and forefinger of the opposite hand are inserted through the circle while trying to force apart the fingers (in an attempt to "break the O ring"). This is a method taught and used by flower-essence developer Machaelle Small Wright of Perelandra Essences. In this technique, you ask yourself mental questions about particular flower essences that can be answered with a yes-or-no response while holding the thumb and forefinger tightly together in the ring formation. A no or negative response is suggested if the fingers can be easily broken apart by the fingers of the opposite hand. Conversely, a positive or yes response to a particular flower remedy is indicated if the fingers remain forcefully closed in a ring.

Wright suggests that you can test the efficacy of this approach for yourself by making the O ring and mentally asking yourself some simple true-and-false questions you already know the answers to. For instance, mentally ask yourself if your name is Rumpelstiltskin. Unless that is your true birth name, you should be able to pull apart your fingers quite easily. Then ask yourself the same question about your real name. This time there should be more resistance to breaking apart the fingers forming the ring. You can then move on to asking questions such as "Is this particular essence one that would be good for me to take?" You may mentally ask if a particular flower essence is correct to take at the present time (or at a later time). If the response to the O-ring test is positive, you can use this same self-kinesiological method to pose mental yes-or-no questions that give feedback on how long to take a remedy, as well as to decide when to switch to a different Bach flower combination. For those so inclined, the pendulum is another intuitive method used by certain flower-essence practitioners to select which essence to prescribe. While mentally holding the question about the flower essence, the pendulum is suspended from the thumb and forefinger and its swinging motion observed. A clockwise rotation usually indicates a positive response, while a counterclockwise rotation of the pendulum suggests a negative response to the mental question posed. There are many books on pendulum dowsing that can help to provide the finer points of this ancient but often accurate intuitive method of information-gathering. Each of the aforementioned approaches, from kinesiology to pendulums, are basically intuitive methods that make use of unconscious muscle movements to amplify the higher psychic perceptions, taking our unconscious perceptions and making them conscious.

After having selected which of the Bach remedies to take, an individual must then go about making a Bach flower combination remedy (if multiple remedies have been chosen). The Bach Flower Remedies can be taken in a variety of different ways. The traditional method is usually the best and simplest. Up to seven different Bach Flower Remedies can be combined into a single customized blend appropriate to an individual's physical, emotional, and mental needs at the time of prescribing. Once the appropriate flower remedies have been selected, two to three drops of each flower essence (which comes as a kind of concentrated mother tincture) are placed in a one-ounce amber dropper bottle. The only exception to this rule is Rescue Remedy, as four to five drops should be added instead of the usual two to three. Even though Rescue Remedy is a combination flower remedy, it still counts as only one when added to the individualized mixture. The bottle that will hold the combination remedy should be sterilized in boiling water for fifteen minutes, then cooled and allowed to dry. After the flower remedies have been added to the sterilized dropper bottle, a teaspoon of preservative—such as brandy, apple-cider vinegar, or vegetable glycerin—needs to be added. The flower essences are very high in life-force energy and thus can easily stimulate bacterial growth. Brandy or alcohol is the traditional preservative. But for those who are sensitive to alcohol, apple-cider vinegar or vegetable glycerin can be used as a substitute. Finally, the bottle is filled the rest of the way with spring water, the cap replaced, and the bottle shaken well. Spring water is the water of choice. Tap water or distilled water should be avoided when making Bach remedies. Bach Flower Remedies can also be taken directly from the main bottles as two to three drops under the tongue, but the diluted flower remedies are usually just as effective and more convenient to use.

Once the personalized Bach Flower Remedy mixture is completed, the remedy is taken as four drops under the tongue a minimum of four times a day and held in the mouth for thirty to sixty seconds. While the remedies can be used more often than this, four times a day is the recommended dosage frequency. Alternately, four drops can be added to a teaspoon of water or mixed in with a small amount of juice or a glass of water and sipped slowly. In between uses, the amber bottle should be kept away from direct sunlight, excessive heat, televisions, microwave ovens, and electrical appliances and stored in a cool, dark place (but not refrigerated). The influence of heat, excessive sunlight, and strong electromagnetic fields can sometimes wipe out the energetic imprinting of the flower remedies. Another issue not often discussed is the shelf life of flower essences. Most bottles of Rescue Remedy come with an expiration date of about five years beyond the manufacture date printed on the side of the bottle. Flower essences older than ten years may have to be discarded, unless they are still producing clinically effective results.

The effects of Bach Flower Remedies are usually perceived gradually (with the exception of Rescue Remedy) over a period of two to four weeks. Essence mixtures should be used for at least six to eight weeks to get the maximum benefit. While the changes are subtle, they can sometimes be quite profound over a period of time. More and more psychiatrists and psychologists are using Bach Flower Remedies in conjunction with, or sometimes as an alternative to, conventional psychotropic medicines such as antidepressants. Some mental-health practitioners find that these remedies can help individuals reconnect with their true selves and their core issues more powerfully and with longer-lasting effects than traditional medications alone. After taking the flower remedies for several months, it is not uncommon for different issues to come to the fore. This is sometimes referred to as the "onion peel" or "artichoke effect." As one begins to peel away the layers of more current energetic and emotional traumas to the personality, older layers of unresolved issues begin to reemerge. As this happens, it is helpful to reassess the need for the specific essences in the combination Bach remedy being taken. Eventually a need will arise to create a new flower-remedy mixture based upon the issues that become of more immediate concern.

Besides the more traditional method of ingesting Bach Flower Remedies, there are additional approaches for using flower essences that are said to affect physical and emotional problems. Bath therapy with Bach Flower Remedies is one such approach. When used in conjunction with visualization exercises, flower essence bath therapy has the added benefits of being both relaxing as well as energetically revitalizing at the same time. A few examples of flower essence bath therapy follow.

After an argument, an emotionally upsetting incident, or an accident, an evening Rescue Remedy bath is said to be quite helpful. Add five drops of Rescue Remedy to bathwater and an additional five drops to a glass of water. Merely sitting in the warm bath and gradually sipping the glass of Rescue Remedy–laced water can aid in relieving acute stress, anxiety, or a recent trauma to the body. After the bath, just bring the water glass of Rescue Remedy with you to bed. Sip a little of the water any time restless thoughts or anxieties recur during the night.

A cleansing Crab Apple bath can be helpful for people who feel abused, put upon, or who merely suffer from low self-esteem and poor self-image. Crab Apple flower remedy combined with healing visualization exercises may help to push energies associated with low self-esteem or abuse out of the physical body, as well as from the spiritual bodies, and into the auric field (where such energies may be partially absorbed by the flower-essence-laced bathwater). To create such an energetically and emotionally cleansing bath treatment, ten drops of Crab Apple flower remedy are added to warm bathwater. While you are soaking and relaxing in the tub, a washcloth should be dunked in the bathwater, folded, and then placed over the eyes. While soaking in the Crab Apple bath, it is helpful to lie back and take stock of your positive qualities. As negative thoughts arise, move your hands through the bathwater away from your body, as if to push negative thought energies away and into the water. Also, as you recall negative thoughts, try to visualize being in an ancient, sacred healing temple. When you breathe in, consciously release any responsibility you might still carry for having caused any abusive or painful incident to occur. Forgive yourself for each aspect of your past, imagining that the divine, healing energies of grace from within the sacred healing temple will assist you in releasing your past. Another helpful visualization exercise to do while soaking in the Crab Apple bath is to imagine being a seed within a large shell. Let the seed represent all your potentials, both tapped and untapped. Envision that the seed is as you were in the very beginning. Within you is the rich potential to manifest anything you can dream; health, happiness, true love, career fulfillment, etc. Then imagine the seed growing and expanding beyond the shell, with energy streams going out in all directions like the spokes of a wheel, each spoke of energy representing all the accomplishments you have ever envisioned for yourself. As you, the seed, continue to expand, begin to erase all the things that blocked your path of growth, including your fears, your emotional pain, and the anxieties you have carried with you for so long. As you erase these darker parts from within, you help release the past and create a life that is fresh, new, and filled with exciting, unlimited potential.

Alternately, a Walnut-essence bath could be of help to individuals who seek to release difficult emotional links with their past. Walnut essence may also help people to let go of difficult emotional ties that may be keeping them from moving forward in life. When trying this bath, remember to visualize and feel love for oneself and to mentally ask for what is best for your "highest good." Add ten drops of Walnut flower remedy to warm bathwater and then soak in the water. As you sit in the tub, think about the person you are trying to release and imagine that person following his or her highest destiny in the present life or the next life. Release any pain or anger attached to this individual. Imagine cutting the emotional energy cords that still connect you with this individual and see yourself free to pursue your own higher happiness, love, and your true soul path.

If flower-essence baths become part of your health regimen, it might be helpful to keep a journal for jotting down any feelings, unusual memories, or profound thoughts that come to you as you sit in the bath. Also, consider recording dreams that come to you in the days and weeks that follow your bath experience, looking for messages from your subconscious mind as well from your higher self that may provide you with symbolic information or direct feedback on the process of change you have just initiated.

One of the most interesting new developments in the use of Bach flower remedies for healing comes out of Germany. Dr. Dietmar Kramer, a German naturopathic physician, has made some rather revolutionary discoveries on the effects of directly applying the Bach Flower Remedies to different parts of the body for both emotional and physically painful conditions. Kramer refers to this approach as Bach flower skin-zone therapy. Dr. Kramer's work stems in part from his revisiting of Bach's original investigative journals. Early on, for example, Dr. Bach made references in his journal to applying flower remedies directly to certain parts of the body in the form of wet compresses for treating acutely painful conditions. One of Bach's case studies mentions using the wet-compress therapy on a man who had sprained his ankle. The patient presented as a high-strung, overly enthusiastic, strong-willed individual who balked at being told his injury would require three weeks of rest to heal fully. Fortunately, Dr. Bach entered the picture with a second opinion on to how to heal the injured ankle. The patient was given the Bach remedy Impatiens orally as well as Vervain in wet-compress form. Vervain had been indicated for individuals who were high-strung and extremely strong-willed (which certainly fit this man's personality). Bach placed two to three drops of Vervain essence into a bowl of warm water and then applied a cloth compress soaked in the flower-essence-laced water to the patient's sore ankle, instructing the patient to add fresh "essence water" to the compress whenever it became dry. By the next day the patient had returned to work, and by that evening he was able to walk normally again, somewhat in disbelief of his remarkable and rapid recovery.

Dr. Kramer looked at this and other clinical case studies reported by Dr. Bach and began to suspect that there might be a connection between certain areas of skin on the body and a rapid responsiveness to particular flower essences within the family of the thirty-eight Bach Flower Remedies. In the beginning, Kramer used Kirlian photos of his patients' hands and feet to assess the bioenergetic effects of flower-essence therapy. The Kirlian photos helped to provide an objective assessment of each patient's energetic state and gave feedback on how a patient's energy field responded to his treatments. In addition to the Kirlian photos, Kramer next added a controversial assessment technique, the intuitive input of clairvoyant observers. The task of the medical intuitives was to give feedback on how the different Bach Flower Remedies interacted with people's energy fields. Kramer's clinical work is perhaps among the most intriguing in the annals of flower-essence research, because he attempted to directly observe and measure the subtle-energy effects of flower essences upon people. His work is impressive because of the inherent difficulties in documenting specific physiological changes in the body produced by flower essences. In contrast, most clinical studies on flower essences have been primarily dependent upon patients' subjective rating of symptom alleviation in order to acquire any kind of meaningful clinical data. Kramer tried to go beyond subjective patient feedback by objectively assessing subtle-energy changes in a patient's multidimensional energy field. The input from the clairvoyants was quite remarkable. This technique of clairvoyant information-gathering goes well beyond the sensitivity of most current physiological measurement devices. Through this innovative approach, Kramer tried to acquire objective feedback on flower-essence-induced changes in his patients' spiritual bodies as well as their auric fields. The use of clairvoyant observers provides important clues to the energetic causes of his patients' illnesses, as well as sensitive feedback on the effects of Bach flower remedies on their physical and spiritual bodies.

Kramer and his clairvoyant observers studied and catalogued energy changes and reactions to different parts of the human body as they placed drops of the thirty-eight different flower essences on various areas of each subject's body. Based upon these observations, Kramer found that different zones of skin throughout the body seemed to respond to only one of the thirty-eight Bach flowers. Furthermore, each Bach skin zone always reacted to the same flower remedy, even in many different subjects. Following many hours of data collection, Kramer and his associates eventually developed a map of the human body that displayed the different Bach-flower skin zones. The team soon discovered that if an individual was experiencing pain in a particular area of the body, they could apply an essence-soaked moist compress containing the zone-specific remedy to the painful region and obtain rapid pain relief in many cases. This rapid response to Bach flowers is quite different from the typically slow patient response normally seen following the oral ingestion of flower essences (with the exception, of course, of Rescue Remedy). One of Kramer's hypotheses that may explain this immediate response to the flower essences is that when the Bach Flower Remedies are applied externally, they may move directly into areas of body tissue that have literally stored the energetic emotional memories of old hurts and traumas. The topical application of flower essences, through salves and creams, through bath therapies, or through direct application of the essence itself to either acupoints (floral acupuncture) or specific Bach flower skin zones, seems to produce the immediate effect of releasing old cellular memories and held-in emotional aftershocks caused by past traumas to the body. Over the course of his research, Kramer soon discovered what he called "active zones" and "silent zones" on the body that would respond to flower-remedy application. Active zones frequently presented as areas of pain, unusual sensation, sensitivity to pressure, or directly observable skin changes. Silent zones represented areas of the body that were energetically disturbed but did not manifest any overlying skin changes, pain, or sensitivity to pressure. Kramer felt that in some cases the silent zone of disturbance might have been due to an emotional conflict that had not yet manifested physically but could be observed clairvoyantly in a patient's auric field.

One unique aspect of the clairvoyants' observation of patients' auric fields was that patients with chronic pain or severe emotional problems frequently displayed dark and muddy colors in their auric fields as well as defects resembling "holes" in their auric fields. The clairvoyants observed that by directly applying the appropriate Bach Flower Remedy to the specific Bach flower skin zone on a patient's body (over which lay the clairvoyantly perceived auric hole), the holes and energy gaps in the aura actually appeared to fill in with new, healthier-looking energy. The clairvoyant observations further indicated that the hole in the aura was filled with a slightly lighter-color energy than the surrounding auric field of the region. Clairvoyant observers reported that over time, the auric field patterns of the flower-essence "responders" seemed to become more normalized.

Based upon his research, Kramer developed an extensive map of the body showing in minute detail how each part of the body seemed to correspond directly with one of the thirty-eight Bach Flower Remedies. The Bach flower skin-zone map developed by Dr. Kramer now serves as a guide for topical Bach flower therapy. If someone experiences pain in a particular area of the body, a practitioner will look on the map to find which Bach Flower Remedy corresponds to that skin zone. Direct topical application of the flower essence (either rubbed on in the form of a flower-remedy-containing salve or applied as a moist compress soaked in warm water mixed with a few drops of the essence) is often advantageous in providing relief from the pain, whether it is of an acute or even chronic nature. Any unusual sensation, such as numbness, tingling, itching, or localized skin changes in a particular Bach flower skin zone, may also respond favorably to a particular flower essence. Depending upon where the pain is, there may be indications for very different Bach Flower Remedies. Remember that Dr. Bach's patient improved with Vervain essence applied topically to the painful ankle. But if we refer to Dr. Kramer's skin map, we would see that Water Violet, White Chestnut, or even Wild Rose could also be indicated for ankle pains in a slightly different location. Another odd coincidence about the pain and the particular correlating Bach remedy is that the indicated remedy often fits the patient's emotional and mental characteristics (to varying degrees). It is fascinating to see how a pain in one part of the body can correlate not only with a specific flower essence but also with a particular negative emotional-energy pattern that needs to be rebalanced by a specific Bach flower essence. It is almost as if the different pains in the body and their specific Bach flower skin zones symbolically provide us with subtle feedback as to specific emotional issues we need to deal with on both an emotional and on a spiritual level. The vibrational energies of the Bach flowers perform a kind of energetic psychotherapy, where the remedies actually seem to rebalance specific negative emotional patterns. Kramer's clairvoyants noted that specific emotional-energy patterns associated with the different Bach remedies tended to gravitate naturally toward certain areas of the human auric field. Based on this clairvoyant data, it appeared to make sense that direct application of a specific flower-energy vibration to its essence-associated body region might actually produce a localized rebalancing of the disturbed auric field. For instance, Kramer noted that when a person "needs" a particular Bach remedy (according to personality and current emotional patterns), the individual will frequently respond much faster to the remedy if, instead of giving the essence by mouth, it is topically applied to that flower's specific Bach flower skin zone.

Based upon his research, Dr. Kramer has reclassified the Bach remedies into groupings of flowers that bear similar clinical applications. He calls the two main groups of essences the "inner" flowers and the "outer" flowers. Kramer recommends that the group of "outer" Bach Flower Remedies be used for treating negative emotional states that arise as reactions to "external" influences. Conversely, the "inner" flowers group is useful for treating negative emotional states produced by "internally generated" problems. Kramer's group of "outer" flowers include Star of Bethlehem, Gorse, Elm, Walnut, and Aspen. He has found that the "outer" flower essences may be indicated for topical use when there is a localized disturbance in the skin and underlying tissues caused by some noxious external influence. The "outer" Bach remedies are often helpful (regardless of which Bach flower skin zone they are applied to) because they possess a kind of generalized regenerative and healing quality for injuries and energy disturbances produced by any type of negative external influence. They appear to be especially helpful in repairing bodily damage caused by physical trauma and external physical forces. In addition, the "outer" essences appear to help the physical and spiritual bodies in energetically processing subtle-energy disturbances that have developed as a result of previous emotional traumas.

When considering specific uses for the various "outer" Bach flowers, Kramer recommends that the Star of Bethlehem essence be applied topically to any skin on the body that has been traumatized as a result of injuries, bruises, sunburns, sun allergies, or even toxic chemical burns. Another "outer" flower, Gorse, can be useful for chronic skin lesions such as wounds or sores that fail

The "Inner" and "Outer" Bach Flower Remedies (According to Kramer)

5 "Outer" Flowers	**Used to treat negative emotional states that are a reaction to "external" influences**
32 "Inner flowers"	**Used to treat negative emotional states produced by "internal" disharmony**
OUTER FLOWERS	
Star of Bethlehem	**For emotional shocks, deep grief, following accidents, also in bone fractures, concussions, and spinal injuries**
Elm	**For feelings of being overwhelmed by responsibility. Also may be useful with difficulty swallowing, lack of concentration, heart palpitations, fainting spells, and nervous breakdowns**
Walnut	**For "new beginnings," helps people adapt to change, eases life transitions such as birth, teething, puberty, moving, career changes, divorces, retirement**
Gorse	**For people who have lost their faith after many failures. All efforts seem useless, people feeling "lost," "stuck," hopelessness**
Aspen	**For people suffering from vague, indefinable, irrational fears, phobias, paranoia, dark premonitions**
INNER FLOWERS	
Made up of 12 "tracts" (which also correspond to 12 astrological signs) each tract consists of a triad of 3 flowers: a "*communication*" flower, a "*compensation*" flower, and a "*decompensation*" flower	
Communication Flower	**Corresponds to our individual characters, how we "communicate" with our environment (both positive and negative emotional aspects)**
Compensation Flower	**If the lesson of the communication flower is not learned, we try to "compensate" for this (usually an artificial, temporary state)**
Decompensation Flower	**Psychopathological end states, they represent obstructions that can block many forms of therapy**
***Example*:**	**(1 tract or triad) *Centaury—Holly—Pine***
Communication Flower: *Centaury*	**Lives only for others, denying own needs, fear of losing love, unable to disagree with others, fear of rejection, can't say no**
Compensation Flower: *Holly*	**Anger, hate, envy, jealousy, suspicion, rage, vengefulness, resentment**
Decompensation Flower: *Pine*	**Guilty conscience, self-blame, self-criticism, difficulty with feeling happy**

to heal in spite of conventional therapies. When applied to the body as a moist compress, Elm flower essence, Kramer has found, can frequently help most kinds of exertional muscle and ligament pains, such as tennis elbow, back pains, and shoulder aches caused by heavy lifting. If overexertion has caused actual physical tissue damage, as is often seen with pulled muscles and ligaments, then Star of Bethlehem–laced moist compresses should provide additional healing benefits. In the case of open wounds and lacerations, direct application of flower essences to the open lesion is not usually recommended. Instead, Kramer suggests sipping from an eight-ounce glass of cool water mixed with Walnut flower remedy (and a few drops of Star of Bethlehem) four to five times a day in order to accelerate open-wound healing and to prevent scar formation.

Dr. Kramer views scars as being not only cosmetically unappealing but also causes of focal-energy disturbances. And because scars can block energy flow (as in the case of postsurgical scars that block meridians), they can create local pain and distant organ and bodily problems as well. With regard to scars and their ability to block energy flow, it is of interest that Dr. Kramer's clairvoyant observers frequently noted deep indentations in patients' auric fields in the regions directly overlying their old scars. Kramer has used Walnut flower essence, applied directly from the stock bottle (mother tincture) to old scars with a Q-tip, to relieve pain or discomfort as well as to improve subtle-energy flow through the body and through the auric field in the location of the scar. As mentioned earlier, there is a medical technique known as neural therapy whereby physicians inject procaine or xylocaine anesthetic directly into scars in order to unblock meridians and to relieve pain caused by scar tissue. Of great interest is Kramer's discovery that Bach Walnut flower remedy, applied directly to old scars, appears to produce the same unblocking effect as neural therapy's anesthetic-injection method when the two approaches are compared by direct clairvoyant observation of patients' energy fields.

Other types of physical problems that may be helped by Bach skin-zone therapy include tension headaches, tension and tightness along the spine, physical complaints due to exhaustion, sore muscles, minor injuries, bruises, abrasions, minor skin rashes, acne, generalized skin sensitivity changes, as well as chronic-fatigue states with no apparent underlying illness. In order to achieve a therapeutic effect, one must apply the appropriate Bach Flower Remedy to the specific zone of the body according to Kramer's Bach skin-zone map. And unlike the traditional oral route of taking Bach Flower Remedies, which usually takes several weeks of usage to produce healing changes, applying the flower remedies topically to their appropriate skin zone results in pain relief and other therapeutic effects, sometimes quite rapidly.

Kramer's use of Bach flowers is actually much more complex than merely applying flower remedies to Bach skin zones. Topical applications form only a portion of the total treatment. Kramer usually prescribes specific Bach Flower

Remedies to be taken by mouth in conjunction with the topically applied essences. Sometimes the diagnosis of the patient's flower-remedy requirements is based upon clairvoyant auric-field diagnosis of the patient. At other times, the choice of flower remedy will be guided by the location of a person's pain, the area of the body affected by localized skin rashes and blemishes, or unusual bodily sensations reported by the patient. Kirlian photographs of patients' hands and feet before and after therapy often provide dramatic feedback to Kramer on the energy-rebalancing effects of selected Bach skin-zone therapies. One unique aspect of using Bach flower maps is that they can be helpful in prescribing Bach remedies for children or individuals who may be poor historians (with regard to their physical- and emotional-health history). Kramer has gone even further to incorporate thirty-two of the Bach remedies into twelve key groups of flower remedies that seem to correspond to the twelve signs of the zodiac.

Needless to say, Kramer's approach to Bach flower therapy is a radical step forward from the older methods of prescribing. His research continues to be based on both clinical feedback and on direct clairvoyant observation of changes in patient's energy fields as they are affected by Bach Flower Remedies, something that has rarely been done so systematically in researching the effects of flower-essence therapy. Also, his methods provide both new ways of prescribing the flower remedies as well as a new approach for obtaining more rapid relief from physical and emotional problems. The use of Bach Flower Remedies forms an entire treatment approach in and of itself. However, since the time of Dr. Bach, there has been tremendous growth in the discovery and development of new flower essences. Next, we'll discuss some of these newer floral healing tools that nature has provided us.

The Many Varieties of Flower Essences: The Blossoming of an Ancient Healing Art

To be sure, the Bach Flower Remedies are unique. Bach felt that the answer to all of humanity's ills could be found in nature and that flowers were one of nature's highest healing tools. While the classical method of prescribing Bach remedies tends mainly to rebalance the emotional, mental, and spiritual bodies, newer approaches, such as Bach skin-zone therapy, may help to relieve pain, heal skin disorders, and indirectly stimulate other forms of healing in the physical body. Because of the positive healing effects of Bach remedies demonstrated over the years, other individuals have gone on to develop flower essences beyond Bach's original thirty-eight. More recently, new flower essences have been developed in England, France, India, Africa, Australia, New Zealand, the United States, Canada, and the tropical rain forests of South America. There

The Preparation of Flower Essences and Their Therapeutic Use

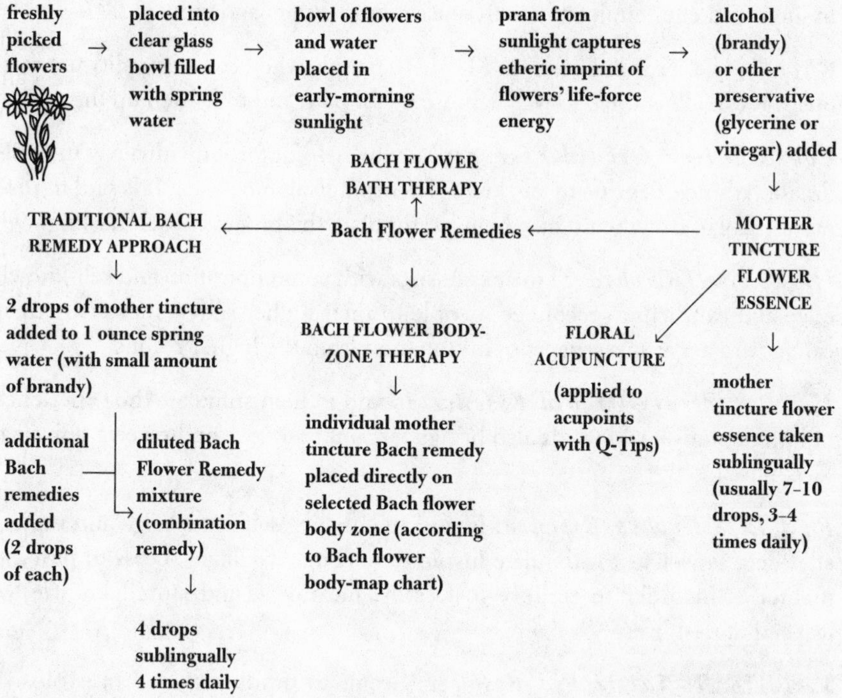

freshly picked flowers → placed into clear glass bowl filled with spring water → bowl of flowers and water placed in early-morning sunlight → prana from sunlight captures etheric imprint of flowers' life-force energy → alcohol (brandy) or other preservative (glycerine or vinegar) added

↓

MOTHER TINCTURE FLOWER ESSENCE

↓

mother tincture flower essence taken sublingually (usually 7–10 drops, 3–4 times daily)

MOTHER TINCTURE FLOWER ESSENCE → FLORAL ACUPUNCTURE (applied to acupoints with Q-Tips)

MOTHER TINCTURE FLOWER ESSENCE → Bach Flower Remedies

Bach Flower Remedies → BACH FLOWER BATH THERAPY

Bach Flower Remedies → BACH FLOWER BODY-ZONE THERAPY → individual mother tincture Bach remedy placed directly on selected Bach flower body zone (according to Bach flower body-map chart)

Bach Flower Remedies → TRADITIONAL BACH REMEDY APPROACH → 2 drops of mother tincture added to 1 ounce spring water (with small amount of brandy) → diluted Bach Flower Remedy mixture (combination remedy) → 4 drops sublingually 4 times daily

additional Bach remedies added (2 drops of each) → diluted Bach Flower Remedy mixture (combination remedy)

has literally been an explosion of growth in the development of flower essences over the past twenty years. Each group of the newer international flower essences possesses a wide range of clinical applications. Since we have been looking primarily at how the Bach remedies can help to rebalance specific negative emotional-energy states, we will see how these newer essences differ from the Bach remedies in rebalancing body, mind, and spirit.

From the Amazon rain forest come special orchid-flower essences developed by Andreas Korte of Africa. Each orchid has a unique name, often given to it by Korte because of the flower's specific and unusual energetic actions. He also produces some very unique flower essences from roses and wild plants. Certain flowers used by Korte to prepare his essences are so rare that the actual names of the flowers are known only by letter and number codes he assigns to each floral species. A few of the more interesting flower essences are listed, along with Korte's suggested applications for the essences:

Aggression Orchid Essence: May be useful for blockages in the lower chakras and can help in releasing basic energy and repressed sexuality. It can also be useful for impulsiveness or aggressive tendencies.

Angel of Protection Essence: May be helpful for fragile, sensitive people who need shelter from hostility and harsh environments. Korte sees it as providing a protective shield for one's sensitivity and vulnerabilities and as an aid to increased communication with one's spirit helpers and guides.

Chocolate Orchid Essence: May help people who take spirituality too seriously. It can allow them to walk a more joyful path and to lighten up their lives.

Coordination Orchid Essence: May be helpful for individuals with coordination difficulties due to any kind of neurological imbalance. It is said to promote cellular growth and may help to stimulate the process of self-healing.

Higher Self Orchid Essence: Assists with communication and self-knowledge and can help to reconnect people to their higher selves, increasing their capacity to act as messengers of information from the higher realms.

Horn of Plenty Orchid Essence: Is said to help stimulate the experience of infinite, universal love. It also brings the ability to give and receive universal love more freely.

Past Life Orchid Essence: Is said to enhance self-knowledge and understanding, as well as to stimulate inspiration. It may facilitate access to past-life memories in order to retrieve lost skills, meanings, and stored knowledge learned in past lives.

Victoria Regia Orchid Essence: Seems to produce a kind of explosive energy and transformation. It appears to bring about a powerful release of life energy and may stimulate the awakening of the kundalini, the subtle energy stored in the base of the spine. It can also be helpful in supporting the transformational process of death and dying.

Apple/Rose Hybrid Essence: May help in achieving unity of body, mind, and spirit. It allows the energy of love to flow throughout the entire system, reconnecting the head with the heart to create a single functional unit of conscious awareness. It helps the energy of love to flow through the entire multidimensional body system, bringing this energy to the crown and even the higher chakras.

Spring Gold Rose Essence: Appears to link the solar plexus chakra with the heart chakra, helping people to accept themselves as they are and to express this same love through the solar plexus chakra.

Arnica Flower Essence: While Arnica produced as a homeopathic remedy from the potentized plant substance is useful for physical trauma, the essence of the Arnica flower may be of help in reestablishing a person's connection with the higher self, especially after shock, trauma, the administering of anesthetics, or in drug addiction.

Arum Lily Essence: The flower essence made from this lily is useful for conflicts concerning sexual identity. It may help people in learning to accept and integrate their male and female aspects.

Dandelion Essence: Made from the common dandelion, it seems to be quite helpful in reducing tension, both emotional and muscular.

Wild Garlic Essence: This is indicated whenever there is any kind of fear operating. It appears to be quite active in strengthening psychological defense mechanisms against any form of fear, promoting a peaceful, calm outlook.

K9 Flower Essence: This flower essence (whose name is protected) is helpful in strengthening the immune system and all the body's natural defenses. It may also have an antiviral effect and is still being actively researched.

From Australia come a number of different native flower essences with unique applications. The Australian Bush Flower Essences were developed by Ian White, a fifth-generation herbalist who has studied the healing properties of flowers and herbs from both family lore as well as from travel and personal research. Some of the more interesting Australian Bush Flower Essences, along with White's suggested uses for them, follow:

Alpine Mint Bush Essence: May be helpful for those who experience mental and emotional exhaustion. Weighted down by a feeling of responsibility, they suffer from a lack of joy in their lives. This essence is said to bring revitalization, joy, and renewal.

Banksi Robur Essence: For people who, while normally energetic, suffer from a temporary loss of drive and enthusiasm due to burnout, frustration, or illness. It may help to restore vital energy and to promote a greater enjoyment and interest in life.

Bush Fuchsia Essence: For people who have a poor ability to learn, such as from dyslexia and left/right brain imbalances. It can also be useful for stuttering. It appears to help balance the left and right hemispheres of the brain, bringing greater clarity to speech and various forms of expression and may also stimulate the development of intuition.

Crowea Essence: Indicated for people who constantly worry, bringing a sense of inner tension and imbalance. It can bring peace, greater vitality, and a sense of balance and centeredness.

Fringed Violet Essence: Useful for shock and trauma. White also sees it as useful for repairing a person's damaged auric field, as well as for protecting personal energy that is being drained by other people. It may also offer some protection against radiation and negative environmental energies. It is said to remove the effects of both recent and old traumas and appears to confer some degree of psychic and energetic protection as well.

Isopogon Essence: May be helpful to those with poor memory, premature senility, and an inability to learn from past experiences. It can sometimes help those with stubborn, controlling personalities as well. Isopogon can bring learning from past experiences and allow one to retrieve forgotten skills and memories.

She Oak Essence: May be of assistance to those having problems with infertility, especially when no physical cause can be found. It may also help women with hormonal imbalances and premenstrual tension. She Oak may restore fertility and hormonal balance, and may reduce tendencies toward fluid retention.

Spinifex Essence: Can be helpful for sufferers of herpes (oral and genital), chlamydia, and surface scrapes and cuts. It can also help to heal skin conditions associated with these conditions when applied topically.

Sturt Desert Pea Essence: For deep hurts, sadness, and emotional pain. It is a powerful essence that allows one to let go of sad memories. It is said to motivate and reenergize.

Sturt Desert Rose Essence: This is indicated by White for guilt and lack of self-esteem that stems from past actions. It helps people to follow their inner convictions and sense of morality.

Radiation Essence: A combination essence made from Bush Fuchsia, Crowea, Fringed Violet, Mulla Mulla, and Waratah essences. White believes that it may negate or reduce the effects of all forms of radiation, including radiation therapy for cancer, solar radiation, electromagnetic fields, nuclear radiation, and negative earth energies. This essence can help to prevent accumulation of negative energies in the body.

Super Learning Essence: A combination essence made from Bush Fuchsia, Isopogon, Paw Paw, and Sundew Essences. This potent flower essence can help to bring about mental clarity, increased focus, and an enhancement of all learning skills and abilities.

From the United States come the Flower Essence Society (FES) essences, which were developed primarily by Richard Katz and Patricia Kaminski. Most of their essences are made from flowers native to California and other parts of North America. Katz and Kaminski have devoted considerable attention to researching flower essences through case studies and clinical reports by flower-essence practitioners. They also conduct seminars and offer practitioner training in the use of their essences. Several FES essences are listed here, along with Katz and Kaminski's suggested application.

Apricot Flower Essence: It is useful for treating mood swings related to blood-sugar imbalances such as hypoglycemia. It appears to be a good lymphatic cleanser and can soothe inflammatory and allergic reactions.

Cedar Flower Essence: May be helpful for disorders of the colon, poor assimilation of nutrients, buildup of toxins in the system, and ulcerous conditions. It may stimulate hair growth as well as increase hair thickness to varying degrees, partly because it can influence sex-hormone balance (of both testosterone and estrogen) on a physical level. For such uses the essence can be applied topically to the scalp as well as taken orally.

Feverfew Essence: The feverfew herb is sometimes prescribed as a migraine-headache preventive. When given as a flower essence made from the same plant, Feverfew can also be helpful in treating headaches, especially those related to hormonal changes of the menstrual cycle.

Jasmine Essence: For various forms of mucous congestion, such as respiratory infections affecting the lungs, throat, sinuses, and nasal passages.

Gooseberry Essence: For menopausal-related problems, such as hot flashes, nervous tension, and the panic attacks that may occur during menopause.

Pansy Essence: Said to have potent antiviral properties. It can help in clearing any bacterial or viral infection from the body. It may also bring relief from painful itching caused by cold sores (which are usually caused by herpes viruses).

Skullcap Flower Essence: Helps to clear the effects of toxins from the body and can ease withdrawal from addictive substances ranging from caffeine to morphine and heroin.

Tobacco Flower Essence: Helps to rebalance the body by clearing the effects of nicotine from the system.

Lavender Essence: For soothing nervousness and overstimulation, especially by spiritual forces that may deplete the physical body. It may be of great aid to those who are high-strung and constantly wound up, often resulting in neck and shoulder tension, chronic muscle-tension headaches, and eye problems. Lavender-flower essence can also be used synergistically with topical application of Lavender essential oil for added healing benefits.

Morning Glory Flower Essence: For those who might be considered "night owls." Often there may be erratic eating and sleeping rhythms and great difficulty in getting up in the morning, sometimes leading to nervous exhaustion and decreased immune functioning. This essence may also be helpful for those with addictive habits.

Mugwort Flower Essence: Indicated for those who are prone to hysterical, overemotional, irrational behavior that may seem out of touch with reality. The essence helps to balance dream experiences, allowing such individuals to gain greater insight from the events of daily life.

Oregon Grape Essence: Indicated for paranoid, fearful people who view others and the outside world in general as being hostile and unfair. It can be helpful for the tension and suspicion that frequently afflict city dwellers. This essence may help to cultivate greater trust in the potential goodness of others.

Scarlet Monkeyflower Essence: Indicated for fear of intense feelings or repression of powerful emotions. Also for those who have an inability to resolve issues of anger or power, or a fear of losing their temper and flying off into a rage. The essence may give the courage to acknowledge and process such feelings, leading to greater emotional depth and honesty as well as clearer communication of emotional feelings. It seems to help integrate our "emotional shadow" or the darker feelings that we sometimes have difficulty acknowledging as an integral part of ourselves.

Sticky Monkeyflower Essence: For those who have a fear of sexual intimacy and contact, frequently compensated for by pursuit of superficial sexual encounters. This flower essence may increase the ability to express greater warmth and deeper feelings of love during sexual relationships.

Tiger Lily Essence: For individuals who are overly aggressive, hostile, competitive types who strive against others rather than working together for a common goal. It is a flower with certain feminine qualities that promote peace and harmony in relationships with others, encouraging greater cooperation.

Another American producer of unique flower essences, gem elixirs, and other vibrational remedies is Pegasus Products out of Loveland, Colorado. Pegasus is run by proprietor/owner Fred Rubenfeld. Rubenfeld has amassed a huge vibrational-remedy bank of over five hundred flower essences, more than three hundred gem elixirs, and a collection of unique vibrational essences known as star elixirs (telescopically charged with the light of individual stars). Rubenfeld's method of flower-essence production is somewhat different from that of other flower-essence manufacturers. The water and flowers used to make Pegasus essences are placed in special quartz bowls, and the sunlight used to charge the water passes through special xenon filters that reputedly remove harmful vibrational qualities that may be carried by sunlight. (Remember, the sunlight of today now passes through our depleted ozone layer, which previously filtered out many harmful and burning rays.) After their production, Pegasus flower essences are energetically charged in pyramidal and other geometric arrays known as Metaforms (developed by subtle-energy researcher Greg Hoag), which are said to enhance and strengthen the vibrational pattern of the essences. The Pegasus essences appear to have specific energy effects upon different levels of human multidimensional anatomy, such as the chakras, meridians, and spiritual bodies. However, unlike most other flower essences, quite a few of the Pegasus essences are thought to work directly at the cellular level upon different cells, organs, and tissues of the body. Here are some suggested uses for a few unique Pegasus flower essences:

Calypso Orchid Essence: Enhances the ability to communicate with the higher self as well as higher-dimensional guides. Calypso is said to do this, in part, by opening and clearing the crown chakra.

Dutchman's Breeches Flower Essence: Enhances the release of old emotional residues into the auric field, where the residue can be more easily cleansed away by movement, water, or air. When sprayed as a mist, it can clear away subtle-energy residues from negative environments.

Banana Flower Essence: Banana is a flower essence that has a wide range of applications. Banana appears to have effects upon a number of physical systems of the body, especially the skeletal system. Banana may be of aid in cases of subluxation affecting any of the vertebrae of the sacrum and spine. It may also strengthen the bony attachment to teeth, helping to heal periodontal diseases that involve a loss of bony support for teeth at the the gum line. With its use, the bone structure of the jaw may be rejuvenated. Thus, Banana flower essence would be appropriate for treating temporomandibular-joint problems like TMJ syndrome. In such cases, the essence can be taken orally, as well as applied topically to the jaw in the form of a salve or cream, and even gargled in the form of a mouthwash. To create a mouthwash, add three to seven drops of Banana flower essence to one to two ounces of spring water and then gargle and swallow the water. The procedure should be repeated three times a day for several weeks, or until symptoms of periodontal disease improve. In addition, this essence has applicability to treating disorders of blood-sugar regulation, such as hypoglycemia. It can be of great aid to people experiencing mood swings caused by blood-sugar fluctuations. It can help with weight loss in overweight individuals whose problem stems from abnormal sugar metabolism. On a subtle-energetic level, Banana essence helps to balance the mental and emotional bodies. Because of this rebalancing effect, Banana flower essence is said to help resolve male sexual conflicts, such as feelings of sexual insecurity. It is interesting to note the obvious relationship beween the Banana flower essence (with its resemblance to the male genitalia) and its influence upon the male sexuality issue, since most Freudians consider the banana to be a phallic symbol.

Bells of Ireland Flower Essence: Has applications in treating infertility and in helping to correct both male and female infertility problems. It can also help with recurrent vaginal dryness, as often occurs in menopause. It does this because it affects parts of the physical and spiritual subtle anatomies that act as binding agents or a connective-tissue-like glue that holds the body together. The affected areas include the connective tissues of the body, such as the cartilage and ligaments of joints, and the ethereal fluidium, a substance that makes up a large part of the etheric body.

Celandine Flower Essence: Said to strongly affect the throat chakra. Thus, Celandine may be ideal for treating disorders relating to energetic imbalances in the throat region. Celandine facilitates transfer of information on many levels, and as such may be helpful for teachers, lecturers, and singers.

Comfrey Flower Essence: From the comfrey plant, the homeopathic remedy Symphytum (also known as bone-knit) is produced. When given as a flower essence, Comfrey appears to invigorate the nervous and muscular systems, helping to release stored nervous stress and muscular tension. It may also be of benefit in treating nervous breakdowns. Comfrey flower essence can also be of aid in treating shingles and possibly in post-herpetic neuralgia (the painful disorder left over when the skin lesions of shingles have long since healed). It has reportedly been used successfully in cases where there has been damage to the nervous system, such as in closed head injuries. The essence does not rejuvenate damaged nerve cells, but may stimulate dormant or atrophied portions of the brain to become more active.

English Hawthorne Flower Essence: On an emotional and subtle-energetic level, Hawthorne may help to rebalance emotional states of extreme stress, grief, and loss. Interestingly, Hawthorne essence also has value in treating distress related to emotional affairs of the heart. Along these same lines, Hawthorne, as both an herb and a flower essence, is helpful for treating various heart and cardiovascular diseases and may also be effective against certain viral diseases.

Khat Flower Essence: Pegasus Products suggests that this essence may help to give greater control over the autonomic nervous system, especially when used in conjunction with meditation and visualization techniques or biofeedback devices. Khat can be a general stimulant to the endocrine and immune systems during times of illness. On a subtle-energetic level, Khat may help to retard the aging process by bringing waves of life-force energy into the cells of the body. It may also help stimulate regeneration of damaged neurological tissues after spinal or brain injuries, since it helps to strengthen the ethereal fluidium, the connective-tissue-like matrix of the etheric body that surrounds and energetically supports the cells of the physical body.

Luffa Flower Essence: This essence, made from the flower of a vegetable-sponge-growing plant, is said to be useful for treating certain skin disorders like eczema and skin ulcers. Luffa can be used both internally as well as applied topically in the form of creams or salves. Pegasus Products also recommends Luffa for certain forms of allergy-related skin disorders. Luffa appears to affect the etheric body, the energetic field closest to the skin of the body, allowing the life force to better penetrate the pores of the skin. As Luffa flower essence affects the skin, a layer that defines the physical boundaries of our "personal space," people who are either too introverted or extroverted may benefit from taking this essence.

Bloodroot Flower Essence: May enhance concentration, meditation, and creative visualization, especially when used in healing work. It activates the heart chakra and is said to strengthen the cells of the body. As the name Bloodroot might suggest, it may have application in helping to treat certain blood disorders, such as blood clots.

Bleeding Heart Flower Essence: Affects the heart chakra and its associated nadi (threadlike "roots" from the chakras) channels. As both a flower essence and in herbal remedies, Bleeding Heart can be useful in treating many different forms of heart disease. To a lesser extent, Bleeding Heart essence can sometimes help in regulating elevated blood pressure and in alleviating circulatory problems caused by vascular disease. On an emotional level, it can bring greater harmony to affairs of the heart and a sense of inner peace.

Pomegranate Flower Essence: Has applications primarily for women. It can treat many female problems, including infertility, difficult childbirth, and ovarian cysts.

Pennyroyal Flower Essence: An unusual essence, thought to be capable of repelling negative thoughtforms because of Pennyroyal's ability to strengthen the etheric body. This makes Pennyroyal potentially beneficial for treating people suffering from certain cases of mental illness. It can ease mental confusion and decrease the barrage of negative thoughts that can sometimes plague such people when taken orally or applied externally. For external use, a Pennyroyal-laced salve is applied to the base of the skull and the upper neck. Pennyroyal essence is especially useful for cleansing crystals and old jewelry of vibrational patterns that have been consciously or unconsciously imprinted upon them by former owners. To clear crystals and jewelry, soak them for ten to fifteen minutes in a bowl of water laced with a few drops of Pennyroyal essence.

Yarrow Flower Essence: Said to provide a form of subtle-energetic protection from radioactive fallout. It may accomplish this by enhancing and strengthening the auric field. White Yarrow essence seems to offer protection from radiation effects, while Pink Yarrow affords protection from negative emotional influences. In this regard, the use of Pink Yarrow may be especially helpful for psychotherapists who work with emotionally disturbed individuals, keeping them from picking up the emotional-energy patterns of clients.

Lotus Flower Essence: This is a most profound and powerful flower essence. It is said to help the entire body more easily assimilate any form of healing. When given in conjunction with other flower essences or vibrational remedies, Lotus acts as an energetic booster to those other therapeutic vibrations. One of the reasons for its ability to amplify other therapies is that Lotus brings all the chakras, nadis, meridians, and subtle or spiritual bodies into a temporary state of alignment, helping the body discharge toxic influences that may block vibrational remedies from working. It may help in the assimilation of nutrients and, taken over a long period of time, may release any miasm (an energetic potential for illness) from a person's physical and subtle bodies. The use of Lotus flower essence can also greatly enhance meditation and creative visualization. Lotus may also have some application in easing confusion and memory problems associated with Alzheimer's disease when taken in conjunction with Pomegranate, Centuary Agave, and Mallow flower essence.

Flower-Essence Therapy: Flower Therapists Versus Self-Treatment with Essences

In a discussion of flower essences for application in various forms of body work, one other flower-therapy product deserves mention. Vita Florum floral products were developed in England by Elizabeth Bellhouse. They are "combination" flower products (containing multiple flower-energy patterns). Vita Florum water and tablets are designed to be ingested. However, the Vita Florum line also includes topically applied body salves, lotions, massage oils, and talcum powders (all charged with floral energies). The Vita Florum water is said to be useful in any form of water therapy, including application to the eyes as well as for vaginal and colonic irrigations. Kirlian photographs, both of Vita Florum water and of patients' fingerprints taken after Vita Florum therapy, suggest that the Vita Florum products have a high vital energy and may help to fill in breaks in an individual's energy field. Vita Florum practitioners often apply the different lotions, salves, or powders to particular skin zones for specific types of ailments. Many of the skin regions used by Vita Florum practitioners correspond to specific chakras that supply energy to the region affected by the particular medical or psychological problem. However, other skin areas deemed to positively influence the dysfunctional organ or body part are used as well. Vita Florum therapy may be helpful for treating a wide variety of disorders, including allergies, anxiety, depression, headaches, eye problems, hyperactivity disorders, TMJ syndrome, arthritis, menstrual disorders, asthma, and stomachaches. While knowledge of the Vita Florum products is not widespread among flower-essence practitioners, information on obtaining these products and literature guiding their use can be found in Appendix 3. With regard to the use of flower essences in bodywork, the Vita Florum massage oil may provide additional therapeutic benefit to massage therapy, chiropractic manipulation, and possibly even acupuncture (applied after needle treatments). As mentioned earlier, some flower-essence practitioners have experimented with dipping a cotton swab in a particular flower essence, which is then applied directly to specific acupoints on the body as a form of "floral acupuncture."

A wide variety of health-care professionals and trained laypeople now work with flower-essence therapy. Even psychiatrists and psychologists are experimenting with Bach flower essences as a way to minimize the number and dosage of psychotropic medications prescribed. Unfortunately, many who use flower-essence therapy with conventional drug therapy rarely advertise this fact to the general public. While antidepressants may rebalance the neurochemistry associated with states of depression, flower-essence therapies provide additional therapeutic benefits by rebalancing the mental- and emotional-energy patterns that may be contributing to an individual's depression. Many holistic physicians, chiropractors, bodyworkers, massage therapists, and psychotherapists are adding flower essences to their therapeutic armamentarium because flower essences are nontoxic yet powerfully transformational in their healing effects at the physical, emotional, mental, and spiritual levels.

If you are uncertain about experimenting with flower essences, you can try to seek out a trained flower-essence practitioner. While most flower-essence therapists are probably familiar with the Bach Flower Remedies and several other varieties, few know about or use every essence within the incredibly huge and still-growing list of floral vibrational remedies. If a local flower-essence practitioner cannot be found, self-treatment with the Bach remedies will sometimes be the only way to get started with flower essences. By studying the Bach Flower Remedy literature, you can often uncover just the right essence that applies to your specific health or psychological needs. After having success with the Bach flowers, you may then move on to experimenting with other varieties of flower essences.

One note worth mentioning regarding Bach Flower Remedies versus flower essences from other manufacturers: The method of adding two drops of mother tincture or "stock bottle" essence to a one-ounce bottle of spring water with brandy (or another preservative) applies mainly to the use of the Bach remedies. In general, if using Pegasus, FES, or any of the other forms of flower essences, it is still advisable to follow the directions on the bottle and to take from three to ten drops of the flower-essence solution directly under the tongue, without diluting or mixing combinations, as is usually done with Bach flowers. If one were to take three different Pegasus remedies, one would, for instance, place under the tongue seven drops from each of the three different remedy bottles three to four times a day.

The books at the end of the chapter can guide you to a vast world of vibrational healing. The gamut of physical and psychological problems that can be potentially helped by flower essences seems as great as the sampling of essences in this chapter would suggest. And the essences mentioned here are but a mere tip of the iceberg. In many cases, flower essences may offer new hope for healing a variety of medical, emotional, and psychospiritual problems for which no known effective conventional treatments yet exist. Also, there are many essences that actually augment the effectiveness of other vibrational and conventional treatments. For example, certain essences can enhance the effectiveness of chiropractic treatments (Banana flower essence), while others may augment the benefits of color therapy (Lemon flower essence). Assuming that we accept the validity of clairvoyant flower-essence investigation, we find that there are a number of essences that may remove energetic blockages to various forms of conventional and alternative therapies.

While the Bach Flower Remedies are available through many health-food stores, some of the more esoteric flower essences—such as the Australian Bush Remedies, Andreas Korte's Orchid Essences, Perelandra Essences, and Findhorn Essences—may be more difficult to locate. However, many of the unusual flower essences can be located through the resources in Appendix 3.

When it comes to scientific research validating their therapeutic effects, flower essences seem to fall into the same category as homeopathic remedies. While there are very few scientific studies that document an essence's physiological effects, there are large numbers of anecdotal and clinical case histories suggesting that extremely positive therapeutic benefits can be produced by taking various flower essences. The Bach remedies, while intended primarily for emotional rebalancing, can sometimes produce very profound physical healing effects, used both orally and topically. Some of the Pegasus and other esoteric flower essences are said to bring about direct cellular healing of specific physical disorders (in selected clinical cases). It must be remembered that much information on the use of flower essences comes through intuitive resources. As such, intuitive diagnosis can often be a guiding factor in their use, especially in the case of pendulum-based and kinesiology-based flower-essence prescribing. In certain situations, electroacupuncture diagnostic equipment can help a health-care practitioner to zero in on the right flower essence of benefit for healing a specific condition. Still, the greater work of scientific validation of flower-essence therapy is yet to be done. That having been said, I truly believe that within the realm of flower essences and herbal medicine may lie the answers to many of humanity's ills, much as the old-time healers have often stated. Flower essences remain one of the most powerful, yet most under-appreciated forms of vibrational medicine in the healer's armamentarium.

Seven

HEALING WITH COLOR AND LIGHT

THE USE OF color and light in healing is actually quite old. The priests of ancient Egypt, Babylonia, and China used color or colored light in many of their healing practices. Sunlight therapy was a common medical practice in historic Greek, Chinese and Roman times for the relief of skin disorders such as psoriasis. In the 1890s, Nobel Prize–winner Dr. Neils Finsen reported that he could heal skin lesions caused by smallpox and German measles by using red- and infrared-light treatments. Even today it is not too difficult to find light therapy used in traditional medical protocols. For example, ultraviolet phototherapy, combined with psoralen-containing creams, is often used by dermatologists to heal severe cases of psoriasis. And the use of lasers in medicine and surgery has become quite commonplace. In recent years, it seems that the use of light and color in healing has become standard practice for a number of specific medical applications. The use of colored light in healing was banned by the Food and Drug Administration (FDA) fifty years ago because it was thought to be nothing more than medical quackery. But today, FDA-approved phototherapy devices like the Lumatron, developed by Dr. John Downing, are being used for the therapeutic application of specific colors and frequencies of light to the body. The use of healing light therapies covers a wide range of approaches and applications, all the way from laser surgery for diabetic eye disease to laser heart revascularization to the use of full-spectrum light boxes for treating Seasonal Affective Disorder (SAD), a form of depression commonly known as the "winter blues."[1] In order to better understand the varied applications of light therapy, or phototherapy as it is sometimes called, we need to examine how light affects the physical body, as well as how it may affect the spiritual bodies and subtle-energy systems that make up the multidimensional human being. Recent medical research has begun to validate some of the many healing applications of light energy. But as you will soon discover, there is an even greater spectrum of color/light healing applications than are known to, or even used by, most conventional physicians.

Can Color and Light Really Heal? Exploring the Theories of How Light Therapies May Work

One of the most basic forms of light, sunlight, is of critical importance to normal human health and wellness. In fact, we might almost consider light to be a kind of vital nutrient to our bodies. Since we really are "beings of light" on a variety of levels, some "enlightened" health professionals, such as psychologist and light therapist Dr. Brian Breiling,[2] refer to human beings not merely as *Homo sapiens* but as *Photo sapiens.* Put quite simply, we absorb and use different components of sunlight through our eyes as well as our skin. The ultraviolet light in sunlight is actually a critical ingredient in the production of vitamin D by our bodies. And without vitamin D, calcium absorption and utilization would not occur as needed in order to maintain healthy bones. Inadequate exposure to sunlight results in weakened bones because of reduced levels of vitamin-D production. Aside from its role in vitamin-D synthesis, the most obvious function of light is to illuminate our environment so that we can see the world around us. Light travels into the eye, through the lens, and onto the retina, the light-sensitive portion of the eye. The retina is made up of special photoreceptor cells known as rods and cones. The color-sensitive cones are responsible for color vision, while the highly sensitive rods help us to see things at night. Once light enters the eye, the retinal cells become activated, sending nerve signals to a number of different areas of the brain. The optic nerves carry information from the eyes in the form of electrical nerve signals that travel to a region of the brain known as the visual cortex (located in the back of the brain). The visual cortex receives light-activated nerve signals from the eyes and creates a three-dimensional image of what we are observing.

Dr. John Downing, a pioneering light-therapy researcher, refers to these light-triggered electrical nerve signals as "photocurrents." Research by Downing suggests that in addition to activating the visual cortex, light-induced photocurrents also stimulate a number of other important brain centers. Photocurrents traveling from the eyes to the visual cortex apparently continue on to stimulate the associative centers of the brain. Without these associative centers, we'd have great difficulty in identifying, interpreting, and understanding the forms and colors of things we are observing. The photocurrents received by the cerebral cortex arrive in such abundance that they are apparently able to act as a general stimulus to the process of thinking, organizing, reasoning, and remembering, as well as enhancing our understanding of the world around us. From the visual cortex, photocurrents travel to the limbic system, a collection of nerve centers that play a role in emotion, learning and memory, sexual behavior, aggression, and even the perception of smells. The ability of photocurrents to stimulate the limbic system has led some researchers to speculate that inadequate photocurrent stimulation of the limbic system might lead to a psychotoxic buildup of painful emotional experiences in our brains. According to this theory, light deprivation might actually be a reason for having anxiety attacks!

In addition to stimulating the cerebral cortex and the limbic system, light-triggered photocurrents also stimulate the hypothalamus. The hypothalamus is an important master control center that helps to regulate many critical life processes. The hypothalamus is linked to both the autonomic nervous system (our body's "autopilot") and the pituitary gland. The pituitary gland, in turn, hormonally regulates all our body's endocrine glands (such as the thyroid, the adrenals, the gonads, and so forth). The hypothalamus also contains a special center (the SCN or suprachiasmatic nucleus) that is believed by many scientists to be the biological clock of the body, the center that tells our brains whether it is really day or night. This SCN center is partly responsible for what are known as "circadian rhythms," internal biorhythm cycles that subtly affect human physiology. We know, for instance, that the production of certain hormones tends to be higher during the daytime and lower at night. Yet other hormones follow the exact opposite cycle of hormone production and release. The circadian rhythms that influence these hormonal patterns are strongly affected by our exposure to cycles of light and darkness, day and night. Another important brain center linked to circadian rhythms is the pineal gland. For hundreds of years, the pineal gland has been the subject of much debate and speculation. Some philosphers have even considered the pineal gland to be the "seat of the soul." In evolutionary biology, the pineal gland has been linked with a true "third eye." This reputation is based upon the fact that some reptilian species (such as the New Zealand tuatara) possess a pineal gland that actually functions as a primitive eye, complete with its own lens. To a certain degree, the pineal still remains a light-sensitive structure, even in human beings. It functions as a specialized hormonal center that not only plays a role in regulating the body's circadian rhythms but also affects sexual maturation, longevity, and immunity. The pineal gland receives both indirect photocurrent stimulation (produced by light entering through the eyes) and direct light stimulation (from low-level light that directly penetrates the skull and gray matter of the brain).

The pineal gland produces an important hormone known as melatonin, which has recently become popular as a nutritional supplement. It appears to play a role in sleep, in proper immune functioning, in the timing of sexual maturation, and perhaps even in promoting longevity by slowing down the aging process. The production of melatonin is directly affected by the amount and timing of our exposure to light. We tend to produce the greatest amount during sleep, the period when we experience the least amount of light (or the greatest degree of darkness). Melatonin has been found to have natural sedative qualities. It can also reduce anxiety. Some studies even suggest that it may increase our natural immunity to illness. In addition, melatonin acts on the pituitary gland to inhibit the release of sexual-maturation hormones until the age of puberty has been reached. The onset of menstruation, for example, is partly determined by an individual's cumulative exposure to light cycles of day and night. Some photobiology researchers have speculated that sexual maturation in teenagers may now be occurring at earlier chronological ages due to the increased exposure to electric light of modern times. This is an entirely feasible proposition. When one considers the widespread use of electric lights during the last century, and the increasing amount of bright light that today's teens are exposed to both day and night (as compared with a century ago), such theories might just be valid, especially when we apply these observations to what we now know about light, hormones, the pituitary, and sexual maturation. Besides inhibiting melatonin production, bright light also appears to boost the levels of a brain neurotransmitter known as serotonin, which functions as a chemical nerve messenger that transmits messages through specific circuits and centers within the brain. Abnormally low serotonin levels in the brain have been linked to a variety of psychological disorders, including Seasonal Affective Disorder, certain forms of eating disorders, and depression. Many antidepressant drugs (like Prozac) treat depression by increasing the effectiveness of the serotonin that is already present, but in lower amounts, within the brains of depression sufferers. Because full-spectrum light boxes and bright-light therapies can increase serotonin levels in the brain, it makes sense that light therapy could be an effective treatment for depression. For the same reason, bright-light therapy has also proved effective for treating certain forms of carbohydrate-craving obesity that affect many people, especially those who put on weight during the light-poor winter months.

Of course, there are other causes of "relative light deprivation." Some individuals appear to suffer from "light-starvation" due to poor illumination or inadequate lighting systems in the home and in the workplace. Health problems thought to be strongly influenced by relative light deprivation include general stress symptoms, fatigue, depression, hyperactivity in children, difficulty concentrating, and weakened bones and teeth.

The timing of cycles of light and darkness can also play a significant role in health and illness. For example, individuals who are exposed to irregular light cycles, such as rotating-shift and night-shift employees, have a higher incidence of heart disease, back pain, respiratory problems, ulcers, and sleep disorders. Chronobiology researchers have discovered that if a person ignores the normal twenty-four hour light/dark cycle by keeping irregular working and sleeping hours, the body's natural circadian rhythms may become disturbed. Some therapists who work with insomnia-afflicted shift workers have had success in rebalancing their patients' disturbed sleep cycles by prescribing both melatonin tablets and daily periods of exposure to carefully timed, high-intensity bright lights (five to ten times brighter than ordinary room lighting).

Based on the knowledge that light-induced photocurrents can activate specific areas of the brain, a number of light-therapy researchers have developed a variety of light-treatment systems that project different colors and intensities of light into the eyes in order to treat certain neurological disorders. Some of these innovative light therapies have proven effective in treating depression, mental and emotional problems, seizure disorders, and memory problems, as well as difficulties with concentration and attention span. The use of light therapy in treating depression heralds the beginning of a revolution in psychiatry and neurology. In the not too distant future, doctors will be using photons, electrons, and even magnetic fields instead of drugs to treat depression. Even now, more and more psychiatrists are prescribing full-spectrum and bright-white-light therapy for treating Seasonal Affective Disorder and other forms of depression. SAD is one of the few health problems actually recognized as a disorder brought on by relative light deprivation. SAD sufferers become depressed mainly during the winter months, when the hours of daylight are much shorter than in the summer or spring months. By starting out each morning sitting in front of a special light box (which emits bright white light or full-spectrum light), most individuals afflicted with SAD can relieve the symptoms of their depression.

As previously mentioned, researchers have only recently concluded that the cells of the body emit light. German researcher Dr. Fritz Popp has confirmed that cells give off what he calls "biophotons" or biologically generated packets of light. This cell-emitted light appears to be in the ultraviolet frequency range. These UV biophotons given off by cells may be part of a "light-based" communication system that relays important biological information between adjacent or nearby cells within the body. With regard to the source of this weak cellular light, surprisingly, Popp's research has led him to conclude that it is actually the cell's DNA double helix that not only emits these ultraviolet photons but also receives them in a manner similar to a spiral antenna exquisitely tuned to transmit and receive ultraweak light signals in the ultraviolet frequency range. Given that the cells of our bodies communicate using light, it seems plausible that human beings could be strongly affected by sunlight as well as by the limited spectrum of light frequencies emitted by the electric lightbulbs and fluorescent tubes used to illuminate our homes and workplaces.

Photobiology researchers in Russia and in the United States have studied the biological effects of different light frequencies on people. Many researchers have found that certain light frequencies produce stronger effects upon cells and tissues than do other frequencies. Specifically, monochromatic (single-frequency) red light, such as the coherent red light of a helium neon laser or even incoherent red and infrared light, has been successfully used to stimulate regeneration of skin and blood cells in areas exposed to the red rays. Red- and infrared-light exposure also appears to increase tissue oxygen levels and to improve local blood flow. Furthermore, Russian light researchers have noted enhanced wound healing, increased nerve stimulation, improved muscle relaxation, and even decreased pain levels in areas of the body directly exposed to red light. As to why such effects might occur, some researchers have theorized that monochromatic light in the blue, red, and far-red regions of the light spectrum may enhance and speed up certain metabolic processes, such as those taking place in mitochondria, the metabolic and electrical power plants of every living cell. They possess certain similarities to batteries, including batterylike electron flow. Russian research into the therapeutic effects of monochromatic light tends to confirm such speculation about monochromatic light's ability to speed up metabolic processes. Furthermore, Russian photobiology research suggests that there are actually minature photoreceptors at the molecular/cellular level. When triggered by certain frequencies of monochromatic light, these intracellular photoreceptors appear to increase DNA and RNA production, accelerate protein and collagen synthesis, and enhance cellular growth and division, all of which leads to rapid regeneration and healing of damaged cells and tissues.

In addition to light's direct effects upon the cells of our physical bodies, different frequencies and colors of light can affect our subtle-energy systems and our spiritual bodies as well. For instance, German naturopathic physician Dr. Peter Mandel has developed an entire healing approach for rebalancing the organ systems of the body by stimulating acupoints and meridians using a technique he calls Colorpuncture. Mandel's approach is distinct from other forms of acupoint stimulation. Prior to his use of colored light for stimulating acupoints, a number of early color and light researchers had advocated shining certain colors of light onto specific areas of the body in order to heal a wide variety of medical and psychological disorders. Dr. Edwin Babbitt and Dinshah Ghadiali, two pioneering color and light therapists, developed whole treatment systems based upon the therapeutic effects of different-colored lights on the body. Later in this chapter, we will explore some of the remarkable research done by these two innovators. But with Dr. Mandel's research, light therapy has been elegantly carried a step beyond the teachings of Babbitt and Ghadiali. Mandel uses a special pen-size illuminator with different-color lenses that focus single wavelengths of colored light into specific acupuncture points on the body. According to Mandel, certain colors (mainly the warmer colors—red, orange, and yellow) seem to stimulate or "tonify" acupoints, while other colors (the cooler colors—green, blue, and violet) "sedate" or disperse the energy of acupoints. Dr. Mandel has also noted that certain acupoints have more of an energy affinity for, or resonance with, certain wavelengths of color. This eventually led him to develop his color-based system of healing using light stimulation of acupoints. Dr. Mandel has postulated that our acupoints may not only absorb ch'i but they may also absorb and transmit light throughout the body's meridian system like a vast fiberoptic network. In fact, Dr. Pankratov of the Institute for Clinical and Experimental Medicine in Moscow has verified that our acupuncture meridians do indeed conduct light, especially light in the white and red spectral range. His research further determined that the light source must either be touching or within one to two millimeters of an acupoint for this effect to occur. Pankratov's discovery confirms Mandel's theory that acupuncture meridians behave like biological fiberoptic light cables. This suggests that light-stimulated acupoints could actually transmit light energy into the organs of the body through the acupuncture-meridian system, just as Mandel has postulated.

Therapists trained in Colorpuncture evaluate each patient's organ-energy status using Kirlian photography of the hands and feet along with a system of interpretation that Mandel calls energy emission analysis (EEA). The EEA technique is based on mapping out specific zones and locations in the ends of the finger and toes that correlate with so-called command acupoints for the body's organ-linked meridians. Kirlian photographic analysis of the corona-discharge patterns of a patients' hands and feet reveals the energetic and health status of the acupuncture meridians as well as the meridian-linked organs. In addition to information gleaned from Kirlian energy-emission analysis, Colorpuncturists also base their acupoint color treatments upon an analysis of meridian energy in terms of the Chinese five-element model. As you may remember, the five-element model describes the energetic interrelationships between the different organs and meridians of the body. In the Chinese five-element model, there are five different organ-linked meridian "circuits." These five circuits are based on meridian connections among the organs of the body. These link-ups include the following: kidney/bladder, liver/gall bladder, lung/large intestine, spleen/pancreas/stomach, and heart/small intestine. By analyzing the different organ circuits from a Kirlian EEA standpoint, a Colorpuncture therapist can determine which organ/meridian system requires acupoint stimulation and rebalancing. The specific frequencies of colored light required for rebalancing are determined by Mandel's specialized system of Colorpuncture.

Until now we have primarily focused on the physiological effects of colored light upon the brain and the cells of body. However, with Colorpuncture and its therapeutic link with acupuncture meridians, we have now moved into the arena of how colored light may affect human subtle-energy systems. As such, we need to examine further not only how different frequencies of light may affect the physical body but also how light can produce powerful vibrational influences upon the chakras as well as upon the etheric, astral/emotional, and mental bodies. Intuitive sources have suggested that the etheric body may actually be the main "connecting link" through which light therapy is able to vibrationally influence and heal the physical body. Thus, different frequencies of colored light may affect the etheric body by directly influencing the etheric fluidium, the matrix or connective tissue of the etheric body that surrounds, energetically nurtures, and invisibly guides the cells of the physical body. With regard to the effects of colored light upon the seven major chakras at the etheric, astral, and higher-energetic levels, a number of esoteric texts have linked each of the seven colors of the rainbow with the seven major chakras of the body. The root chakra resonates with red, the sacral chakra with orange, the solar plexus with yellow, the heart center with green, the throat chakra with blue, the brow center with indigo, and the crown chakra with the color violet. If a particular unbalanced chakra is therapeutically stimulated (and rebalanced) by its resonant (or complementary) color, the rebalanced chakra might then assist in healing damaged or diseased cells and tissues via an increase of supportive pranic-energy flow to the local chakra-fed tissues.

The Many Different Forms of Color and Light Therapy

As pioneering light therapist Dr. Jacob Liberman says, light really is the medicine of the future.[3] It is perhaps the least invasive form of therapy in use today. Short of looking directly down the barrel of a high-intensity laser and accidentally burning one's retina, the vast majority of light therapies produce very few side effects. Of course, high-powered surgical lasers have the potential to do damage if not used in a responsible manner by a skilled surgeon. But under most circumstances, light has a greater potential to heal than to harm. Laser therapies have become increasingly popular over the last few decades as many new types of lasers have been developed. The selection is amazing, varying from tunable color lasers to solid-state lasers no bigger than an electronic chip.

Without doubt, a number of medical and surgical procedures have become more efficient because of lasers. For instance, laser scalpels offer the advantages of cleanly cutting through tissue while simultaneously sealing and cauterizing cut blood vessels. And because different cellular structures absorb various colors of light differently, doctors have been able to develop very color-specific lasers for different treatment applications. For example, green-argon lasers are often used for delicate diabetic eye surgery because the green-light tends to affect only blood vessels, leaving other sensitive portions of the eye undamaged.

Not only can the light of the laser beam cut through tissue, it can also activate light-sensitive chemicals that are able to selectively kill certain types of cancer cells. There are a number of anticancer therapies that use specific frequencies of either coherent laser light or normal incoherent light to activate the cancer-killing properties of certain types of medicines. These light-activated medicines can be administered topically or intravenously to heal different forms of cancer. In photodynamic therapy (PDT), certain light-sensitive dyes (which tend to be absorbed primarily by abnormal cells) are administered to patients in order to selectively kill cancer cells while leaving the surrounding normal tissues relatively unharmed. Many of these organic, light-sensitive dyes are derived from chemicals known as porphyrins, the ringlike chemical structures that lie at the very heart of every oxygen-carrying hemoglobin molecule in our bloodstream. It just so happens that the porphyrin molecule is easily excited by certain frequencies of light. When the porphyrin ring is chemically modified and combined with other therapeutic medicines, a special dye known as photofrin is created. This photofrin dye becomes highly reactive to cells in the presence of certain frequencies of light. Such porphyrin-based dyes are known as photosensitizing agents. This means that exposure to specific frequencies of light will activate the dyes' disease-fighting properties. One unique property that makes such dyes ideal for photodynamic therapy is that the porphyrin molecules in the dye tend to bind tightly to cancer cells. This property makes the dyes useful in both diagnosis and treatment of cancer. When exposed to ultraviolet light, the cancer-cell-bound dyes actively fluoresce, causing the cancerous cells to be easily identified by their characteristic glow. From a therapeutic standpoint, when the porphyrin dyes become activated by exposure to light in the red frequency band, the light-activated dye destroys the identified cancer cells.

Bronchial tumors of the lung are one form of cancer sometimes treated using photodynamic therapy. Bronchial-cancer patients are given an intravenous infusion of a light-sensitive, porphyrin-based chemical dye that selectively binds to their cancer cells. After several days, most of the dye has left the normal cells but remains bound to the cancerous cells. The patient then undergoes an endoscopic procedure in which a flexible fiberoptic scope is inserted into the bronchial airways to make it possible to directly observe the tumor.

When the lung tumor comes into view through the bronchoscope, the appropriate frequency of light is transmitted through a thin fiberoptic cable in the scope. The intense light beam (in the appropriate dye-activating frequency) bathes the tumor in light. The light-activated, cell-destroying properties of the cancer-fighting photochemical agent are brought to bear against the bronchial tumor cells. In China, photodynamic therapy has been used for more than twenty years with reports of great success in treating certain types of tumors that are easily reachable by flexible endoscopes or are easily accessible to direct light exposure. Photodynamic therapy is also being investigated for its ability to heal skin cancers like basal-cell and squamous-cell carcinomas. Isn't it ironic that the very thing that can lead to skin cancer, overexposure to light, may also be effective in curing it? While photodynamic therapy is proving to be very effective against certain forms of cancer, it must be noted that there is a price to pay with such light-energy treatments. Aside from healing the cancer, this procedure can make patients overly sensitive to sunlight for long periods of time.

In addition to healing cancer with light therapy, recent research has led to the development of photodynamic therapy for treating even precancerous skin lesions. A photosensitizing agent known as Levulan (based upon 5-aminolevulinic acid, a natural compound found in all human cells), used in conjunction with blue-light therapy, has been found to be effective in healing actinic keratoses, a form of sun-related skin damage said to be a precursor to skin cancer. The drug is applied topically to the skin lesions followed by blue-light treatment. Pending FDA approval of the drug Levulan, dermatologists may soon adopt this form of phototherapy as a safe and effective treatment for this all too common skin problem.

Another form of light therapy that makes use of light-sensitized agents is known as PUVA therapy. PUVA therapy refers to the use of the drug psoralen given in combination with UV-A ultraviolet-light therapy. This is one of a variety of ultraviolet-light therapies currently in therapeutic use or undergoing some form of scientific investigation. Ultraviolet light is divided into three types of rays: UV-A, UV-B, and UV-C. UV-A, considered the least harmful, is of a longer wavelength, has less skin-penetrating power, and is reponsible for slow tanning. UV-C is considered the most harmful, being of a shorter wavelength and thus more potent. However, very little of this form of UV light penetrates the earth's ozone layer. It is UV-B that poses the greatest threat to human health. Ultraviolet-light energy in the UV-B wavelength is the one responsible for sunburn. UV-B overexposure may also damage the eyes by causing corneal disease, retinal damage, and early cataract formation.

While UV-B and UV-C may be bad for the body, ultraviolet light in the UV-A frequency range seems to possess some interesting therapeutic properties. In PUVA therapy, UV-A ultraviolet light is combined with the light-sensitive drug psoralen, in order to treat skin disorders such as psoriasis and vitiligo (the disease that causes patchy, lightened, pigment-free spots to appear throughout the skin). In psoriasis, a disease of scaling and crusting skin lesions caused by rapidly dividing skin cells, the ultraviolet light in the UV-A frequency band actually appears to inhibit diseased psoriatic skin cells from dividing, especially when given in combination with psoralen therapy. Studies of PUVA therapy have shown that up to 90 to 95 percent of psoriasis sufferers will respond favorably to a series of thirty PUVA sessions spaced over the course of ten weeks. Some researchers, notably Dr. Hugh McGrath of Louisiana State University Medical Center, have also experimented with using UVA-1 light for healing other types of skin disorders. UVA-1 is a specific subfrequency of UV-A ultraviolet light. This subfrequency of UV light is used to treat diseases such as lupus or systemic lupus erythematosis (SLE). SLE is an autoimmune disorder that can cause inflammation and deterioration of the skin, blood vessels, nervous system, and even the heart. An investigational study conducted by Dr. McGrath found that when patients with lupus were given UVA-1 light treatments, they experienced decreased joint pains, fewer headaches and skin rashes, and improved energy levels.

In addition to healing lupus and various skin disorders, there appear to be other therapeutic benefits associated with controlled exposure to certain wavelengths of ultraviolet light. As early as the 1900s, medical researchers were using ultraviolet radiation (from sunlight) as a means of lowering blood pressure in both normal and hypertensive individuals. In some cases, just a single treatment of ultraviolet light dramatically lowered elevated blood pressure levels for up to six days. More recently, in research carried out at the Tulane School of Medicine, Dr. Raymond Johnson exposed 20 individuals to ultraviolet-light therapy and found that their cardiac outputs (the efficiency of the heart's pumping action) had increased by an average of 39 percent for 18 of the 20 patients. Ultraviolet-light therapy has also proven to be helpful in unclogging arteries affected by atherosclerosis. One Russian study used ultraviolet therapy on 169 patients with cerebrovascular disease (narrowing of the arteries that feed oxygen and nutrients to the brain). The entire patient population showed improved cerebral circulation even at their one-year follow-up reevaluations. In addition, all UV-treated patients reported feeling better, and many of the (formerly disabled) patients were happily back at work. UV-light therapy even appears to affect cholesterol and other blood-fat levels that have been implicated in heart and vascular disease. In studies on patients with hypertension and related circulatory problems, there was a nearly 13 percent decrease in serum cholesterol within two hours of exposure to ultraviolet light. The majority of patients in the studies maintained lower serum levels for up to twenty-four hours. There is also a suggestion that ultraviolet light may stimulate the thyroid gland, causing the body to increase its metabolism and burn an increased number of calories. Experiments with animals show that treatment with ultraviolet light does seem to stimulate weight loss, probably through thyroid stimulation. Ultraviolet light is known to be effective in killing certain types of disease-producing bacteria as well, including several forms of the tuberculosis bacillus. Interestingly, Russian and German doctors have routinely used ultraviolet light to treat simple infectious diseases in schools and workplaces for many years. Furthermore, both Russian and German practitioners use UV treatments to combat black-lung disease in coal miners. Russian doctors believe that UV may help the body to more easily eliminate inhaled dust particles via the bloodstream. There is also evidence that certain wavelengths of ultraviolet light taken in through the eyes appear to boost the immune response. However, all procedures involving UV therapy do have some risk. The important thing to remember about ultraviolet light is that certain wavelengths are beneficial to health while others may be detrimental. Extreme caution must be taken in using this type of phototherapy, and only under the supervision of a UV-therapy-trained specialist.

Having discussed some of the more conventional uses of light therapy for health problems, let us return to the innovative work of light researcher Dr. John Downing. As mentioned earlier, some of the most convincing clinical evidence for the effectiveness of specific frequencies of light in healing comes from Dr. Downing's research. Downing spent more than twenty-seven years studying and developing a technique he calls Ocular Light Therapy (the therapeutic projection of different colors and wavelengths of light through the eyes in order to improve various types of brain dysfunction). We discussed how the projection of light through the eyes results in photocurrents that stimulate critical areas of the brain involved in vision, emotional expression, thinking, memory, and even coordination and balance. Through his research, Dr. Downing developed a device known as the Lumatron Ocular Light Stimulator. This device has the distinction of being one of the first colored-light-therapy devices to be approved by the FDA. Downing has developed a unique system able to determine the correct color needed for light stimulation of the eyes based upon a patient's neurological pattern of response. This neurological typing system analyzes patients' symptoms and emotional patterns using both a clinical examination and a constitutional profile questionnaire filled out by patients seeking Ocular Light Therapy. After the examination and profile are complete, the patient is classified as one of two neurological types. There is a "slow neurological type" (associated with a parasympathetic-dominant autonomic nervous system response) and a "fast neurological type" (associated with a sympathetic-dominant response to stimulation). While the parasympathetic nervous system is more active during rest, digestion, and meditation, the sympathetic nervous system is responsible for the "fight or flight" adrenaline-producing response. "Fast" neurological types are usually treated with the blue side of the color spectrum in order to slow down and bring their nervous system into balance. Conversely, patients categorized as "slow" neurological types are usually treated with the red side of the color spectrum in order to speed up and rebalance their nervous system. The Lumatron device, as it is used for colored-light therapy, looks somewhat like a psychiatrist's couch with a large light-box device mounted on it at about eye level. The patient merely lies on the couch and looks into the colored light box. The colored light is broadcast into the eyes at different rates of flashing, depending upon which color is chosen and what therapeutic effect is needed for rebalancing. The flashing rates also correspond to different frequencies of brainwave activity, with red colors pulsing more in the faster "beta" range (13 to 15 cycles per second) and violet down in the "theta" or "delta" range (1 to 7 cycles per second). Extremely "fast" or "slow" neurological types sometimes require faster or slower flashing rates than are routinely used.

A variety of different neurological disorders have been found to respond favorably to Ocular Light Therapy. For instance, patients with poorly controlled epilepsy have responded favorably. In one case, Dr. Downing treated a patient with epilepsy who was experiencing typical grand mal seizures every two weeks in spite of adequate doses of seizure medication. Because the patient would often fall, lose consciousness, and shake uncontrollably without warning, he began to experience brain damage caused by his numerous seizure-induced falls. Following an examination, Dr. Downing determined that he was a "slow" neurological type with a correspondingly slow brain-wave pattern on his EEG recordings. Dr. Downing tried red-light stimulation though the patient's eyes using the Lumatron over a series of twenty-minute sessions. He followed these sessions with a month off from light therapy. Next, the patient received a second series of twenty-minute Lumatron sessions using orange-light stimulation. Following Ocular Light Therapy, the patient did continue to experience seizures. However, instead of occurring every two weeks, his seizures were now occurring every two months. In time the seizures decreased to every three months, and eventually they were a year apart. At a seven-year follow-up, the patient was experiencing much milder seizures about once every nine months. In addition, the patient reported that he was functioning much better, both mentally and emotionally, and had noticed a significant reduction in the slurring of his speech.

Another patient with epilepsy treated with the Lumatron and Ocular Light Therapy required a different approach. Based upon his examination and patient history, Dr. Downing determined that this second patient was a "fast" neurological type. The patient was fairly hyperactive, had an aggressive Type-A personality, and complained of petit mal absence seizures every morning. The seizures would cause him to "blank out" in the middle of normal daily routines. This created much fear and emotional distress for the young man. He had previously tried biofeedback therapy in an attempt to control his abnormal brainwave patterns. But the patient's biofeedback therapist reported that the young man's eyes-closed brain-wave pattern would not slow down from his persistent rapid beta pattern of 17 cycles per second. Dr. Downing decided to use violet-light Ocular Light Therapy on the patient via the Lumatron while the patient was simultaneously hooked up to a brain-wave-analyzing device. After twenty minutes of violet-light stimulation, Dr. Downing and the attending biofeedback specialist examined their patient's brain-wave patterns and discovered that they had dropped into the slow theta range at 4 cycles per second. Such a decrease was something the patient had never been able to do on his own during previous attempts. Eventually he underwent a total of twenty light sessions using violet-, then indigo-, and finally blue-light treatments. Following the completion of light therapy, the young man's EEG revealed a normal apha brain-wave pattern of 10 cycles per second with a corresponding drop in seizure activity from one every morning to one every other morning. Although further light therapy was recommended to decrease the seizures even further, the patient declined, reporting that he was satisfied with the success and was feeling more relaxed and happier with the results of the light therapy he had already undergone. Dr. Downing's two cases illustrate how neurological problems such as seizure disorders may be therapeutically influenced by Ocular Light Therapy. These cases also show how different colors of light were able to produce quantifiable, lasting changes in the brain's electrical activity.

Another interesting case treated by Dr. Downing highlights the benefits of colored-light therapy in treating depression, panic attacks, anxiety, insomnia, and muscle pain. In this case, a sixty-four-year-old woman had been referred to Dr. Downing by her osteopathic physician. She had proven resistant to conventional medical and osteopathic therapy. The patient suffered from chronic muscle pain caused by fibromyalgia, a disorder often associated with disturbed sleep patterns. She suffered from sleep-onset insomnia and was unable to fall asleep until at least one o' clock on most nights. Her thoughts were bothered by incessant, unwanted ideas, which often kept her awake. The patient was taking Xanax, a Valium-type anxiety-relieving medication, frequently used for treating panic attacks. She also complained of severe fatigue. Upon examination of the patient, Dr. Downing determined that she was a "fast" neurological type and immediately began violet-light therapy on the Lumatron. Dr. Downing used a flashing rate of 6 cycles per second. After one session with violet-light treatment, the patient's depression dramatically lifted. After five sessions, she reported that her fibromyalgic muscle pain had decreased and that she was sleeping better with fewer disturbing thoughts prior to falling asleep. The back pain she had suffered from had decreased as well. Furthermore, the patient testified to being a bit more relaxed and clear-headed. In fact, she started to have greater insight into some of her marital problems. Dr. Downing then decreased her violet-light flashing rate from 6 to 4 cycles per second, after which the patient reported she no longer needed to take Xanax for panic attacks. Her depression also had completely cleared.

As mentioned previously, Dr. Downing had been aware that light-induced photocurrents stimulated the visual cortex, as well as other areas of the brain. He had also heard reports that some patients exposed to light therapy had experienced significant improvements in their visual-field perception. Based on this knowledge, he did visual-field testing on his fibromyalgia patient before and after she had completed violet-light therapy. The results indicated a dramatic increase in size of the woman's visual-field perception after Ocular Light Therapy. Thus, the woman's ability to see more things in her visual field had clearly improved (something that rarely happens on its own). Many other optometrists using Ocular Light Therapy (as well as other forms of colored-light therapy) have also reported similar improvements in their patients' visual fields following colored-light treatments (even in some patients who were nearly blind prior to therapy).

Strokes are another type of neurological disorder that may respond favorably to Ocular Light Therapy. One of Dr. Downing's Norwegian light-therapy students has reported great success in treating stroke victims in Norway. One particular case of interest was a fifty-five-year-old businesswoman who had to close her small retail business after she was disabled by a stroke. Following her stroke, she had lost the use of her left hand and had to drag her left foot behind her when she walked. Her stroke had occurred in the right hemisphere of her brain. At the time she sought out light therapy, a year and a half had already elapsed since the stroke. She had made no further progress in recovering any lost function in her left arm and leg. However, after her tenth session of Ocular Light Therapy on the Lumatron, the woman began to move the little finger on her left hand. Within two weeks of further light treatments she could move her whole left hand, and she was no longer dragging her left foot. After another twenty sessions, the patient was basically back to her normal pre-stroke state! This is quite impressive, since most stroke-induced neurological problems that have not responded to physical and occupational therapy rarely improve on their own by the time a year and a half has passed.

Dr. Downing has found that Ocular Light Therapy can also improve a variety of other physiological imbalances as well. The long list of problems that have been shown to improve with Ocular Light Therapy includes fatigue, stress symptoms, headaches, various types of pain syndromes, hormonal imbalances, premenstrual syndrome, mood swings, insomnia, depression, fear and anxiety, hyperactivity, memory and concentration problems, learning disabilities, and also selected vision problems. As with UV therapy, though, there can be some small risk with Lumatron flashing-light therapy when used on certain individuals. For example, some patients with epilepsy may actually have seizures triggered by flashing lights or a flickering TV set (so-called photo-convulsive seizures). Patients with seizure disorders who are candidates for Ocular Light Therapy should see a neurologist for EEG brainwave testing with a flashing strobe light prior to beginning light treatments. This test will determine whether or not a patient with a history of seizures may be susceptible to photoconvulsive seizures. In cases where patients are susceptible to any type of flickering-light-induced seizures, they can still be treated with the appropriate wavelength of light on the Lumatron but without the added light pulsation. Dr. Downing has noted that flashing-light therapy is usually more effective with the pulsation. But either form of light therapy has been found to be effective in many cases.

Dr. Downing's use of Ocular Light Therapy to heal disorders of the nervous sytem is quite remarkable. But actually, this type of therapy dates back to the 1930s and 1940s, when devices for delivering Ocular Light treatments were developed by physician/optometrist Dr. Harry Riley Spitler. In 1941, Dr. Spitler wrote about what he called "the syntonic principle" of light therapy. The word "syntonic" is derived from the Greek word meaning "to bring to balance." Spitler was one of the first physicians to speculate that if a person's autonomic nervous system was "too dominated" by either their sympathetic (fight-or-flight) or their parasympathetic (relaxation-response) system, they would end up with health problems. Through his research, Dr. Spitler discovered a number of basic light-therapy principles. He found a relationship between the frequency of light to which tissues are exposed and the rate of growth and cell division in those tissues. Spitler carried out experimental light research on rabbits by exposing them to environmental conditions illuminated with different gel-filtered frequencies of light. After analyzing his data, Spitler determined that rabbits grew differently depending on the wavelengths of light that illuminated their living environments. Within three to eighteen months of living in altered lighting conditions, some of the rabbits began to develop both partial and complete loss of their fur, toxic symptoms, digestive problems, abnormal body weight, sterility, abnormal bone development, and even cataracts. Dr. Spitler concluded that some of these health problems might be related to imbalances in the rabbits' autonomic nervous systems arising from the varied light frequencies used in his experiment. Such experiments ultimately led Dr. Spitler to search for ways in which different frequencies and colors of light therapy might be used to heal the "sick" rabbits.

Since certain colors of light seemed to induce illness (and presumed autonomic instability), he speculated that other frequencies and colors of light might actually rebalance the rabbits' disturbed sympathetic and parasympathetic nervous systems. Through his research, Spitler learned that different light frequencies seemed to trigger the release of specific neurochemicals and hormones from the pituitary and endocrine glands. Based on his studies, Spitler began to develop the light-therapy principles for what came to be called syntonic optometric phototherapy. The basis for syntonic phototherapy was his observation that specific frequencies of light, directed through the eyes, could actually bring about a rebalancing of the nervous and endocrine systems of the body. In 1951, Dr. S. V. Krakov, a Russian scientist researching the relationship between color vision and the activity of the autonomic nervous system, confirmed Spitler's theories about the effects of different colors upon the sympathetic and parasympathetic nervous systems. Krakov found that the color red stimulated the sympathetic nervous system, while the color blue stimulated the parasympathetic system. His findings were later confirmed by other light researchers.

Today, the College of Syntonic Optometry continues to offer postdoctoral training for eye-care professionals in the principles of colored-light therapy first established by Dr. Spitler back in the 1930s. Throughout the United States, a significant number of optometrists are trained in syntonic phototherapy techniques and continue to use syntonic light therapy in their practices. The "syntonic method" of colored light therapy produces a number of other interesting neurological changes besides the rebalancing of the autonomic nervous system. As with Ocular Light Therapy patients, many individuals who undergo syntonic colored-light therapy also experience an increase in the size of their visual fields. This increase in visual fields occurs even in patients who may be experiencing a loss in their visual fields due to eye diseases like glaucoma and retinitis pigmentosa. Also, syntonic light therapy has been shown to improve visual attention span (the ability to focus upon and concentrate on a visual target) and visual memory (the ability to remember what one sees), as well as auditory memory.

In addition to altering seizure activity and improving vision, a number of health-care professionals have successfully used colored light (directed through the eyes) to induce states of profound relaxation and increased creativity. A sunglasseslike device called the Shealy Relaxmate (developed by energy-medicine pioneer Dr. Norman Shealy) uses red light-emitting diodes (LEDs) to stimulate the eyes with red light as a form of relaxation therapy. In the Relaxmate, the pulsing red light of the light-emitting diodes is used to stimulate the brain via the eyes. Relaxation-inducing systems like the Relaxmate operate according to the principle of "bioentrainment." We know that the human brain usually responds to oscillating light, sound, or magnetic-field energies by becoming "entrained" to the frequency at which the energy is pulsing. The pulsing red light of the Relaxmate can be adjusted to the frequency of relaxing "alpha" brainwaves (in the range of 8 to 10 cycles per second). The flashing red LEDs entrain the electrical activity of the brain by producing photocurrents in the brain that oscillate at the alpha frequency. As the pulsing alpha photocurrents entrain brain-wave activity into the relaxation-associated alpha range, the Relaxmate wearer goes into a calm, relaxed state. Similar versions of LED-based brain-wave-entrainment devices will often use different-colored filters placed over the red LEDs to produce different frequencies of light to stimulate the brain. Many LED-type relaxation devices are tunable to different frequency ranges associated with not only alpha for relaxation but also theta brain activity (6 to 8 cycles per second), a state associated with creative insight. A number of these relaxation devices also come with earphones that produce an auditory tone that oscillates at the same frequency as the LEDs in the eyepieces, providing synchronized light- and sound-energy stimulation able to entrain the brain-wave activity of the user into a slower, more relaxed brain-wave rhythm.

Red light is also used in laser-therapy systems that are used to project light onto highly specific areas of the body other than the eyes (such as acupoints and reflexology points) in order to assist the body in healing itself. Some of these light-therapy devices come in the form of miniature soft-laser devices, which project a beam similar to the red-light of laser pointers. In addition to the classic helium-neon red-laser systems, there has been a proliferation of tiny red-LED light-therapy devices, such as the Light Shaker and the Tri-Light, which also produce an intense beam of red light that can be easily used to therapeutically stimulate acupoints. These small, portable, battery-powered appliances are commonly used by acupuncturists, chiropractors, physiotherapists, massage workers, and naturopaths. Acupuncturists and other health-care professionals familiar with auricular therapy will frequently use red-LED devices to stimulate acupuncture points on the ear. In auricular medicine, there is a point on the ear, commonly known as the "Valium" point, which, when stimulated, can produce deep relaxation. Patients who are being treated for various pain syndromes can sometimes benefit from monochromatic red light focused directly into the Valium point. Following red-light stimulation of the Valium point with the red-LED device, a patient will often report a sense of calming and a wave of progressive relaxation. Anecdotal reports from pain patients also attest to decreases in pain, tightness, and spasm in the affected painful areas of the body following red-light stimulation of the Valium point.

Another use of monochromatic red-light LED therapy is for stimulating reflexology points located upon the hands and the feet. Reflexology is a method of healing based upon stimulating specific points on the hands and feet that correspond to different parts of the body. Dr. William Fitzgerald, an American physician, is credited with the discovery of reflexology. However, the knowledge of treating reflex points on the feet actually goes back much further. A number of American Indian tribes had used reflex-point stimulation to heal the internal organs of the body for many years before Fitzgerald's remarkable discovery. The stimulation of reflex points on the hands and feet is actually quite similar to needling of body acupoints on the ears in auricular acupuncture. If you remember, in auricular medicine, ear acupuncturists use a special map of the ear showing which points correlate with each of the body's different organ systems and skeletal structures. Similar body maps exist in reflexology for the hands and feet. Each reflexology map is a kind of miniature hologram of the whole body. The reflex zones on the foot or the hand that correspond to the painful or diseased body system will often be tender to the touch.

Reflexologists believe there is a basic energy polarity to the body, with the two opposite energetic poles consisting of the head and the feet. They also subscribe to the theory that there are ten separate life-energy currents that flow along longitudinal zones going from the top of the head down to the feet. The body's different organs are said to lie within one of these ten energy zones. While each of these ten energy streams is said to end in the feet, there are also outlets that end in the hands. The theory behind reflexology is that various types of manipulation of the reflex zones on the hands and feet can assist in releasing energy blocks (in one or more of these ten body-energy currents) that may be affecting the health of the body's organs. For example, using deep massage to stimulate the knee reflex point on your foot might help to relieve recurrent pain in your sore knee. Recently a number of reflexology therapists have found that stimulating the appropriate reflexology point on the hands and feet with monochromatic red light from red-LED devices often provides relief from various health problems in a manner similar to the therapeutic effects of deep massage upon the same reflex point. In such cases, a red-LED device such as the Light Shaker is pressed directly upon the reflex points and rotated slowly in a a clockwise direction for several minutes. Reflexologists claim to have had success in using localized deep massage and red-LED stimulation of hand and foot reflex points to improve a variety of health ailments, including weight problems, menstrual disorders and PMS, kidney and bladder problems, constipation, sluggish digestion, and a variety of pain syndromes.

Still another variation of reflexology combined with light therapy involves the use of the reflexology crystal torch developed by Pauline Willis.[4] Willis is a trained color therapist and reflexologist who works in London, England. The torch device she uses consists of a high-intensity flashlight that has a special nylon head mounted over the flashlight lens. The nylon head is removable and has a slot for inserting circular colored-light filters made of stained glass in front of the projected light beam. On the projecting part of the nylon head is a single-terminated quartz crystal that focuses the colored light onto the appropriate reflex zone. The color filters used by Willis include the usual red, orange, yellow, green, blue, indigo, and violet, but also magenta, turquoise, and gold. She selects a specific color or colors for treating a particular illness according to guidelines established by color therapists in England. She then focuses the appropriate therapeutic color into the foot reflex zone (which corresponds to the problem area of the body) for sixty to ninety seconds. In addition, Willis focuses the selected therapeutic color or colors into the foot-reflexology chakra points corresponding to the chakra that "rules" that affected organ system. In reflexology, it is said that there are seven chakra-reflex points on the feet that connect to and somehow energetically affect each of the seven major chakras.

During a typical reflexology-treatment session, Willis uses colored light from the crystal torch to help rebalance the seven major chakras. The color selected for stimulation of a chakra-reflex point depends upon whether or not specific physical problems are already present. If there are no physical ailments, red is used to rebalance the root chakra, orange for the sacral, yellow for solar plexus, green for the heart center, blue for the throat chakra, indigo for the brow center, and violet for the crown chakra. But when treating a specific physical disorder, Willis uses the reflexology crystal torch to administer the therapeutic color most appropriate for the diseased organ or health condition. The crystal torch is used to stimulate the chakra-reflex points on the feet for the chakras that govern the affected organ systems. For instance, if a patient is suffering from tension related to stress-induced hypertension, Willis directs blue light, a cooling and relaxing color, into the solar plexus chakra-reflex points on both the right and left feet. The solar plexus chakra points are treated because it is the solar plexus center that supplies pranic energy to the adrenals, the endocrine glands that release stress hormones such as adrenaline and cortisol into the general circulation. This would be in contrast to a general reflexology-treatment session for an individual who has only minor imbalances in the solar plexus chakra and no active health problems. In this second example, Willis would use yellow light (the usual color associated with this chakra) to rebalance the solar plexus chakra-reflex points.

When treating a particular condition with color reflexology, Willis sometimes uses not only the selected therapeutic color but also its complementary (or opposite) color. For example, when treating a localized infection, such as a painful abscess, Willis will use the color red on the appropriate reflexology body point to stimulate local circulation to the area. Next she will apply turquoise light, the complementary color to red, in order to reduce local inflammation. According to Willis, green light can also be very helpful for treating infections. Thus, for infections, not only red and turquoise but also green light would be sequentially directed into the foot reflex points that correspond to the body regions affected by infection.

While color reflexologists work primarily with colored light to stimulate body-reflex zones, there are other color therapists who focus colored light upon the entire body in an attempt to heal illness. The use of colored light focused upon the body to bring about healing has a unique history. Here again we go back to the work of Dr. Edwin Babbitt and Dinshah Ghadiali. Dr. Babbitt did his work on color therapy during the late 1800s, while Dinshah worked in the early part of the 1900s. Babbitt built a colored-light projection device known as the Chromodisc, which he used to heal illness by radiating certain areas of the body with specific colors of light energy. He also recommended the use of colored "solar elixirs" or sun-charged water for healing as well. Babbitt created solarized elixirs by placing water in either a colored bottle or a clear bottle wrapped in a particular color cloth or transparent gel. The bottles were then placed in direct sunlight for "energizing." According to Babbitt, the water became solarized or imprinted with the healing color vibration of the color of the bottle (in a fashion similar to "sun-charged" flower essences). He would have people make their own blue-solarized water by charging it in blue milk of magnesia bottles. The cooling blue color was often used for treating sore throats and burns.

In 1920, Dinshah Ghadiali, from India, developed a system of color healing he called Spectro-Chrome. Ghadiali's system was based upon three main color-therapy principles: 1) The human body reacts to light, 2) different colors of light relate to specific physiological functions in the body, and 3) color "tonation" (exposing the body to specific colors of light) aids those bodily functions. In addition, Ghadiali determined that there were specific zones throughout the body where different-colored light should be directed in order to heal specific organs. In 1926, Dr. Kate Baldwin, a physician with thirty-seven years of experience in medicine and surgery, trained with Ghadiali in color therapy. After working with Ghadiali's color-healing methods, she became convinced of the benefits of colored light for healing illness. Baldwin found that sprains, bruises, and traumas to the body frequently responded better to color treatments than to other therapies of the day. Conditions such as sepsis, in which a patient suffers from overwhelming infection, also seemed to respond favorably to specific colors of light directed onto the appropriate body zones. In addition, Baldwin reported that heart problems, asthma, hay fever, pneumonia, corneal ulcers, glaucoma, and even cataracts improved with Ghadiali's Spectro-Chrome method of color therapy. Unfortunately, the FDA considered Ghadiali's techniques to be little more than medical quackery. It issued injunctions against Ghadiali forbidding him to ship his instruments across state lines. Every effort was made to suppress any interest in color therapy. But due to the efforts of Dinshah's sons, his work has been kept alive to this day.

In the section that follows, we will examine the therapeutic attributes of colors as described by Dinshah Ghadiali. We will also compare Ghadiali's color system with Pauline Willis's modern color-therapy interpretations of colored lights' effects upon the body (as used in color reflexology). The colors listed below are used for either reflex-point stimulation (via the reflexology crystal torch) or for whole- or partial-body exposure using special color-filtered electric lamps. In addition, where applicable, specific auric-field color associations are also mentioned.

Red: Stimulates the sensory nervous system. To Dinshah, red was considered a liver stimulant that helped to regenerate liver function. It was thought to increase red-blood-cell production and to improve circulation. Dinshah also believed that the color red was able to stimulate the excretion of toxins from the body through the skin. Modern light research also suggests that red light tends to speed up the nervous system, partly because it stimulates the sympathetic nervous system, our adrenaline-charged fight-or-flight response. Willis agrees with Dinshah on red's ability to increase red-blood-cell production in that it may aid the body's uptake of iron, which is used to manufacture hemoglobin, the source of our blood's red color. This makes red a good color for treating anemias, especially those due to iron deficiency. Red light directed to the root chakra or to the root-chakra reflex points on the feet is said to help rebalance and to clear blockages from this center. Because red is a stimulating color, Willis advises against using it on people who suffer from asthma, high blood pressure, heart disease, or epilepsy. Esoterically, red is considered the symbol of life, strength, and vitality. An excess of red in a person's auric field is thought to indicate indicate strong physical propensities, while darker red may sometimes be due to strong emotions, ranging all the way from passion and courage to anger and hatred. Red is the color usually associated with the first or root chakra, which is our grounding point to the earth.

Orange: Was thought to accelerate regeneration of lung tissue while improving the health of the respiratory system. Dinshah used orange light to stimulate the thyroid gland (in hypothyroidism), as well as to inhibit the parathyroid gland (such as might be used in cases of overactive parathyroid glands or hyperparathyroidism). Orange light was felt to have decongestant properties as well. According to Willis, orange is the color of vital energy in the auric field, where bright, clear orange often denotes health and vitality. An excess of orange in the aura is said to denote abundant vital energy. Orange is the dominant color that affects the second or sacral chakra. But orange is said to have a strong affinity for the splenic chakra (a minor but important chakra) as well. According to clairvoyant studies of the chakras, the splenic chakra is one of the main intake points for absorbing etheric energy (prana) into the body. Some clairvoyant investigators have noted that the splenic chakra absorbs and then refracts full-spectrum prana, like a prism refracting white light, into its various component-color frequencies and then distributes those colored streams of etheric energy to the other chakras of the body. Color therapist Willis believes that orange light (from a color-reflexology standpoint) is useful for treating kidney stones and gallstones and may also be of benefit in cases of chronic bronchitis. This perspective complements Dinshah's claim that orange stimulates the respiratory system. Willis believes that orange may also have benefit as an antispasmodic, and as such can be beneficial for treating muscle spasms, including abdominal cramps. Because orange is said to affect the sacral chakra, it may also be helpful in treating ovarian cysts, uterine fibroids, and prostate diseases (afflictions of organs linked to the sacral center). Orange light is frequently administered in conjunction with its complementary color of blue.

Yellow: Dinshah believed that yellow was able to stimulate the motor nervous system (the neuromuscular system). Yellow was said to help regenerate damaged nerves in both the sensory and motor nervous systems. Another suggested use was as a stimulant for the lymphatic system and the intestinal tract. To clairvoyants, yellow in the auric field is the symbol of the mind and the intellect. It is also the dominant color associated with the solar plexus chakra. Willis agrees with Dinshah's use of yellow light for strengthening the nerves and for stimulating higher intellectual functioning. Yellow is said to activate the motor nerves and thus generate energy in the muscles of the body. Because of this property, yellow was deemed useful in treating any form of paralysis. In cases of paralysis or dysfunction of the nervous system, yellow light, followed by its complementary color of violet, is often used by color reflexologists to stimulate the spinal-reflex points on the feet. Dinshah believed that yellow light was useful as a cleanser of the skin and as a healing therapy for skin disorders ranging from scars to eczema. Yellow-light therapy, when used to treat arthritis, was seen as a way to help break down calcium deposits that had formed in the joints. Since the pancreas is energetically linked to the solar plexus chakra, color reflexologists will sometimes use yellow light (in conjunction with its complementary color, violet) to stimulate the solar plexus chakra reflex points on the feet in an attempt to help regulate blood-sugar levels in patients with certain forms of diabetes. Similarly, indigestion caused by stomach problems may also respond to yellow and violet light directed to either the body or the stomach and solar plexus reflex points on the feet.

Lemon: A color created by Dinshah through combining yellow and green gel filters in a projection lamp. Dinshah felt that yellow light was the great "restorer" and was used for people who were being treated for chronic conditions that had left them feeling depleted. Similarly, lemon-colored light was thought to promote healing in persistent or chronic medical disorders. Lemon light energy was also said to dissolve blood clots. In addition, Dinshah used lemon light as an expectorant and as a stimulant to the brain and to the immune system. Finally, lemon was thought to be a mild stimulant for sluggish digestive systems.

Green: Dinshah considered green to be an "equilibrator" of brain function and of the physical body in general. Green light was also thought to stimulate the rebuilding of damaged muscles and body tissues. Additionally, green was said to stimulate and rebalance the pituitary gland in healing endocrine disorders caused by pituitary dysfunction. Dinshah classified green light as a disinfectant, an antiseptic, and a germicide that could be helpful for treating infections. Green is also the dominant color associated with the heart chakra, making it a useful color (to use with its complementary color, magenta) in treating certain forms of heart disease. However, if the heart disorder is felt to be of an emotional origin, color reflexologist Willis advocates using rose pink and pale violet instead of green. Willis agrees with Dinshah in her clinical observation that green light does indeed help in treating infections. She also suggests that green light cleanses the etheric body (when adminstered through the throat-chakra reflex point on the feet). Green used in conjunction with magenta may also be helpful in treating toxicity caused by problems in the colon, liver, or kidneys, which are the major organs of detoxification and waste excretion. In patients requiring detoxification, Willis suggests that both the throat-chakra reflex points and the key organ reflex points (such as those for the liver, colon, or kidney) should be treated with green as well as magenta. In general, green is said to be a color that balances body, mind, and spirit, and one that clairvoyants frequently observe emanating from the hands of healers. Green in the auric field can often be associated with spiritual growth and development.

Turquoise: This color is obtained by combining green and blue gel filters in a projection lamp. While green is a great rebuilder and cleanser, "cooling" blue is said to have the ability to reduce pain and bring down fevers. Therefore turquoise carries some of the attributes of both colors. Turquoise is classified by Willis as a healing aid for acute illnesses and as a brain depressant. It may also be helpful in healing damaged skin in cases of serious burns. Turquoise is a color not normally associated with the seven major chakras. However, Willis feels that turquoise is associated with a minor chakra just above the heart chakra known as the midchest or thymus chakra, which, as its name implies, feeds pranic energy to the thymus gland. Willis suggests that this minor chakra is enlarging and becoming more significant as humans proceed along the evolutionary curve of psychospiritual development. Since turquoise may stimulate the thymus chakra, it also affects the thymus gland, the major immune center of the body. Willis notes that turquoise may be helpful in boosting thymic and immune function in cases of chronic immune dysfunction ranging from chronic fatigue syndrome to AIDS. She also uses turquoise for treating inflammatory problems. In cases of inflammatory disease in a particular area of the body, turquoise is administered in conjunction with its complementary color red to the associated body-reflex points on the feet. It is thought that red light stimulates local blood flow in the area that is inflamed, while turquoise light reduces the inflammation. In patients who are recovering from a bacterial or viral infection, Willis will sometimes treat the lymph-reflex points on the feet with turquoise to enhance immune response so that the infection may be rapidly cleared from the body. With regard to treating infections, turquoise light, in conjunction with red, may be of help in treating acute or recurrent sore throats, ear infections, nephritis (an inflammatory condition of the kidneys frequently caused by infections), as well as bladder infections.

Blue: Blue was a color frequently used by Dinshah and his associate Dr. Kate Baldwin for treating burns, for relieving itching, and to decrease painful irritation caused by skin abrasions. Because it was considered to be a "cooling" color, blue was said to decrease fevers and to reduce inflammation. It was thought to be a stimulant to the pineal gland as well. More recent photobiology research has confirmed that blue light has a calming influence upon the nervous system, especially for someone who is overly tense. Blue is also the predominant color associated with the throat chakra. Many clairvoyant observers view an excess of blue in the auric field as possibly indicating artistic tendencies, a harmonious nature, and spiritual understanding. Willis agrees with Dinshah in his claim that blue light has a calming effect, and often uses blue for treating tension, fear, palpitations, and rapid heartbeat, as well as for relieving insomnia. Because of its cooling, anti-inflammatory properties, blue can also be used for treating inflammatory problems (by directing blue light into the appropriate body-reflex points on the hands or feet). In cases of gastric and duodenal ulcers, blue, along with its complementary color orange, is directed to the appropriate stomach- or small-intestine-reflex points on the feet to assist in healing the ulcer. Color reflexologists will sometimes treat asthma and pleurisy (often associated with lung inflammation) by directing blue and orange light into the lung-reflex points on the feet. Other inflammatory disorders that Dinshah felt may be helped by blue light (and its complementary color, orange) include cystic mastitis (a painful inflammation of the breasts), diarrhea (associated with inflammation of the colon), and certain cases of prostate enlargement (where there may be subclinical prostatitis). Since blue is considered the major color of the throat chakra, it is also thought to be useful in treating disorders of the throat region, such as laryngitis, sore throats, and possibly certain thyroid disorders, such as goiters and overactive thyroid glands. Color reflexologists will sometimes use blue light for treating jaundice via the liver-reflex point. Even mainstream medicine has recognized the beneficial effects that blue light has in relieving jaundice. In fact, for many years hospitals have used intense blue light to treat newborn infants who are jaundiced.

Indigo: This color was thought by Dinshah to be stimulating to the parathyroid glands (situated on the back of the thyroid gland), especially in cases where there may be inadequate production of parathyroid hormone, a calcium-regulating hormone in the body. Conversely, indigo was said to inhibit thyroid function, especially in cases of overactive thyroid glands. It was also thought to have certain sedative qualities, along with helping to increase white-blood-cell counts. In addition, indigo has been used for treating problems related to closed head injuries. From an auric-field standpoint, some clairvoyants consider indigo to be associated with deep devotion carried with a sense of dignity and unconditional love. Because of indigo's link with the sixth or third-eye chakra, it is also a color associated with intuition and transcendant vision. Willis feels that indigo, a combination of blue and violet light, stimulates the pituitary gland, the major endocrine center associated with the sixth chakra. Since the eyes and ears (energetically connected to the sixth chakra) are stimulated by indigo light, color reflexologists find indigo useful for treating disorders such as cataracts and hearing loss. Indigo may also have the ability to induce anesthesia in the physical body, making it a powerful painkiller for treating various types of pain syndromes. Thus, a color-reflexology practitioner will sometimes use indigo light, along with its complementary color, gold, for treating painful sinus infections, spine pain (such as low-back pain and sciatica), angina (heart pain), shoulder pain (due to muscle strains or tension), and even headaches. Willis believes that indigo may have certain blood-purifying properties as well. It might also be beneficial for treating cases of hepatitis or inflammation of the small intestine (such as ileitis caused by Crohn's disease) with color reflexology. In each of these disorders, the appropriate organ-reflex point as well as the chakra-reflex point associated with that organ would be stimulated with indigo light.

Violet: Dinshah believed that violet light was a stimulant to the spleen, one of the immune-system-related organs of the body. It was also thought to decrease muscular activity, including depressing the activity of the heart muscle. Dinshah also felt that violet inhibited the activity of lymphatic glands and depressed the function of the pancreas as well. Like indigo (made up of blue and violet light), violet was said to increase production of white blood cells and to have a tranquilizing effect upon the nervous system. Along these same lines, patients with anxiety and depression who are considered neurological "fast" types by Dr. John Downing's classification scheme (sympathetic dominant) have found significant therapeutic benefit from pulsed-violet-light therapy to the eyes delivered by the Lumatron device. From an auric-field perspective, clairvoyants describe violet as a color associated with spirituality as well as self-respect and dignity. A deep violet (or purple) color in the aura often denotes a high level of spiritual attainment and a "higher love." It is also a color linked with the insight and wisdom of the higher self. Violet is considered to be the dominant color associated with the crown chakra. Willis finds violet a good color for treating psychological problems, including manic-depressive illness. She also notes that violet light may be helpful for treating disorders of the nervous system as well as painful sciatica. It has even been suggested as a treatment for hair loss, since violet affects the crown chakra, which energetically feeds the hair and scalp on top of the head. Violet is frequently used in conjunction with its complementary color, yellow.

Purple: Dinshah distinguished purple from violet via the use of specific colored gel filters (listed at the end of the book in Appendix 4, the resource section). According to Dinshah, purple was said to induce relaxation and promote sleep. Purple was also thought to depress activity in the adrenal glands and in the kidneys. According to Dinshah, purple light was able to lower body temperature, to decrease blood pressure, and to slow down the heart rate (in cases of rapid heartbeat).

Magenta: As a blend of red and violet light, magenta was thought to be an "emotional equilibrator" that could help in times of emotional upsets. According to Willis, pale magenta is the color of the first chakra above the head, sometimes referred to as the eighth or transpersonal chakra. Willis suggests that from an auric-field perspective, magenta is the color of "letting go." On a physical and mental level, this can mean letting go of ideas and thought patterns that no longer serve us. On an emotional level, magenta represents letting go of old emotional patterns and feelings no longer relevant to our current situation. When magenta becomes pale pink, it is said to transform into the color of spiritual love. As such, pale pink could be useful in healing "broken hearts" caused by the loss of a loved one or the ending of a romantic relationship. Willis believes that magenta may also be helpful for treating malignant tumors (in conjunction with other therapies). When used as an adjunctive therapy in treating malignant tumors of the pituitary gland, lung, breast, stomach, colon, kidney, uterus, or testicles, magenta would be applied (in conjunction with its complementary color, green) to the organ-reflex points and the appropriate chakra-linked reflex points on the foot. Additionally, Willis suggests that color reflexology using magenta light (along with green) may be useful for healing breast cysts, detached retinas, tinnitus (ringing of the ears), and water retention.

Gold: Although gold is not one of the colors mentioned by Dinshah, it is a color used by Willis in color reflexology. Gold is made up of orange and yellow light and is thought to be the complementary color of indigo. From an auric-field perspective, gold in the aura is said to be the color of wisdom and "grace," and is often linked with mysticism. In the "combined" color gold, yellow, the color of the intellect, becomes tranformed into a color more associated with cosmic consciousness. Gold is said to have the healing quality of increasing vitality and vital energy. When used in color reflexology, gold and indigo (its complementary color) may be useful for revitalizing the nervous system by treating foot-reflex points connected with the spine. Gold light, directed to the spleen-reflex point on the foot, is said to have an energizing effect similar to the color orange (especially when used with its complementary color, blue).

According to experienced color therapists, some conditions may respond better than others to color and light therapy. However, color therapy is likely to be of greatest benefit when used in combination with other vibrational and medical therapies. For instance, it would not be wise to treat a malignant tumor solely using colored-light therapy. On the other hand, a person undergoing conventional chemotherapy for a malignant tumor could also take a 1M homeopathic dilution of the chemotherapy drug (to decrease drug side effects) along with magenta- and-green-light therapy directed to the appropriate organ and chakra reflexology points on the hands and feet (or even to the cancer-affected body region itself). The key to deriving maximum benefit from using all the many forms of vibrational medicine is the correct combining of modalities that are complementary, synergistic, and mutually reinforcing in their healing effects upon body, mind, and spirit.

A slightly different form of color therapy involves the use of different-colored crystals for healing. Various types of crystals and gemstones are said to have special subtle-energy and healing properties and thus may sometimes be used by certain color therapists and healers to bring additional healing forces to bear upon various ailments. Certain crystals, most notably quartz, are thought to be subtle-energy amplifiers. Therefore, healers might employ particular crystals in an attempt to amplify healing energies that they are projecting into the body. Another aspect of healing with gemstones that is relevant to color healing involves the use of specific colored gemstones that correspond to the seven main colors of the visible spectrum. Gemstone-color therapists believe that each of these gemstone colors in turn corresponds to and resonates with one of the seven main chakras. Colored gemstones frequently used by color therapists include

Red: ruby and garnet

Orange: carnelian and orange jasper

Yellow: amber, citrine, and yellow topaz

Green: emerald and malachite

Blue: sapphire and lapis lazuli

Violet: amethyst and violet fluorite

Magenta: rose quartz

If a color therapist (or crystal healer) uses gemstones, the color-energy frequencies contained within the gemstones can be transferred to the body in a number of different ways. The simplest method of using gemstones for color healing is probably the "chakra layout pattern," which involves having the patient lie on his or her back while the healer places seven different-colored gemstones directly upon the seven corresponding chakra regions of the body. The gemstones are left in place on top of the chakra zones for a period of twenty to forty-five minutes. Another method of assimilating gemstone-color energies is through the direct ingestion of gemstone-powders, a technique that has been used by Ayurvedic and gemstone therapists in India for many years. Many gemstone color therapists subscribe to the theory that different-colored gemstones are repositories of what might be called "the cosmic colors of the rainbow." In reality, patients do not actually ingest polished gems. The gemstones are first burned to fine ash, which is then ingested. Instead of consuming the actual gemstones, a more subtle method of taking in the energetic frequency of the various colored gems is through the use of "gem elixirs." Like flower essences, gem elixirs are prepared using the "sun method," whereby gemstones are placed in water in direct sunlight in order to transfer an energetic imprint of the gem's vibrational-energy pattern into the water. The water is ingested as drops under the tongue from a dropper bottle. Gem elixirs made from the gems mentioned above appear to resonate with specific chakras as well as with the organs associated with those chakras. However these unique gem-based remedies are said to have many other, more subtle vibrational-healing properties as well.

The Inner Path to Color Healing: The Use of the Mind and Spirit in Healing with Color

Until now, we have primarily looked at color therapy from the perspective of light-therapy techniques. These techniques allow practitioners to direct various colored-light energies into the eyes, the hands and feet, and to the entire body by way of colored-light devices. But there are also a number of "inner" color-healing techniques that mainly use the power of the mind and spirit to produce healing effects. While we have discussed the use of colored light directed to the foot-reflexology points, a number of color-reflexology practitioners, including Pauline Willis, sometimes merely visualize sending streams of the color energy from their hands into the appropriate reflex points on the patients' feet. Joseph Corvo and Lilian Verner-Bonds, a healer and a clairvoyant, resrectively, who work with color therapy, deserve mention when one looks at the power of the mind and color therapy. Corvo and Verner-Bonds developed a system based upon the ten body-energy zones used in "zone therapy" and in reflexology. Zone therapy, first developed in the early part of this century by ear, nose, and throat surgeon Dr. William Fitzgerald, is considered by many reflexology therapists to be the "mother" of reflexology. Corvo and Verner-Bonds call their unique color healing system "color-zone therapy." Their approach consists of localized deep massage of reflexology points, as well as visualization techniques that involve sending color energy into the appropriate organ-associated reflex points on the feet. The colors used are merely visualized ones. There is no actual colored-light source. Corvo and Verner-Bonds feel that the physical manipulation of organ reflex points on the feet helps to directly stimulate the illness-affected organs of the physical body while the psychic color energy directed through the reflex points works at the level of the emotional and spiritual bodies to heal and rebalance the emotional patterns that may have contributed to a patient's physical illness. In addition to sending visualized streams of color into the appropriate reflex points, color-zone therapists will often instruct a patient to simultaneously visualize taking in that same stream of color and directing it to the painful or diseased area of the body via a guided color-imagery meditation.

There are other types of body workers besides reflexologists who work with color-imagery techniques to channel different-colored energies into specific regions of the body. The reason given for the effectiveness of color visualization is rather interesting. The act of visualizing streams of colored light going out through the healer's hands actually is believed to create strong mental thoughtforms that open up psychic-energy pathways within the healer. This may allow a healer to become a frequency-specific channel for directing different types or "colors" of healing energy out through his or her hands. When healers are visualizing individual color streams emanating from their hands and moving into patients' bodies, simultaneous observations by clairvoyants have often validated that this is indeed what is occurring at a subtle-energy level. In many cases, the clairvoyants have actually been able to accurately describe the "intended" visualized colors coming from the hands of the healer.

Another fascinating approach along this same line of healer-directed color energy healing relates to a specialized technique called pranic healing as taught by Choa Kok Sui, a Philippine pranic healer and teacher. Pranic healing is a form of psychic or spiritual healing that involves the transmission of specific colors of pranic energy from the healer to the patient. Patients are initially assessed using hand-scanning techniques (similar to those used in therapeutic touch), where the healer literally feels for "energy-field disturbances" over the chakras and within the auric fields of patients. Pranic healers then visualize taking in prana—a nutritive, environmental subtle energy—and focus it in highly specialized ways through minor chakras in their hands. The focused prana is used to correct perceived energy disturbances in patients' physical and etheric bodies as well as in their auric fields. By treating illness at the level of distorted etheric-information patterns, pranic healers try to correct the higher-energetic causes behind a person's physical health problems. Pranic healers learn to use visualization and specialized breathing techniques to take in prana from the earth through their feet, from the air through their breath, and even from nearby living plants. This absorbed prana is then directed through the major chakras of the pranic healers' bodies as well as out through their hand and finger chakras directly into patients' bodies. Again, pranic healers use controlled breathing and special color-visualization techniques to send specific colors of prana into patients' bodies, depending upon their particular illnesses. There are specialized methods pranic healers use to cleanse and remove "diseased energy" from the body. One technique involves healers' sweeping their hands through congested areas of the auric field, like a brush or broom sweeping away dirt. Following the removal of "diseased energy," the pranic healer then charges weakened energy zones of the patient's energy field, the chakra system, and the etheric body, using specific streams of colored pranic energy. The color of prana the healer sends will often depend upon the nature of the patient's ailment as well as which chakra region of the body seems to be out of balance.

Pranic healers, such as Choa Kok Sui, have made extensive clairvoyant studies of the chakra systems and the auric fields of hundreds of patients. Sui has carefully observed the healing influence of specific visualized colors of prana upon patients' chakras and auric fields, and also clairvoyantly studied the normal "metabolism" of prana by the body. According to clairvoyant observations dating back nearly a century by the Reverend Charles Leadbeater, which were similar to Sui's modern intuitive research, naturally occurring white prana is normally absorbed by specific chakras and then "digested" and broken down into its component colors. In agreement with Leadbeaters's clairvoyant studies, Sui notes that the spleen chakra (a minor chakra located in the upper-left abdomen over the physical spleen) absorbs a substantial amount of white airborne prana. White prana is said to be highly charged with the subtle energy of sunlight that clairvoyants refer to as "vitality globules." As the spleen absorbs white prana, it prismatically breaks it down into six major rainbow colors and then redistributes individual colored pranic streams to the other chakras of the body. The major pranic colors are red, orange, yellow, green, blue, and violet. Red, blue, and green pranic energy are the colors most frequently used by pranic healers in their healing work. Pranic healers learn to use their breathing, combined with imagery techniques, to direct specific colors into various chakras and energetically imbalanced areas of a patient's body, depending upon the illness and energy imbalance diagnosed. The healing properties of colored prana, as described by Choa Kok Sui, are described in the following paragraphs. It should be noted that Sui's color recommendations frequently concur with the uses of specific color energies as directed by other, more "traditional" color-therapy approaches. However, in some cases, discrepancies may be noted.

Red prana: Has the energetic qualities of strengthening, warming, increasing circulation, aiding in rapid tissue healing or cellular repair. Red prana sustains the physical body and vitalizes the blood, bodily tissues, and the skeletal system. As in other color-healing systems, red prana is recognized as being stimulating and activating in its effects. Red prana is used by pranic healers for rapidly healing internal and external wounds, broken bones, allergy attacks, sluggish or weakened organs, general fatigue, paralysis, loss of consciousness, and for healing measles.

Orange prana: Has the energy qualities of being expelling, eliminative, decongestive, cleansing, melting, extracting, splitting, as well as "exploding" in nature. Pranic healers use orange prana for treating ailments that require a great deal of elimination of waste, germs, and toxins from the body, as well as for removing diseased etheric matter. As such, it is considered useful for healing kidney and bladder ailments and constipation. In addition, orange prana is used for treating menstrual disorders, for eliminating blood clots from the body, and for shrinking cysts. It is also said to have therapeutic value in healing painful arthritic joints, as well as for treating lung problems, including coughs and colds. Because it is believed to have such a potent effect, intense energizing with orange prana may actually produce tissue damage and is not generally used for treating delicate organs such as the eyes, the brain, or the heart.

Yellow prana: Has the energetic properties of aiding with cohesion, cementing, and "gluing" together tissues that have been disrupted and traumatized, as in the case of skin wounds and broken bones. It can be helpful in stimulating the repair of damaged cells, because yellow prana tends to produce increased cellular multiplication and growth. Yellow prana is thought to be stimulating to the nerves as well.

Green prana: Has the energetic characteristics of digesting and breaking things down. Green prana is also disinfecting, detoxifying, dissolving, and decongesting in nature. According to Sui, green (along with orange) prana is useful for decongesting and loosening up "stubborn diseased etheric matter" from the etheric body so that it may be more easily removed using the special "sweeping hands" cleansing method. After removal of "diseased" etheric matter, the pranic healer reenergizes the same area of the etheric body and the auric field with fresh, clean prana, which helps in restoring health to the physical body. Green prana is also used to break down blood clots, as a general disinfectant, and for treating fevers and colds. It is thought to be milder and safer to use than orange prana. In certain cases of treating delicate internal organs, light-green prana is used as a precautionary pretreatment to energize the illness-affected organs before the more potent orange, red, or even violet form of prana is used. The use of light-green prana followed by light-orange prana on a specific region of the body can effectively decongest and cleanse both the physical and etheric aspects of that body region or organ system. When darker shades of orange and green are sent into the body simultaneously, destructive effects can be produced. Thus, a pranic healer may sometimes use dark-green prana when attempting to shrink a tumor.

Blue prana: Has the energetic properties of being inhibitory, contracting, and localizing in nature. For instance, when dark-orange/green is used for treating tumors, pranic healers often use blue pranic energy first to localize and limit the destructive effects to only the tumor-afflicted region of the body. Blue prana tends to be cooling (like blue light in general) and may have a soothing and mildly anesthetic effect upon the body. As such, blue prana is useful for treating painful, inflamed areas of the body and to bring relief from any other kind of painful condition. The cooling effects of blue prana are said to reduce fevers as well. Like blue light, which is known to bring on a relaxed, parasympathetic-dominant state within the nervous system, blue prana tends to produce restful states of awareness and can even induce sleep. Because blue prana has such strong disinfectant properties, pranic healers will commonly use it for healing infection-related health problems. Pranic healers also use blue prana to stop bleeding from wounds or internal hemorrhaging caused by trauma.

Violet prana: This uniquely colored prana possesses the energy qualities of all of the other colored pranic energies and is considered extremely potent. Because it is so potent, violet prana tends to be used only for more severe health problems, and its use is generally avoided when treating milder cases of illness. Light-violet or green-violet prana may be used to promote rapid healing of partly damaged internal organs, as in the case of severe stomach ulcers. Pranic healers might also use light-violet prana in treating severe infections, such as pneumonia and advanced syphilis.

White prana: Many different kinds of healers visualize sending white light into patients through their hands when they do healing work. In pranic healing, white prana (like white light) is thought to contain all the other colors of the pranic rainbow spectrum. Pranic healers consider white prana to be milder in intensity than colored prana. Specific colors of pranic energy are said to be more potent and specialized in their use than white prana. Pranic healers will often use white prana to treat infants, very young children, and older patients, because it is milder in its energetic effects.

In pranic healing, as in most other forms of psychic or spiritual healing, the "intention" of the healer holds the key as to how the psychically directed color energy is sent and used. When visualizing streams of color going from their hands into a patient's body, pranic healers will use their intention to powerfully activate the specific qualities of the energy they are sending, as well as to direct the colored pranic energy to the parts of the body where it may be of greatest benefit in promoting the healing process. Pranic healers (who follow the teachings of Sui) do not use indigo. Choa Kok Sui explains that, to his clairvoyant vision and that of his teachers, there is no observable indigo-colored energy in either the chakras or the etheric body. This observation differs somewhat from what has traditionally been considered the color of the third-eye chakra by many color therapists and some clairvoyants. Such minor variations in color interpretation are often found in very specialized vibrational-medicine practices. But for the most part, there is a great deal of similarity between different schools of color healing.

With regard to the use of visualized streams of color energy by healers, a few of the similarities as well as differences between how different healers interpret and use specific colors for healing are interesting to note. Barbara Brennan, a scientist, teacher, and healer who founded the Barbara Brennan School of Healing in New York, has a slightly different color-healing scheme from Choa Kok Sui's approach. One area of difference is that Brennan uses "channeled" indigo light to help to open spiritual perception as well as to induce a feeling of ecstasy. Brennan states that she may intentionally send specific colors of healing energy to patients or she may merely observe the color of the energy she is channeling from her healing spirit guides. It is important to remember that pranic-healing techniques are not necessarily the same as spiritual-healing techniques (which Brennan teaches) and may involve a different kind of healing energy. Specifically, pranic healers feel they work primarily on healing the chakras and the etheric body, while Brennan teaches her healing students techniques thought to bring about healing changes not only in the chakras and the etheric body but also in the higher spiritual bodies and the entire auric field. In her outstanding book, *Hands of Light,* Brennan also discusses the meaning of auric-field colors on a soul task level. Brennan's interpretation of auric colors comes from information channeled by colleague Pat Rodegast. Brennan feels that the meaning of the auric-field colors applies mainly to observations of the auric field at the "soul level." Her schema of color-energy healing properties along with Rodegast's channeled auric-field color interpretations are as follows:

Red: Increases one's connection to the earth and helps to strengthen one's basic life-force energy, stimulating the will to live (red is associated with the root chakra, the survival-instinct chakra). Red energy is also used to charge the auric field, to protect and to shield certain areas, and is thought to be of benefit to all organs fed energy by the root chakra. Red energy is also used in warming "cold" areas of an individual's energy field as well as for "burning out" tumors. Red seen in the aura can have different meanings, depending upon its shade and tones. For instance a rose-red color can indicate love and passion. Red-orange is especially associated with sexual passion. Dark red, on the other hand is associated with stagnated anger (as clairvoyantly observed by Brennan). Maroon, a red-purple color in the aura, indicates moving into one's soul task or life purpose.

Orange: Orange projected during healing is also used for charging the auric field with energy. Because of orange's resonance with the sacral chakra, orange healing light is useful for treating sexual disorders and increasing sexual potency. It can also be helpful for disorders of any organ fed by the sacral chakra, such as the uterus, ovaries, male genitals, and lower intestinal region. In addition, orange is used for boosting immunity. Orange energy, when seen naturally in the aura, may be associated with ambition.

Yellow: Sending this color energy brings increased mental clarity and can clear a foggy head. Again, yellow's association is with the mind and intellect, as in most color-healing schemes.

Green: This is a balancing color and brings a sense of being "okay" with the world. Green energy is good for healing all organs connected with the heart chakra and may be useful in treating diseases of the heart and lungs. Green is a color used in general healing and for charging the auric field with energy. As mentioned previously, clairvoyants have observed that many natural healers tend to have green energy coming from their hands. Green energy in a person's aura frequently indicates a healer and nurturer.

Blue: Blue energy has a calming, cooling effect (as is recognized in most other systems of color healing). It is also used for restructuring distorted energy patterns at the etheric-body level. Brennan teaches that the color blue helps to increase sensitivity and to strengthen one's inner spiritual teacher. Blue energy sent during healing is used for treating ailments of any organs in the throat-chakra region, such as the mouth, throat, larynx, and thyroid gland. As in pranic healing, blue energy may sometimes be used for the energetic shielding of certain areas of the body and the auric field. It is also used for cauterizing wounds during spiritual surgery, a technique Brennan teaches that uses psychically created beams of color and energy to "surgically" operate upon and to energetically correct health problems in the etheric and higher-spiritual bodies. Blue as an auric-field color is said to indicate sensitivity and someone who is a natural teacher.

Indigo: Indigo is used to open the third-eye or brow chakra, activating a person's clairvoyant abilities. Indigo, like yellow, may also be useful for clearing a foggy head. It can also bring about a feeling of ecstasy. Indigo is a color that helps an individual to connect with the deeper mysteries of spiritual life, and is good for treating dysfunction of any structures near the brow chakra (such as the eyes, ears, and sinuses, as well as the pituitary gland and parts of the brain). Indigo in the auric field often indicates that a person is moving toward a deeper connection with spirit.

Purple: When used in healing, purple energy helps a person to connect with spirit. The color purple in one's environment may also increase a sense of leadership and respect. Purple in the aura usually indicates a strong connection with spirit and things of a spiritual nature.

Purple-blue: This is a color that Brennan uses in taking away pain when doing deep-tissue work or work on bone cells. It can also help to expand a patient's auric field in order to make a deeper connection with one's lifetime "task" or soul purpose.

Lavender: This light-violet-colored energy is used to purge the auric field of negative or dysfunctional energy patterns. Lavender in the aura is a color Brennan associates with the energy of spirit.

Gold: Brennan uses gold light for restructuring the highest levels of human spiritual anatomy, what she calls the seventh layer of the field, an energy structure that may be analogous to the causal body (literally, the soul level). Gold light is also useful for charging the auric field. Gold light seen in a person's auric field tends to indicate a strong connection with God and a loving dedication to the service of humanity.

White: White light is used to charge the auric field, to take away pain, and to bring peace and comfort to someone in distress. White light in the aura is a color frequently associated with truth.

With such a vast array of color-energy modalities available, it seems obvious that the average person would easily require the assistance of a healer or reflexology therapist trained in a particular color therapy or technique to get started. But of all mentioned techniques, color-zone therapy and color reflexology (using visualized colors) may be the color-therapy approaches most easily learned by a layperson. Instructional books with specific guidelines for these color-healing systems are listed in Appendix 4. There are also a variety of very simple approaches to self-healing using color visualization that most individuals can easily try on themselves.

One of the simplest approaches to self-healing with color is known as "color breathing." In color breathing, an individual imagines or visualizes breathing in specific colors through each of the chakras in successive order. A simple color-breathing meditation taught by color therapist Pauline Willis can be useful for energizing and rebalancing the entire body and is easily performed by following these directions: Willis advises using different-colored flowers or leaves to more easily imagine the healing colors in your mind's eye. Sit comfortably in a straight-backed chair and breathe in and out through your nose slowly, paying attention to the movement of your chest and the rhythm of your breathing as you feel your body relax. Imagine all the stress and tension in your body melting away as you slowly breathe in and out. As you relax, imagine a bright red rose in your mind's eye. When you next inhale, visualize drawing up streams of the same rose-red color from the earth. Draw it up through the soles of your feet and up into your root chakra (located in your perineal region). As you imagine drawing up red energy into your root chakra on the inward breath, with your next exhalation imagine sending that rose-red energy out from your root center into the entire auric field that surrounds your body. Inhalation and exhalation of each color is done three times before moving on to the next highest chakra. Then imagine the rose becoming an orange chrysanthemum. Inhale this orange-colored energy up through your feet and into your sacral chakra, located just below your navel. With the out breath, imagine sending this colored energy out through your sacral chakra and into your auric field until you are bathed in the orange light. Moving up to the next energy center, imagine a bright yellow daffodil, and then breathe in yellow energy up through your feet and into your solar plexus chakra, located in the pit of your stomach area. Again, on the out breath, send this yellow energy out through your solar plexus center and into your auric field. As before, repeat this visualization three times. Next, imagine a bright green leaf and visualize drawing in this bright green energy horizontally and directly into your heart chakra in the middle of your chest. As you breathe out, send this green energy into your aura, bathing your entire energy field in brilliant green light. Next, imagine a blue cornflower. Imagine that you are breathing in this blue energy through the top of your head and directly into your throat chakra before breathing it out into your auric field on the out breath. Repeat this visualization three times. Next, replace the blue flower with an indigo iris in your mind's eye. Imagine breathing indigo light in through the top of your head and into your brow chakra, in the middle of your forehead, just above your eyes. Then send the indigo light out through your brow center as you breathe outward. Again, repeat for a total of three times. Finally, visualize tiny violets and imagine breathing in violet energy through the top of your head, your crown chakra, and then breathe out violet light through your crown center and into your auric field. Again, repeat three times. Then relax and imagine all these bright colors swirling around you while moving into and through your physical body. When you feel comfortable, breathe in and raise your hands above your head, stretching your entire body. As you breathe out, bring your arms back down to the sides of your body. The entire cycle should be repeated twice more for the greatest effectiveness.

Other simple imagery exercises also involve visualizing taking specific colors into your body to deal with specific problems. For example, when dealing with insomnia, you can visualize breathing the color blue into your entire body, imagining the blue light penetrating every cell and organ of your body. Imagine that your body becomes a container filled with blue until you drift off to sleep. If first attempts at falling asleep fail, try again, as repeated practice may often result in success. In addition, try sleeping under blue bedsheets, wearing blue pajamas or nightgowns, and even using a soft blue night-light near the bed to help encourage sleep (since blue is known to be a calming, relaxing color). For treating depression, some color therapists advocate using the same color-visualization exercise to take in orange-colored light, a more stimulating color. Because of its stimulating nature, this exercise should not be done at night, as it might actually interfere with falling asleep. For dealing with painful arthritis, imagine breathing in yellow light throughout your body, especially concentrating it in joints afflicted with pain, stiffness, or swelling.

Some color practitioners also believe that every day we take in color

The Therapeutic Actions and Subtle-Energy Associations of Color

COLOR	ACTIONS (ACCORDING TO DINSHAH/WILLIS)	ASSOCIATED CHAKRA
Red	**Stimulates the sensory system** **Liver stimulant** **Increases red-blood-cell production** **Stimulates sympathetic nervous system**	**Root chakra**
Orange	**Stimulates regeneration of lung tissue** **Stimulates respiratory system** **Inhibits parathyroid gland** **Antispasmodic** **In reflexology, may decrease kidney stones and gallstones**	**Sacral chakra**
Yellow	**Stimulates higher intellectual functioning** **Stimulates nerve regeneration** **Activates motor nerves** **Stimulates lymphatic system** **Stimulates intestinal tract** **Cleanses and heals the skin**	**Solar plexus chakra**
Lemon (yellow + green)	**Promotes healing in chronic medical conditions** **Dissolves blood clots** **Stimulates the brain** **Stimulates the immune system** **Stimulates sluggish digestion**	
Green	**Equilibrates brain and physical body** **Disinfectant, antiseptic, and germicidal** **May help to treat heart disorders** **May cleanse etheric body (reflexology)** **Balances body, mind, and spirit** **May relieve toxicity related to liver, kidney, and colon diseases**	**Heart chakra**
Turquoise (green + blue)	**Great rebuilder and cleanser** **Reduces pain and fever** **May accelerate burn healing** **Boosts immune function (stimulates thymus gland)** **In reflexology, reduces inflammation** **Relieves infections (reflexology)**	**Midchest or Thymus chakra**
Blue	**Relieves painful burns** **Relieves itching and skin irritation** **Calming influence** **Stimulates parasympathetic nervous system** **Relieves inflammation** **Treats jaundice (especially in infants)**	**Throat chakra**

AURIC FIELD ASSOCIATIONS	COMPLEMENTARY COLOR
Strong physical energy **Passion** **Courage** **Anger (Dark Red)**	**Turquoise**
Color of vital energy **Energy of health and vitality** **Color of imagination** **Purposeful spiritual direction** **Optimism**	**Blue**
Color of mind and intellect **Clear thinking** **Ability to think on a philosophical level**	**Violet**
	Violet-Magenta
Color of healing **Color of growth (emotional/spiritual)** **Color of harmony** **Inner tranquillity** **Adaptability**	**Magenta**
	Red
Religious and devotional nature **Objectivity** **Loyalty** > **(Bright Blue)** **Confidence** **Indicates spiritual nature**	**Orange**

(continued)

COLOR	ACTIONS (ACCORDING TO DINSHAH/WILLIS)	ASSOCIATED CHAKRA
Indigo (blue + violet)	**Stimulates parathyroid glands May inhibit overactive thyroid gland Mild sedative qualities May enhance white-blood-cell activity May be helpful in closed head injuries Can reduce brain swelling May stimulate pituitary gland In reflexology, stimulates pituitary May relieve pain, induces anesthesia With gold, may treat angina (reflexology)**	**Third-eye chakra**
Violet	**Stimulates spleen Decreases muscular activity Depresses heart action May shrink enlarged lymph nodes May increase white-blood-cell production May treat anxiety and depression (Lumatron therapy) May relieve sciatica**	**Crown chakra**
Purple	**Induces relaxation Promotes sleep Decreases adrenal and kidney activity May lower blood pressure, decrease heart rate and body temperature**	
Magenta (red + violet)	**Emotional equilibrator**	
Gold (yellow + orange)	**Increases vitality and vital energy In reflexology, revitalizes nervous system**	

energy in various forms through the clothes we wear, from the color schemes of our homes and offices, and even by the color of the foods we eat. Dr. Gabriel Cousens, a holistic physician who works with various forms of vibrational medicine, is the author of the book *Spiritual Nutrition and the Rainbow Diet.* Cousens feels that the colors of the food we eat resonate with, and feed subtle color energy to, our chakras. As his book states, red food feeds the root chakra, green foods are good for the heart center, and purple and violet foods feed the brain along with the brow and crown chakras, and so on. Cousens advocates starting the day on the lower red, orange, and yellow spectrum of foods. Next,

AURIC FIELD ASSOCIATIONS	COMPLEMENTARY COLOR
Color of deep devotion **Intuition** **Transcendant vision** **Purposefulness** **Spiritual will** **Seeking after truth**	**Gold**
Color of spirituality **Spiritual attainment (deep violet)** **Wisdom of the higher self**	**Yellow**
Priestly nature **Color of mysticism** **Color of the seeker**	
Color of letting go (of old emotional and mental patterns) **Color of spiritual love (pale pink-magenta)**	**Green**
Color of wisdom **Color of spiritual "grace"**	**Indigo**

he suggests that an individual move up to the higher-frequency food colors, like green, blue, and violet with the later meals of the day. Cousens classifies foods in the following color schemes:

> *Red*: Tomatoes, red cabbage, radishes, beetroot, raspberries, red cherries, red peppers, red currants, red plums, and red apples.
>
> *Orange*: Oranges and orange juice, carrots and carrot juice, mangoes, tangerines, pumpkins, apricots, and cheddar cheeses.

Yellow: Bananas, pineapples, yellow cheeses, yellow peppers, sweet corn, golden plums, parsnips, lemons, grapefruits, honeydew melons, butter, and egg yolks.

Green: Spinach, lettuce, cabbage, green apples, green beans, peas, lentils, kiwi fruit, watercress, green peppers, and any green leafy vegetables.

Blue: Blueberries, blue plums, grapes, and bilberries.

Violet: Purple grapes and grape juice, plums, purple broccoli, and eggplant.

Magenta: Strawberries and any pale-pink fruits or vegetables.

Besides eating various colored foods for their potential healing and rebalancing effects, color therapist Pauline Willis also suggests wearing specific colored clothes in the colors that you think you might need to help in your healing process. Willis recommends wearing specific colors of clothing depending upon both your mood and your health problem. For instance, people with high blood pressure might benefit from wearing blue clothes to help them lower it. For people who experience cold hands and feet, especially during the winter months, she suggests wearing red socks or mittens as well as red flannel underwear (similar to the kind our grandparents used to wear). According to Willis, when selecting colored clothing to wear for its therapeutic effects, cotton or silk (colored with natural and not synthetic dyes) is the best choice. In fact, for dealing with hypertension, Willis even advocates lying under a body-length piece of naturally dyed blue silk the first thing in the morning and the last thing at night. If cotton is used, it needs to be 100 percent cotton, not a cotton-synthetic blend. When wearing a particular color for color therapy, any undergarments worn need to be made of white material.

Whether one uses color visualization, color reflexology, Colorpuncture, colored-light therapy, or decides to go to a healer who channels specific colors of healing energy, the modalities of color and light are finding new applications and increased uses beyond their ancient roots. Color can be a powerful yet simple vibrational-healing tool, easily accessible to all. While color energy cannot heal everything, it appears that color healing may ultimately prove to be of great benefit in treating a wide variety of physical, psychological, and even spiritual problems.

FOOTNOTES

1. "Electrons, Photons May Battle Depression in Future," *Internal Medicine News*, August 15, 1996, p. 22.
2. David Olszewski and Brian Breiling, "Getting into Light: The Use of Phototherapy in Everyday Life," in *Light Years Ahead: The Illustrated Guide to Full Spectrum*

and Colored Light in Mindbody Healing (Berkeley, Calif.: Celestial Arts, 1996), p. 286.

3. Jacob Liberman, *Light: Medicine of the Future* (Santa Fe, N.M.: Bear & Company Publishing, 1991).
4. Pauline Willis, *Reflexology and Color Therapy: A Practical Introduction* (Boston: Element Books, Inc., 1998), p. 93.

E i g h t

MAGNETOBIOLOGY AND THE ART AND SCIENCE OF MAGNETIC HEALING

THE USE OF magnets for healing is nothing new. For centuries, healers have attempted to use permanent magnets in healing a variety of health problems that could not be helped by the medical therapies of the day. Magnets, as healing tools, have suddenly become a topic of great interest, as evidenced by the increasing number of books on magnetic healing, magnetic products, and television news stories focusing attention upon magnets as a form of alternative healing therapy. Just look at the bevy of mail-order catalogs suddenly offering magnetic belts or braces specifically for back pain and arthritic joint pains. Is this just a case of another health fad that will come and go like so many others, or is there really a scientific basis for the healing benefits of magnetic therapy? We will better understand this diverse and somewhat controversial area of vibrational medicine after examining the history of magnetic healing, as well as some of the theories behind why magnets may have a beneficial effect upon human health.

The History of Magnetic Therapies

According to legend, the properties of magnets were accidentally discovered by a Greek shepherd names Magnes, who, while tending his sheep, found that the iron nails in his sandals were mysteriously attracted to a large rock because of some unknown and invisible force. So strong was the attraction between the rock and his iron-laden sandals that Magnes had to use all his strength to pull his foot away from the rock's strange grip. This rock became known as the Magnes stone, which still bears his name, and which today we refer to as the lodestone. According to the same legend, Magnes's amazement at the force of this great magnetic rock led him to insert small pieces of the rock in his sandals, enabling him to walk great distances without feeling fatigued. Although this story may or may not be true, it is likely that similar magnetic rock fragments had been used for their curative powers many centuries before the ancient Greeks. A variety of ancient cultures—including the Chinese, Eastern Indians, Arabic peoples, Hebrews, and the early-dynasty Egyptians—all used magnets for their healing properties. Legend has it that Cleopatra, seeking to retard the aging process, slept with a lodestone on her forehead. In the third century B.C., Aristotle wrote about the healing properties of natural magnets, which he referred to as "white magnets." In the first century A.D., Pliny the Elder, a Roman historian, talked about the use of magnets in healing disorders of the eyes. Also during the first century A.D., a number of ancient Chinese geomancers began to document the subtle effects of the earth's magnetic field on human health and illness after using very sensitive compasses to monitor geomagnetic conditions. In the third century A.D., the great physician Galen recommended the use of magnets in treating constipation and a number of painful conditions. By the fourth century A.D., Marcel, the French philosopher and physician, recommended wearing a magnet around the neck to relieve headaches. In the sixth century, Alexander of Tralles used magnets for treating painful joints. Then, during the ninth century, Islamic physician Ibn Sina, also known as Avicenna, reported on how he was able to help treat depression through the use of magnetic therapy. Around A.D. 1000, a Persian physician documented the use of magnets to relieve disorders such as gout and muscle spasms.

Magnets were used for healing a number of different medical problems by various physicians and healers right until the sixteenth century, when the great physician Paracelsus not only advocated magnets for healing specific disorders but also detailed the different healing effects of the north and south magnetic polarities on living systems. Paracelsus was among the first to postulate that the earth itself was one great magnet. In his writings about magnetic therapy, Paracelsus commented that the "magnet is the king of all secrets." In 1777, the French Royal Society of Medicine examined the magnetic-healing studies of a

French abbot by the name of Le Noble. Their reports on the effects of magnetic treatments were so favorable, they concluded that the magnet seemed destined to play as large a role in future medical practice and theory as it was just beginning to play in the field of experimental physics. Interestingly enough, a few years later, the same Royal Society of Medicine condemned the work on "animal magnetism" done by Franz Anton Mesmer, who used "magnetic passes" over patients while employing the energy of "human magnetism" as opposed to magnetic lodestones.

Mesmer looked at magnetic healing via an astrological theory that saw the sun, the moon, and even the earth as possessing subtle magnetic energies that could influence the human nervous system and energize the body. Mesmer's theories were similar to those of Paracelsus before him. Both men postulated that a subtle, invisible magnetic fluid or force of nature was exchanged between heaven and earth, and that this magnetic force could heal and energize living things. During Mesmer's early work with magnetic healing, he attempted to heal a woman with a seizure disorder by placing pieces of magnetic lodestone over specific areas of her body. Mesmer hypothesized that the magnets would artificially reproduce the subtle magnetic influence of the earth's geomagnetic field upon her nervous system and somehow rebalance the irregular patterns he suspected were creating difficulties in her brain and nerves. Mesmer placed permanent magnets on the woman's abdomen and legs. As soon as he had placed them, the woman immediately began to describe a sensation of painful internal currents flowing from the lower part of her body down into her legs. After six hours the painful sensations disappeared, as did the woman's seizures, delirium, and fainting spells. Mesmer used both permanent magnets and certain hand gestures on the body. The gestures were known as "magnetic passes." Sometimes Mesmer's magnetic passes worked just as well for healing as did the permanent magnets.

Mesmer eventually became convinced that we humans possess a unique form of magnetism which he named "animal magnetism" to distinguish it from the iron-filings-attracting form of "ferromagnetism." He went on to develop techniques for capturing and utilizing this form of human subtle energy in healing patients. Mesmer often substituted his own animal magnetism in place of the energy of the permanent magnets his early work had depended upon. Although he was misunderstood during his own time, twentieth-century experimenters have recently found evidence that Mesmer might not have been far from the truth with his claims about animal magnetism. Studies in the last thirty years have demonstrated a remarkable similarity between the beneficial biological effects of a healer's hands and the therapeutic effects of permanent magnets on living systems. Such confirmatory research also suggests that part of the reason for Mesmer's therapeutic successes was that he might have actually been a powerful hands-on healer. Unfortunately, Mesmer's work, as intriguing and controversial as it was, ultimately did little to advance scientific interest in magnetic healing during his lifetime.

Just about eighty years later, the famous French chemist Louis Pasteur wrote about discoveries he had made concerning the effects of magnets on the process of fermentation. Pasteur was the same chemist who developed a treatment for rabies as well as the process of sterilizing milk (which still bears his name in the word "pasteurization"). He noted that placing a magnet near a fermentation vat containing fruit (as is used in the production of alcoholic beverages) seemed to speed up the fermentation process. During the same time period, Samuel Hahnemann, the developer of homeopathy, also experimented with magnets for healing. Hahnemann eventually advocated the use of magnets for treating a variety of health disorders. While the use of magnets for healing began to grow throughout Europe during Hahnemann's era, interest in magnetic therapies began to rise in popularity in the United States as well. European and American businessmen soon began patenting magnetic-healing devices, seeing a rich market to be tapped. The enterprising fellows often made extraordinary healing claims for their products. Of note were ads in the post–Civil War–era Sears Roebuck catalog, which advertised a wide range of magnetic jewelry and apparel, purportedly able to heal ailments ranging from menstrual cramps and sexual impotence to baldness.

Magnets continued to enjoy popularity even into the 1920s, at a time when magnetic clothing and similar items were said to cure almost any disease. The greatest magnetic-therapy advocate of them all was Dr. C. J. Thacher. Sometimes referred to as the "king of magnetic healers," Thacher, of the Chicago Magnet Company, promised health without the use of medicine by way of magnets placed upon the body, sewed into clothing, and even put in the soles of shoes. Thacher explained that the energy of life came from the magnetic force of the sun (prana?) and was conducted through the bloodstream due to its rich iron content. Nearly a century later, in 1954, Linus Pauling received the Nobel Prize in chemistry for his discoveries of the magnetic properties of iron-rich hemoglobin in the bloodstream. Pauling thus confirmed at least some aspects of Thacher's theories. While Thacher's claims for magnets might have been extravagant, his use of magnetic insoles was insightful in the light of recent research into the analgesic benefits of magnets for healing painful diabetic neuropathy. This notoriously difficult-to-treat diabetic nerve disorder causes its sufferers a great deal of foot and leg pain. Magnetic insoles placed in the shoes of neuropathy sufferers were found to significantly decrease or eliminate the disease-related extremity pains. But preliminary results have suggested that lasting pain relief comes only with continued use of the special magnetic shoe inserts.

By the mid-twentieth century, interest in magnetic healing was growing rapidly in countries like India, Russia, and Japan. India especially has had a number of physicians and researchers actively working in the field of magnetotherapy for at least the last thirty to forty years. In the late 1950s, research on magnetic therapies was also being published in Japan. While some magnetic products were promoted and used in the United States during that time period, the remarkable stateside resurgence in magnetic therapies has really just begun to grow over the last ten to twenty years. Even though magnets have been successfully used for treating a variety of health problems through the centuries, one must ultimately attempt to answer the question "Why do magnets even heal at all?" What is their kingly secret, hinted at hundreds of years ago by Paracelsus? The question of how magnetic healing works is not so easily answered, for there are a variety of competing theories that attempt to explain their effectiveness.

Magnetism and the Role of the Geomagnetic Field in Health and Illness

Before we can answer the question of why magnets heal, we must understand the relationship between living systems and magnetic fields in general. What exactly is the role of magnetic fields in health and illness? Are there bad as well as good magnetic fields, fields that can produce harmful as well as helpful effects on human health? Research, for instance, continues in the effort to try to determine whether living close to power transformers and the powerful magnetic fields they generate might actually increase the incidence of certain types of cancers. While this question has not been definitively answered, there are indications that certain types of man-made pulsed magnetic fields may do so. A growing awareness of the subtle biological effects of magnetic and electromagnetic fields has led to the redesign of computer monitors to reduce the amount of electromagnetic radiation they emit. There are a variety of different types and characters of magnetism that exist in the spectrum of natural and man-made magnetic fields. Each of these types of magnetic fields has the potential to overtly or silently influence human biology. It seems that north- versus south-pole fields, static versus pulsed magnetic fields, weak versus strong magnetic fields, and slow versus fast magnetic fields all produce different biological effects upon individual cells as well as upon whole living organisms.

To understand the critical role that magnetic fields play in human health, we must look again at human beings from the vibrational-medicine model of the multidimensional human being. Start with the premise that the body's auric field is a composite of biomagnetic energies generated by the cellular activity of the physical body, as well as by the subtle magnetic energies related to chakra activity and from our spiritual bodies' energy fields. Add to this natural orchestra of biomagnetic and subtle magnetic energies a cacophony of electromagnetic pollution and the background magnetic-energy field of the earth itself. Soon the theory that magnetic fields influence the human body gets complicated.

It has only recently become apparent that we as human beings, as well as all earthly life forms, are somehow energetically attuned to the weak background magnetic field of our planet. We now know this because science has shown that if we deprive our body of this background geomagnetic energy, ill health can eventually develop. If we are isolated from the earth's geomagnetic field for long periods (either by living in magnetically shielded structures or through prolonged space flight), our bodies can develop fatigue, aches and pains, and a variety of other symptoms. It was only some thirty-odd years ago that NASA began to send humans into space, an environment extremely low in gravity and earth magnetism. Only after the early longer space flights did space scientists become aware of the astronauts' dependency upon not only gravity but also the earth's magnetic field for maintaining optimal health. During those early space missions, NASA scientists observed a new, previously unreported illness experienced by a number of the first astronauts. This illness came to be known as "space sickness." Eventually, a decreased exposure to the earth's magnetic field was found to be responsible for space sickness. But either by placing small magnets in space suits or by incorporating certain types of weak magnetic material into space suits and living quarters, space sickness could be prevented. Although it is not widely known, our NASA space shuttles have magnetic-field generators on board. These are the so-called earth-field generators that reproduce the slowly oscillating magnetic fields of the earth's normal 7.8 cycles per second. The earthfield generators act to simulate the effect of the geomagnetic field so as to keep the astronauts healthy during long space flights. The Russian space program has also had to include magnetic devices in their spacecraft, following observations made by Russian physicians that cosmonauts who spent nearly a year in space on long missions had lost nearly 80 percent of their bone density. The bone-density loss was a kind of premature osteoporosis caused by a lack of gravitational and geomagnetic-field exposure. While the Russians have further investigated magnetic fields for their potential healing influence, NASA's studies of the beneficial influence of weak magnetic fields upon human health were not continued with regard to finding any potential implications for magnetotherapy as a form of medical treatment.

If we go back to the late 1950s and early 1960s, while American space efforts were still in their infancy, a number of Japanese researchers began intensively studying the effects of weak magnetic fields on human health and illness. The Japanese rapidly came to conclusions similar to NASA's research results about human beings' dependence upon exposure to the magnetic influence of the earth's geomagnetic field. The Japanese research, led by Dr. Kyoichi Nakagawa, suggested that certain forms of illness that had begun to appear in Japanese city dwellers in the late 1950s (characterized by symptoms ranging from weakness, malaise, chronic fatigue, and insomnia) were triggered in part by what was called "magnetic-field deficiency syndrome" (or MFDS). Nakagawa and fellow researchers discovered that the iron and steel girders commonly used to build the substructures of large modern buildings tended to keep the flow of the earth's background magnetic field from penetrating the interior of certain buildings. Thus, people who spent a lot of time in iron- and steel-based structures did not get the same daily exposure to the natural geomagnetic field as people who lived in houses constructed of wood and other natural, organic building materials. Japanese researchers who studied the biological effects of the earth's magnetic field found it to be an important but silent ingredient for maintaining good health. Nakagawa's research suggests that if people artificially shield themselves from the natural geomagnetic field over long periods of time, a "magnetic-field deficiency" can result that will eventually produce negative health consequences for field-deprived individuals. It seems that just as it's crucial for us to receive a certain minimum exposure to sunlight for maintaining optimal health, so it may be equally important for each of us to maintain a certain daily nonshielded exposure to the magnetic field of the earth itself. Interestingly enough, this same research team also discovered that the application of magnetic fields to the bodies of magnetically deficient patients, through various magnetic appliances or magnetic pads, seemed to provide relief from the varied symptoms of magnetic-field deficiency.

You may ask why we need a daily exposure to the earth's energy for optimal health or, more specifically, to the geomagnetic field? A discussion about carrier pigeons and rocks will help to explain. Consider, if you will, that we all live in an energy cocoon generated by the earth's magnetic field, the so-called magnetosphere. This big cocoon influences many living and nonliving components of the planet. Geologists believe that the magnetosphere is generated by molten iron currents constantly flowing at the heated core of our planet. When molten rock or magma flows up through volcanoes and cools down to become hardened rock, the volcanic rock becomes permanently imprinted with the weak magnetic energy of the geomagnetic field. All living organisms, from single-celled amoebas to multicellular human beings, also become imprinted by and attuned to this same geomagnetic field. Like the cooling magma that captures its magnetic orientation within the planetary field, certain animals, such as carrier pigeons, become magnetically imprinted by the highly specific geomagnetic field patterns of the place where they are born. By following an innate magnetic sense that allows them to orient themselves magnetically in space, carrier pigeons can follow the earth's invisible lines of magnetic force to find their way home unerringly from distant locations. This was confirmed when biologists found they could disorient carrier pigeons by magnetically interfering with their natural homing sense. After placing a tiny battery-powered magnetic coil over pigeons' heads (and a small blindfold over their eyes to prevent them from using visual cues), a magnetic interference was created that disrupted the birds' natural geomagnetic homing sense. Sure enough, the magnetic interference caused the pigeons to veer from their usual homeward-bound course.

The pigeons' magnetic sense appears to be related to the presence of small amounts of magnetite, a naturally magnetic mineral, within the brains of the birds. Recent research has confirmed that even human beings possess a subtle magnetic sense. More specifically, we each have tiny deposits of magnetite both in our brains and in our adrenal glands. Furthermore, it is possible that the subtle magnetic sense possessed by humans is the basis for dowsing skills. Experimenters have successfully blocked the dowsing skills of master dowsers by covering their heads and adrenal regions with magnetically shielding material. This suggests that there may indeed be some relationship between dowsing and the human ability to sense weak magnetic fields.

The polarity and directionality of the earth's magnetic field appears to produce different influences upon cellular growth and metabolism. This is evidenced by an unusual plant-growth-enhancing effect attributed to a specific north-south geomagnetic orientation. One group of researchers experimented with rye seeds. The experimental trials showed that when rye seeds were planted so the long axes of the seeds were oriented north-south, the seeds sprouted more often and grew into taller plants. Repeated experimentation confirmed that seed placement, with orientation of the seeds' long axes along the north-south magnetic flow of the earth's field, tended to promote increased plant growth. The results showed plants can be affected by geomagnetic fields.

It should be noted that the intensity of the earth's magnetic field is not the same everywhere on the planet. It varies with the location, depending on the topography of the internal layers of the earth as well as on the presence of magnetic ore and mineral deposits that conduct or block the transmission of magnetism. The geomagnetic field, for example, is twice as strong over the United States and Russia as it is over Brazil. Also, field strength varies with time of day. It is always stronger on the night side of the planet, which, of course, faces away from the sun. An explanation of the field-strength variation involves the magnetosphere of the earth, which constantly interacts with the magnetic solar wind. During the daytime hours, when a landmass on the earth's globe is closer to the sun, the sun's magnetic influence compresses the magnetosphere so its field intensity is slightly weaker at the landmass. At night, when that same landmass is now rotated away from the sun, the magnetosphere expands farther outward into space, producing a slightly greater magnetic-field intensity at the same landmass. This geomagnetic-field variation associated with night and day may be one of the reasons shift workers find it more difficult to sleep during the daytime hours. Perhaps their bodies are used to "recharging" when the geomagnetic field is at a higher strength, i.e., during the night. And maybe the daylight sleep time doesn't provide enough geomagnetic-high-field recharge time.

The geomagnetic field also pulsates with a peak frequency of oscillation at about 7.8 cycles per second, a figure sometimes known as the Schumann resonance of geomagnetic activity. Some researchers have found that there are even micropulsations upon this background geomagnetic oscillation, with frequencies of pulsation ranging from less than 1 cycle per second all the way up to 25 cycles per second. By studying the earth's geomagnetic imprint on different types and ages of rocks around the planet, geologists have made an interesting discovery. The earth's magnetic field seems to have a kind of slow energy metabolism that causes the geomagnetic field to build in strength and then diminish in intensity in cycles of five hundred thousand years. Current magnetometer assessments of magnetically imprinted rock layers have revealed a decrease by about 90 percent in the strength of this magnetic field. The field strength is down from a figure of 4 gauss (a unit of magnetic measurement) to about 0.4 to 0.5 gauss. Since the earth's magnetic field is already weaker than it was a few thousand years ago, and since all living things are dependent upon daily exposure to the geomagnetic field, it looks as if anything that blocks or further weakens earth magnetism could be potentially deleterious in its effects upon human health. Conversely, anything that can help maintain an appropriate level of exposure to earth's geomagnetic field should be advantageous to humans.

We have already mentioned the magnetism-blocking effect of apartment buildings constructed using large amounts of iron and steel. Living in such magnetically shielded structures tends to diminish an apartment dweller's exposure to the geomagnetic field, possibly contributing to certain forms of ill health over time. While some building materials can block geomagnetic fields, other materials have the ability to enhance and focus this background earth energy in beneficial ways. Just as feng shui, the ancient Chinese art of architectural design and landscaping, follows certain principles of design and object placement that promote good ch'i-energy flow, there are also ancient geomagnetic principles of building that can selectively focus and enhance the flow of natural earth energies. Ch'i is described as a form of naturally occurring (but difficult to measure) "subtle" magnetic energy that encourages life processes, as opposed to geomagnetic energy, its ferromagnetic cousin, which can be easily measured using sensitive magnetic detectors.

The geomagnetic principles of architectural design are based in part on constructing buildings that incorporate certain unique geometric patterns and building materials with different geomagnetism-conducting properties. Within nature there are certain nonmetallic materials, such as specific types of rocks and minerals, that demonstrate the peculiar magnetic properties of diamagnetism and paramagnetism. Diamagnetism refers to the tendency of a substance to be repelled by a strong magnetic field. Conversely, paramagnetism refers to the tendency of different materials to become easily attracted to magnetic fields. For example, granite, clay, and sandstone are common building materials possessing varying degrees of paramagnetism. These materials display magnetic properties markedly different from those of iron or cobalt. Not only are iron and cobalt attracted to magnetic fields, they can also hold on to a magnetic charge, becoming magnetized themselves. For example, a steel pin that is easily attracted to a small magnet can become magnetized merely by being stroked against the magnet several times. In contrast to iron or steel, paramagnetic materials are only weakly attracted by magnetic fields, and they will not become magnetized themselves. The question becomes, can paramagnetism play a role in the way geomagnetic energies influence living systems? The answer to that question appears to be yes. It appears that different materials with varying paramagnetic qualities, ranging from the soil covering the surrounding land to the stone blocks used to construct buildings, have the potential to influence plant and animal life positively and negatively because of their ability to block, enhance, or even focus the magnetic energy of the earth's field in both potentially beneficial and harmful ways. While the awareness of geomagnetic effects upon human and plant life seems to be relatively recent to modern science, the knowledge of how to work consciously with natural magnetic (and other subtle earth energies) may actually be quite old.

Geomagnetic design principles appear to have been part of the secret knowledge possessed by a number of ancient architects who designed and built many sacred monuments and buildings throughout the world, ranging from the pyramids and obelisks of Egypt to the famous Round Towers of Ireland. Recently we have begun to comprehend that some of our so-called primitive forebears might have possessed more knowledge about the constructive use of geomagnetism than we might previously have suspected. The key to creating structures that geomagnetically enhance certain life processes is in focusing the local geomagnetic field in highly specific and beneficial ways. It appears that certain geometric forms, such as the obelisk, the pyramid, and other specific geometric shapes, somehow resonate with and focus the magnetic energy of the planet, almost like an antenna or a very large capacitor (a holder of energy charge). For instance, buildings constructed of paramagnetic materials that have been patterned after certain geometric shapes (such as the Round Towers of Ireland) have been found to energetically enhance the growth of surrounding plant life. In other words, the towers might not have served merely a housing function but might also have been natural geomagnetic and cosmic-energy antennae designed and constructed to enhance the growth of surrounding crops and fields. And perhaps that subtle geomagnetic influence even affected the consciousness of the building dwellers themselves.

Dr. Phillip Callahan, a professor of entomology (the biology of insects) at the University of Florida, has worked as a consultant for the U.S. Department of Agriculture. Because of his insect studies, Callahan has come to appreciate the role geometry plays in attuning to and focusing certain types of electromagnetic energies. Callahan has done much research on the geometric shape of the antennae and sensor arrays used by moths in sensing different frequencies of electromagnetic radiation. Callahan's appreciation of the insect world's use of particular geometric forms to naturally resonate with and focus specific types of energy led him to wonder about how similar geometric patterns, consciously incorporated into specific sacred monuments by geomagnetically knowledgeable ancient builders, might have somehow endowed such structures with certain unique energetic properties. Callahan sought to confirm experimentally whether certain forms of geometry combined with paramagnetic building materials could somehow influence living systems beneficially. Callahan built a small-scale model of one of the famous Round Towers of Ireland. Around the actual towers, plant growth is especially lush and fertile. The Round Towers themselves are tall, circular, columnlike buildings capped by a cone. Since the original towers were made from weakly paramagnetic materials such as sandstone and clay, Callahan constructed his scale model out of thick paper impregnated with the weakly paramagnetic material carborundum (a component of sandpaper used for abrasive cleaning).

Callahan planted seeds in soil at various locations and distances from the Round Tower model, which had also been placed firmly into the same soil. He discovered there was an increased sprouting and growth of plant seeds on the north-facing side of his Round Tower model. Callahan attributed this enhanced growth to what seems to many an unlikely source. Specifically, his experiments with the carborundum Round Tower model showed that it functioned as a radiofrequency and magnetic-energy antenna and possibly as a geomagnetic-energy "accumulator" that could intensify and radiate magnetic energy outward in a highly specific pattern. Callahan believed the plants thrived because of the geomagnetic energy focused by the tower model. Callahan's association of enhanced plant growth with magnetic fields has also been confirmed in a slightly different way by other researchers. More precisely, water that has been exposed to a magnetic field appears to increase plant growth. Now, just suppose that the intensified geomagnetic energy radiating outward from the tower in Callahan's experiment was then picked up by water in the nearby soil and relayed to the plant life. Here again would be even more exposure to magnetic fields for Callahan's thriving tower plants! One certainly begins to wonder if it is possible that the original builders of the Round Towers had this geomagnetic, plant-growth-enhancing effect in mind when they constructed them. In other words, did these early builders have knowledge of the principles of so-called sacred geometry to help in focusing earth energies in ways that would enhance the growth of nearby crops?

Aside from the possible magnetic influence of the towers upon nearby plant life, Callahan speculates that the towers might even have had an effect upon the humans who inhabited these buildings. The towers seem to have been used by various religious orders at one time. So isn't it feasible that perhaps priests and holy men prayed or meditated at or near windows in the towers? This is not just some whimsical thought by any means. You see, Callahan found that in his model tower, there were different levels of highly focused magnetic activity corresponding with window locations in the original Round Towers! Could such a magnetic enhancement in those locations possibly have been intended for their ability to focus geomagnetic energy purposefully to induce certain altered states of consciousness, such as deeper meditation or perhaps even visionary states? The evidence, albeit tentative, considered in conjunction with other geobiology research, suggests that what we had previously thought of as primitive cultures might actually have possessed a knowledge of tapping into Earth energies in ways we are only now beginning to rediscover and understand.

Since it may be possible to use earth magnetism and geometric principles for positively focusing geomagnetic energy in life-enhancing ways, might there not be other unsuspected geomagnetic effects caused by our ignorance of these principles in modern architecture and land management? You may here again recall Japanese research on city buildings that block geomagnetism. Dr. Callahan speculates even further about blocked geomagnetism in the cities where those steel buildings rise. He has put forth a theory that even the paramagnetic quality of the soil throughout the areas we live in may also block or enhance the amount of geomagnetism that reaches city dwellers. In order to confirm his theories, Callahan traveled around the world making measurements of the paramagnetic qualities of soil in various cities in different countries. His preliminary work suggests a number of remarkable conclusions. First of all, the places having soil of high paramagnetic quality seem also to be the most beneficial regions for plant growth. This may relate to the theory that the more paramagnetic the soil is, the better it transmits and distributes the geomagnetic field to the crops planted in it. Second—and this is still in the realm of speculation—Callahan found that there were higher levels of violent crimes and civil unrest in areas where the soil had poorly paramagnetic qualities. In other words, it almost seems as if a diminished penetration of the earth's magnetic field into certain geographic regions might actually produce mental and behavioral disturbances in susceptible individuals. As Callahan speculated that certain forms of focused geomagnetism (as in the Round Towers) possibly enhanced the meditation of holy men, he also began to speculate upon how other alterations in the geomagnetic field might just as easily induce anxious or agitated states of consciousness. There are certain areas of geomagnetic-field research that support the notion that human behavior and even states of illness may somehow be connected to alterations in the local geomagnetic field. Historically speaking, it is interesting to note that both Paracelsus and Mesmer claimed that the sun, the moon, and even the stars exerted magnetic influences upon the subtle magnetic field of the earth, as well as upon human beings. We now know solar magnetic activity affects the geomagnetic field, and the moon, especially the full moon, reflects more of this solar magnetic activity to the earth's field. People have long speculated about the influences of the sun and the moon upon human health. Aside from observing the influences of the visible spectrum of light upon vitamin-D synthesis, most modern scientists have never taken seriously the idea that subtle magnetic influences from the earth and heavens might produce such effects. On the other hand, a few researchers have actively sought out evidence supporting the idea of a possible subtle magnetic connection to health and illness.

For example, French researchers have previously shown a strong correlation between an increased incidence of heart attacks and periods of strong solar-flare activity. Heavy solar activity, such as the solar-energy bursts associated with sunspots, frequently produces changes in the earth's magnetic field or magnetosphere. European geobiology researchers have confirmed that sunspots and solar flares, both connected with unusually intense solar magnetic activity, can actually influence blood clotting, and even the whole cardiovascular system. It is thought that sunspots do this through the secondary effects of the magnetic solar wind on earth's geomagnetic field. In 1959, Dr. Marcel Poumailloux, a French physician, and Dr. R. Viart, a meteorologist, announced that they had found a striking correlation between increases in heart failure, solar activity peaks, and periods of agitated geomagnetic activity. The French researchers found six specific periods between 1957 and 1958 (one of the years of greatest solar activity ever recorded) when heart attacks rose exactly at times of peak solar activity. They also recorded a drop in hospital admissions for heart failure during a four-month period of relatively calm solar and geomagnetic activity. Both researchers pointed to an increased tendency toward blood-clot formation during times of maximal solar and earth magnetic activity. Could the more intense magnetic fields have produced some type of increased coagulability of blood due to blood's iron-rich hemoglobin? Such an increase in blood coagulability or blood viscosity could certainly produce health problems in people with preexisting vascular problems, such as coronary artery disease.

Is heightened geomagnetic and solar activity an accurate predictor of the timing of heart attacks and heart failure? While few American cardiac researchers have pursued this line of questioning, one early study, conducted in three Dallas hospitals between 1946 and 1951, found an increased frequency of heart attacks during periods of sudden weather changes, (which might have been indirectly related to heightened geomagnetic and solar magnetic activity during those same time periods). Epidemiologic studies carried out in Russia and Eastern Europe as far back as the late 1950s and early 1960s have con-

The Many Dimensions of Magnetism: A Glossary of Terms and Energy Effects

CATEGORY OF MAGNETISM	DEFINITION	ASSOCIATED ENERGY EFFECTS
Ferromagnetism	**Iron-filing-type magnetic effects associated with permanent magnets**	**May enhance healing of certain health problems** **North-pole versus south-pole bioeffects**
Electromagnetism	**Magnetic fields produced by electron flow/electrical currents**	**Has both positive and negative health effects** **Largely frequency-dependent in its bioeffects** **Can entrain bioelectric rhythms in brain and heart**
Biomagnetism	**Weak magnetic fields generated by ionic currents and cellular activity**	**Human DNA oscillates in gigahertz range of biomagnetic activity** **Heart and brain magnetic rhythms can reveal disease patterns**
Animal Magnetism	**Subtle magnetic currents of ch'i and prana**	**Life force is subtle magnetic in nature** **Possibly etheric level of energy**
Subtle Magnetism	**Subtle magnetic energies of chakras, etheric, astral, mental bodies, and auric field**	**Thoughtforms follow "subtle" law of magnetic attraction (like attracts like)**
Paramagnetism	**Weakly attracted to strong magnetic fields**	**Highly paramagnetic soil may enhance plant growth by allowing greater penetration of geomagnetic field**
Diamagnetism	**Weakly repelled by strong magnetic fields**	**Biological effects unknown**
Geomagnetism	**Earth-generated magnetic fields** **Key frequency (Schumann resonance) = 7.8Hz cycles/second**	**May act as carrier energy for spiritual healing** **All life is dependent upon minimal level of geomagnetic energy to maintain health** **Steel and concrete buildings may shield out geomagnetic-field—magnetic-field deficiency syndrome (MFDS)** **Abnormal geomagnetic energy gradients may be associated with "geopathic stress"**
Solar Magnetism	**Conventional and subtle magnetic fields generated by solar activity. Includes gigahertz frequencies and solar prana**	**Solar magnetism affects geomagnetic field** **Solar flares and solar magnetic storms associated with increased heart attacks and heart failure** **All life is dependent upon both sunlight, (gigahertz energy utilization by DNA?) and solar prana**
Cosmic and Stellar Magnetism	**Subtle magnetic currents emanating from stars, galaxies, and celestial bodies**	**Astral and higher forms of subtle magnetism may influence astral body's chakras** **Possible energetic basis for astrology**

firmed the French findings of strong correlations between high solar magnetic and geomagnetic activity and an increased incidence of certain heart and blood disorders.[1]

Earlier we touched on Dr. Callahan's speculations about the connection between geomagnetic fields and human behavior. The idea that magnetic changes in the earth's field produced significant behavioral changes in certain susceptible individuals was also researched by Dr. Robert Becker. Dr. Becker, co-author of *The Body Electric* and pioneer of electromagnetic medicine, examined the medical charts for psychiatric admissions to a number of state hospitals. He looked for a pattern that would confirm his suspicion that there might be an increased level of schizophrenic and psychotic behavior associated with peaks of abnormal geomagnetic activity. Working with psychiatrist Dr. Howard Friedman, Becker analyzed the number of psychiatric admissions as well as the behavioral activity of inpatient schizophrenics to see if there was any association between increased psychotic behavior and periods when there were active magnetic storms, a phenomenon frequently associated with abnormal geomagnetic patterns. Becker found that there were indeed more admissions to psychiatric wards and more psychotic behavior among inpatient schizophrenics and manic-depressives during such periods. Later Becker recognized that this ability of magnetic fields to alter activity within the nervous system might translate into potential therapeutic benefits. In his own experimental work, Becker was eventually able to induce complete anesthesia in a salamander using only magnetic fields. Interestingly, the magnetically anesthetized salamanders usually remained "groggy" for up to a half hour after the magnetic field had been turned off, almost as if they had been given potent anesthetic drugs!

Not only can abnormal geomagnetic activity produce behavioral changes and pathological alterations in cardiovascular function, it may even contribute to certain degenerative and inflammatory diseases, such as arthritis and cancer. Earlier we discussed a phenomenon known as geopathic stress, a kind of "bad" earth energy associated with an increased incidence of cancer, certain forms of arthritis, and a variety of other health problems. One important correlate of geopathic stress in the home is an abnormal magnetic-field pattern in the area of a person's sleeping quarters. We have talked about how different building materials such as iron and steel, can block the geomagnetic energies from penetrating fully. But even homes made of natural organic materials can be affected by the abnormal magnetic energy fields associated with geopathic stress.

Ludger Meersman, a Florida geobiology researcher, uses sensitive magnetometer equipment to map out the magnetic-energy levels directly over beds and in the sleeping quarters of individuals who have been suspected of living in geopathic-stress zones. He noted that the beds of most healthy individuals often have a homogeneous magnetic-field intensity across the entire bed. However, in the bedrooms of sick individuals who were thought to live

The Many Varieties of Geomagnetic and Geopathic Stress

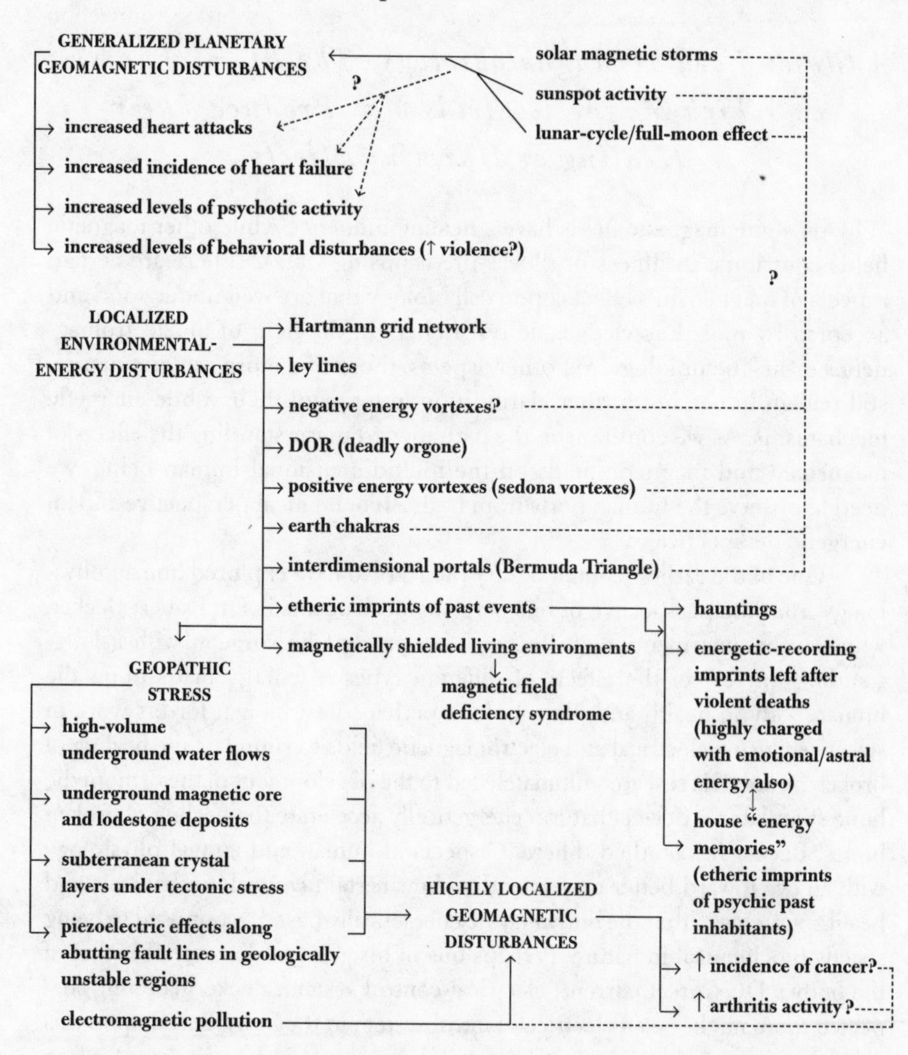

in geopathic-stress zones, Meersman discovered that there were small areas of high- and low-intensity geomagnetic fields distributed along the length of the beds. Often the region of the bed with the most anomalous magnetic readings corresponded to the part of the patient's body afflicted with arthritis or cancer. These anomalous areas showed abnormal magnetic-field gradients, with zones of high to low intensity distributed over a very small space. Given the theory that low geomagnetic-field intensity (as in the case of magnetically shielded apartments) or even abnormal geomagnetic-field gradients (as in the case of geopathic stress) can be associated with illness, it seems logical that the application of magnetic fields might have therapeutic health benefits.

A Closer Look at Magnetobiology: The Biomechanisms by Which Magnetic Fields May Produce Their Healing or Harmful Effects

Why do some magnetic fields have a healing influence while other magnetic fields contribute to illness or illness-predisposing states? There are certain aspects of magnetism's effects upon cell biology that are well understood and accepted by most knowledgeable researchers in the field of bioelectromagnetics or magnetobiology. Yet other aspects, those of a more esoteric nature, still remain in need of further clarification in terms of their subtle-energetic mechanisms. As we continue on the path toward understanding the effects of magnetism and magnetic fields on the multidimensional human being, we need to observe the human body from both a biochemical perspective and an energetic perspective.

One of the earliest contemporary pioneers to have explored human physiology from the perspective of bioenergetic mechanisms is Dr. Robert Becker, whom we've just mentioned. Becker is an orthopedic surgeon with a longstanding interest in the effects of different types of energy fields upon the human body in health and illness. He is perhaps best known for his work in selectively using electrical and electromagnetic fields to stimulate the healing of broken bones. His research ultimately led to the development of the orthopedic bone stimulator, a device that can energetically accelerate the healing of broken bones. Becker has studied different aspects of human and animal physiology with an eye toward better understanding how certain control mechanisms and healing systems within the body may be energetically based as opposed to being strictly biochemical in nature. Perhaps one of his greatest discoveries is that of the body's DC (direct current) electrical-control system. Becker had long suspected there might be an electrical-control system in the body. Based upon this suspicion, he was able to use weak electrical currents to induce an amputee frog to grow an entirely new arm and hand! Also, Becker had read with interest the work of another earlier bioenergetic pioneer, Harold S. Burr. Burr was a neuroanatomist at Yale University who had mapped out electrical-field patterns around various living organisms. The energy fields he mapped came to be called L-fields (or life fields). L-fields seemed to energetically direct growth and development in plants and animals. Following his own studies in the area of tissue regeneration and bone healing and the clues left by Dr. Burr, Becker began to postulate the existence of a unique electrical-control system within the body that played a role not only in healing and repair of cellular damage but also in the regulation of states of consciousness.

Becker's work led him to discover that anesthetized humans showed a marked alteration in the DC current around their heads. This was especially true in test subjects who had been hypnotically conditioned to temporarily feel no pain through hypnotic anesthesia. But these hypnotically anesthetized subjects showed changes in their DC-current potentials similar to individuals given actual chemical anesthesia. This meant that an altered state of consciousness (hypnosis) could produce measurable electrical changes within the body. Becker had shown that there was a direct correlation between states of consciousness and alterations in the body's DC electrical-control system. According to Dr. Becker's hypothesis, the DC-current system may be an alternative pathway for sending electrical messages (including pain impulses) to the brain and local tissue-healing systems during times of tissue injury. One aspect of the DC current itself is that it may provide a local electrical stimulus that remains "turned on" until healing of local tissue damage is complete. Once healing of local tissue damage is complete, the electrically active cells return to normal, and the abnormal DC signal is "turned off." But aside from the importance of DC currents in local tissue healing caused by injury, these currents seem also to be part of a major bioinformation network that strongly influences both body and mind. One important carrier of the DC current signal appears to be a network of nerve-support cells known as glial or perineural cells. Prior to Becker's work, it was never suspected that the glial cells might have a purpose in the transmission of bioinformation throughout the body.

Becker came to discover that the DC-current system tended to transmit information by slowly varying the electrical charge or voltage of the membranes of glial cells. He also noted that this slow voltage change in the glial cells could be influenced by the application of external energy fields, especially magnetic fields. Becker was eventually able to make use of his newfound information by producing complete anesthesia in animals using only an externally applied magnetic field. The use of magnetic and electromagnetic fields to induce sleep is a phenomenon first studied in the 1960s by Russian researchers who developed some of the first electrosleep devices. During typical electrosleep, a device generates an oscillating magnetic or electromagnetic field that pulsates at very low frequencies. These slow frequencies are in the same frequency range as the slow-wave electrical rhythms the human brain produces when an individual is deeply asleep. Electrosleep devices produce magnetic and electrical fields and currents that lull the brain into a rhythmic pattern normally associated with human sleep.

Most of the time, human cells oscillate and resonate with the background 7.8-cycle-per-second hum of the earth's magnetic field (which is located in the mid-alpha-theta brainwave range, a brainwave frequency normally associated with relaxed waking consciousness and creative reverie or daydreaming). Our cells are somehow attuned to this magnetic background rhythm. Perhaps the "earth hum" acts as some kind of an "inner timing" rhythm to the electrical-control systems of the body. Becker even noted that the DC-current system had its own oscillating circadian rhythm, which seemed to have been tied to planetary geomagnetic rhythms. In other words, the DC system seemed to be responsive to changes in the earth's magnetic field.

Unfortunately, in today's modern world we live in an ocean of different types of magnetic and electromagnetic fields that literally bathe our planet in radiowaves, 60-cycle-per-second electrical and magnetic fields, microwave energies, and a variety of other energy-field by-products of our new electronic technologies. These new background energies might possibly be providing confusing electromagnetic messages to the body's DC electrical-control system discovered by Dr. Becker. And perhaps the confusing electromagnetic messages could eventually lead to illness. Becker's DC control system seems to be part of a larger, whole-body electrical circulatory system postulated by Dr. Bjorn Nordenstrom. In Nordenstrom's model of electrical tissue healing, bioelectrical currents are thought to be generated not only by the DC control system but also by the flow of ions between damaged cells and across broken blood-vessel walls. The movement of positively and negatively charged ions in the body's tissues creates batterylike electrical effects in inflamed, infected, cancerous, or damaged tissues. According to Nordenstrom, these batterylike electrical effects produce localized electrical microcurrents that electrically attract repair cells and healing cells to exactly where they are most needed to carry out repairs to damaged bodily tissues. This phenomenon had been previously known to biologists as the so-called current of injury.

Dr. Becker had initially become interested in the DC current/field system and the current of injury via his studies of tissue regeneration in salamanders and frogs. Through his amphibian research, Becker was able to confirm something he had long suspected. Most cells of the body possessed unique semiconductor properties, the same properties as can be found in electronic and integrated circuits. One of the special capabilities of biological semiconductors, though, is that they can carry weak electrical currents over long distances througout the body, as well as within individual cells.

The finding of biological semiconductors was something predicted by the famous physiologist Dr. Albert Szent-Gyorgi. Becker had read Szent-Gyorgi's speculations on biological semiconductors and bioelectronics in his early years. The information inspired many of Becker's eventual discoveries into the energetic nature of human and animal biology. If indeed the human body possesses unique bioelectronic-control systems, this may provide one of the mechanisms by which magnets can influence and heal the body.

Of course, to achieve healing in specific medical conditions, it seems that certain types of magnetic-field delivery systems work better than others. If weak magnetic fields do influence the bioelectrical and bioelectronic control and communication systems of the body, then the magnetic influence could be either positive or even negative in nature, depending upon the characteristics of the magnetic field. As we shall soon discuss, the north polarity of the magnet seems to produce different bioenergetic and physiological changes in body tissues than does the the south pole. Also, certain frequenices of oscillating magnetic fields appear to accelerate tissue healing, while other frequencies appear to inhibit healing or may actually promote inappropriate cell growth. These are just some of the finer points of magnetic-field treatment that will become clearer as we examine specific applications of magnetotherapy. But first we'll discuss the mechanisms by which magnets actually cause physiological changes in the cells and tissues of the body.

Consider a magnet from its most basic level. Magnets are an everlasting source of an invisible, poorly understood force. This magnetic force is able to attract metals like iron and steel and can also affect the movement of electrically charged particles. Modern science has established that there exists a unique relationship between magnetic fields and charged particles like electrons. Electrons are, of course, the charged particles making up electricity. The very word "electromagnetism" seems to imply some kind of connection between the flow of electrons (i.e., electricity) and magnetic fields. One way in which this relationship can be seen happens when someone passes a metallic wire through the field of a permanent magnet. The magnetic field will cause the electrons in the wire to move, creating an electrical current in the wire. This is one of the principles by which electrical generators work. A second example of the electron-magnetism relationship occurs if you connect that same wire to the two poles of a battery. By doing this, you complete an electrical circuit so that current can flow through the wire. In turn, the flow of electrons produces a magnetic field around the wire. Magnetic fields affect not only electrons but also other electrically charged particles. Some of the greatest sources of electrically charged particles in the human body are the various chemical ions like sodium, potassium, calcium, and magnesium. These ions help to maintain the appropriate electrical charge within nerve-cell membranes. The movement of ions also produces bioelectrical currents throughout the nervous system, as well as within the cells and tissues of the body. The cells of the body maintain their electrical charge via a delicate balance of high potassium and low sodium within cells and a high sodium with low potassium in the surrounding fluids that bathe the cells. There are tiny sodium-potassium pumps within all cell membranes working constantly to maintain the critical balance of sodium to potassium inside and outside cells. Medical conditions that cause a person's sodium or potassium blood levels to become too low can upset this delicate balance and interfere with nerve- and muscle-cell function by inhibiting the natural bioelectrical equilibrium. But other forces, specifically magnetic forces, can also change this delicate biolectrical equilibrium in both therapeutic and deleterious ways. If the normal flow of ions across cell membranes becomes altered even slightly (as occurs with exposure to strong magnetic fields), then nerve cell function, message transmissions throughout the brain and nervous system, and even the DC-current control system can all be adversely affected.

To illustrate, remember Dr. Becker's amphibian research? Well, the doctor discovered during the course of his work that he was able to use a powerful magnetic field to induce complete anesthesia in salamanders. The magnetic field produced simultaneous bioelectrical changes in both the amphibian's nervous system and its DC field system. Becker induced a slower brainwave activity in the salamander's EEG, similar to what was normally seen with chemical anesthesia. The explanation for the magnet's effects in inducing anesthesia was that the magnetic field produced a change in the flow of ions across the cell membranes of nerves within the brain and spinal cord, as well as bioelectrical changes within the DC control system. These magnetically induced changes resulted in a reduction of pain sensitivity throughout the entire nervous system and an altered state of consciousness in the salamander. This ability of magnetic fields to interfere with the flow of electrons moving through a semiconductor (in this case, the biological semiconductor of nerve- and glial-cell membranes) is known in physics as the Hall effect. Becker saw that this magnetic anesthetizing effect occurred only when the externally applied magnetic field was at least 3,000 gauss (a magnetic intensity about 6,000 times stronger than the earth's magnetic field). If the field strength was reduced below 3,000 gauss, the salamander immediately began to wake up.

The capacity of magnetic fields to alter nerve cells and DC bioelectrical current flow to produce anesthesia and decreased pain sensitivity may provide an important clue as to why magnets can relieve different kinds of pain. Here, too, there is an explanation for why continued magnetic applications are sometimes needed to maintain relief of chronic pain, as when magnets are used to relieve chronic low-back pain or the foot pain of diabetic neuropathy. Of course, to achieve pain relief, you do not want to use magnets to produce general anesthesia and complete sleep, as in Becker's salamander subjects. It is necessary only to magnetically affect the flow of local biocurrents in the painful area of the body.

Becker worked mainly with static or nonoscillating magnetic fields in his experiments. Oscillating magnetic fields produce a variety of similar as well as different biological effects, compared with static or nonoscillating fields. One important magnetic field that is always oscillating is the geomagnetic field of the earth that bathes us in magnetic energy twenty-four hours a day. Perhaps the oscillation of the earth's magnetic field provides some kind of silent magnetic

The Hall Effect: How Magnetic Fields May Affect Biological Currents

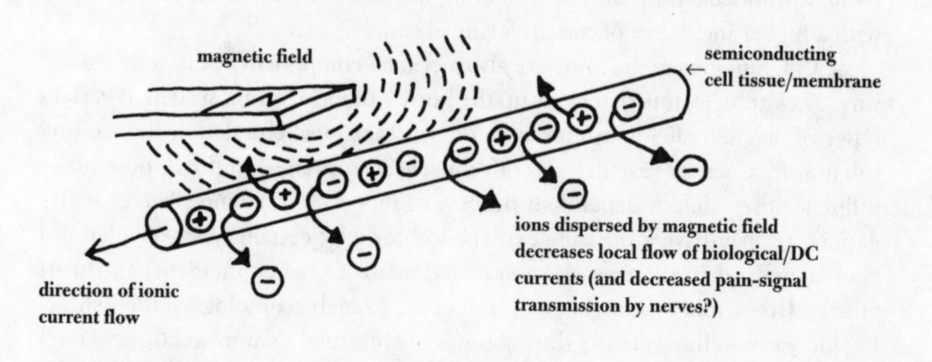

timing signal necessary to maintain optimal health. Furthermore, some have suggested that the earth's magnetic field causes the ions of the body to vibrate and move more energetically, almost as if they had a greater "atomic zing." The unique frequency of the earth's magnetic field, the 7.8-cycle-per-second background rhythm, seems to be a frequency at which all living cells like to vibrate. On the other hand, other frequencies of magnetic fields besides the 7.8-Hz (7.8 cycles per second) geomagnetic signal can influence the body in both healthy and unhealthy ways.

One focal point of magnetic influence may be the movement of certain ions across cell membranes. The calcium ion is unique in its function as an important "second messenger" within many of the body's cells. The movement of calcium ions across cell membranes is used to transmit different types of signals that may turn on or off a variety of critical cellular functions, depending upon which type of cell the calcium is moving into or out of. For instance, in smooth-muscle cells (the unique muscle cells that wrap around many of the blood vessels of the body), calcium movement into the cell may produce vasoconstriction of blood vessels and reduced localized blood flow. Alternatively, in blood platelets, this same calcium flow across platelet membranes can lead to platelet aggregation, contributing to blood-clot formation and platelet "clumps" that may block the flow of blood through small blood vessels. Some bioelectromagnetic researchers have reported that a pulsating magnetic field oscillating at 60 cycles per second (the frequency of alternating current that flows through the U.S. power grid) may influence the flow of calcium ions across cell membranes in ways that may ultimately be harmful to the body. Specifically, magnetic fields pulsating at the frequency of typical house current not only affected calcium flux, they also seemed to enhance the growth rate of existing tumor cells, depending upon the strength of the magnetic-field exposure. This negative influence of 60-Hz pulsed magnetic fields on calcium movement in the body could be one of the reasons high-voltage power transformers (which produce strong 60-Hz oscillating magnetic fields) might be associated with a higher incidence of certain forms of cancer.

Calcium ions in the body are also a critical component in the activation of various clotting factors that make up the body's blood-clotting system. If certain types of magnetic fields can increase or decrease local circulation (by altering calcium flux across vascular smooth-muscle membranes) and can potentially influence the calcium-dependent process of blood clotting and platelet aggregation, we may have a possible explanation for why certain types of solar and geomagnetic disturbances are associated with a greater incidence of heart attacks. (Remember the earlier mention of the French geobiology studies showing increases in heart attacks during times of abnormal solar magnetic activity?) A heart attack is a medical condition in which heart-muscle cells are literally starved to death because of a critical blockage in one of the coronary arteries that feeds oxygen to the heart muscle itself. Most often the blockage is due to thrombosis or blood-clot formation in a narrowed or constricted coronary artery. One possible energetic mechanism behind the increase in heart attacks during heavy sunspot and solar magnetic activity might be a magnetic-field effect that produces an imbalanced flow of calcium ions across the cell membranes of blood platelets and vascular smooth-muscle cells. This critical combination could lead to a greater tendency toward coronary artery spasm and blood-clot formation (coronary thrombosis), along with increased platelet aggregation, causing a complete blockage in already narrowed coronary arteries of susceptible individuals. Of course, this hypothesis remains a matter of speculation until further scientific research can elaborate upon the actual mechanism behind the unusual, little-known association between heart attacks and abnormal solar and geomagnetic activity.

The two phenomena—the increased tumor-cell growth in 60-Hz magnetic fields and increased heart attacks at times of major geomagnetic aberrations—illustrate, biologically speaking, the harmful side of a two-edged sword. Yet, magnets can also heal. A greater knowledge of magnetobiology will ultimately be our guide to either wielding magnets as a healing tool or to allowing ourselves to be unwittingly sliced by the harmful edge of the magnetic sword.

We definitely need to better understand the nature of magnetobiology, not only to explore the untapped healing potentials of magnets but also to keep from accidentally making ourselves sick with the unintended magnetic and electromagnetic fields produced by our wondrous new electronic technologies. Also, a greater understanding of solar and geomagnetic biological effects might eventually lead to the development of an early-geomagnetic-warning system that could alert physicians about critical time periods when heart-disease patients would be at greater risk of developing heart attacks. Perhaps a greater understanding of the subtle magnetic forces at work will even lead to new ways of preventing heart attacks through using magnetic-healing methods. In addition, a greater understanding of paramagnetism and its relationship to plant growth could eventually lead to new biomagnetic agricultural methods that would see farmers constructing certain geometric forms and paramagnetic structures to focus geomagnetic fields in an effort to signficantly increase the production of plant-based foods.

But just as positive benefits may result from being bathed in and entrained to the normal background magnetic earthfield rhythm, distortion of that rhythm through excessive sunspot activity, solar magnetic storms, living in geomagnetic-shielded housing, or even living over the abnormal magnetic-field gradients of geopathic-stress sites can all result in a host of negative health influences. And if the cells of our bodies do become disturbed because of distorted geomagnetic-field patterns, shouldn't it be theoretically possible to apply slightly stronger homogeneous, balanced magnetic fields (as in the form of various magnetic-healing appliances and devices) to restore a level of magnetic balance (as well as electrical balance) within the body's electrical-control systems? We'll now take a look at how feasible this theory is.

Let's start by getting back to the ability of magnetic fields to affect ionic flow across cell membranes. There is another unique phenomenon seen in magnet therapy that may also be explained by the ionic-flow effect. Magnet therapists have noted that placing magnets on the body seems to increase blood circulation. An increase in local circulation may be one of the mechanisms by which magnet therapies promote healing of bodily tissues and relief from various forms of chronic pain. An increase in the local circulation around and through a painful or inflamed body region brings more life-giving oxygen, as well as an increased count of various healing and repair cells. A number of clinical studies on magnetotherapy's effects have confirmed that magnets applied to the body do increase local circulation of blood. One reason for this increase may have to do with the flow of calcium ions across the membranes of smooth-muscle cells surrounding the arteries and smaller arterioles of the body. Normally, when calcium ions flow across the membranes of smooth-muscle cells, they send a signal that can trigger spasms and constriction of the arterioles, leading to a decrease in local blood flow. Abnormal geomagnetic and solar magnetic activity might actually enhance such spasms and blood-vessel constrictions, perhaps by increasing calcium flow into vascular smooth-muscle cells. But if one could inhibit that vascular constriction with a different type of magnetic field, it might be possible to use magnets to increase regional blood flow by altering calcium flux across the vascular muscle cells, but in the opposite direction. Since research has shown that certain types of magnetic fields can alter the natural flow of calcium ions across cell membranes, it seems likely that some magnets (or even artificially generated magnetic fields) may inhibit the flow of calcium ions into the smooth-muscle cells of blood vessel walls, causing the arterioles to relax and dilate, bringing increased blood flow to the nearby tissues.

A few magnetotherapy researchers have speculated that a magnet's ability to increase local blood circulation is due partly to the magnetic field's ability to attract iron-rich red blood cells to the region of the body where the magnet has been placed. Red blood cells are filled with hemoglobin molecules, of which iron is a key component. The iron atoms in hemoglobin are primarily responsible for hemoglobin's ability to carry oxygen to the tissues of the body. People take iron tablets when they are anemic because oral iron increases the body's production of iron-rich hemoglobin. Iron, as we know, is strongly attracted to and influenced by magnetic fields. So it would seem to make sense that placing a magnet on the body might attract iron-containing red blood cells to the area adjacent to the magnet. Critics of this theory of red blood cell/magnetic field attraction state that because the iron atoms in hemoglobin are bound to protein molecules, they do not behave the same way as aggregates of iron atoms do, as in the case of iron filings drawn to a piece of lodestone. Also, critics suggest, if magnets attracted iron-rich hemoglobin to a particular area on the body, it would cause a clogging of local blood vessels beneath the magnets due to aggregates of red blood cells behaving like blood clots. It turns out that there is a form of hemoglobin in red blood cells that would make them respond strongly to magnetic fields. That magnetically sensitive molecule is deoxygenated hemoglobin. Deoxygenated hemoglobin is the form that the hemoglobin molecule takes after its "oxygen cargo" has been "delivered" to the body's tissues and it is awaiting oxygen recharge in the lungs. Of interest to our magnetic discussion is the fact that the deoxygenated-hemoglobin molecule has an electron configuation that makes it more susceptible to the influence of magnetic fields. Despite criticism of the red blood cell/magnetic field attraction theory, confirmation of the idea that magnets might affect movement of red blood cells comes from a number of magnetobiology researchers who have actually observed microscopic changes in the movement of red cells on a glass slide when they were exposed to an external magnetic field.

In one case, magnet researcher Dr. Albert Roy Davis observed red blood cells under a microscope before and during exposure to the north or south poles of a bar magnet. When a magnet was placed beneath the slide, Davis saw the red blood cells begin to spin actively. If the polarity of the magnet facing the microscope slide was switched, the spinning motion of the red blood cells switched its direction of rotation. (Alteration of the spin of charged particles is another one of the many effects produced by magnetic fields.) In addition, other researchers discovered that if red blood cells were exposed to a magnetic field, their outer membranes seemed to become more electrically charged and less sticky compared to other nearby red blood cells. The change in stickiness of red blood cells may be yet another of the magnetobiological effects partly responsible for increased capillary blood flow. The magnetic alteration in the charge of red blood cell membranes might be another reason some magnetic therapists have reported success in slowly dissolving blood clots in the legs after placing permanent magnets directly over the affected blood vessels.

Besides stimulating local circulation, magnetotherapy appears to increase the delivery of oxygen to the body's tissues, suggesting another possible mechanism for magnetic enhancement of healing. In studying the effects of magnets upon blood flow, German physician Ulrich Warnke used pulsing magnetic fields applied to the head region (5-gauss intensity at 10 to 20 cycles per second) of fifty-eight subjects and then measured the blood flow in the arms and hands of the subjects. He noted markedly increased circulation within only two minutes of magnetic stimulation. However, he was surprised to find that the amount of dissolved oxygen in the body tissues had also increased by 200 percent, something that could not be explained entirely by a mere increase in blood flow. The maximal effect on improved circulation and increased oxygenation was seen at magnetic-field pulse rates of about 15 cycles per second. Further research on the magnetic field's ability to increase blood flow was carried out by Dr. Benjamin Lau of Loma Linda University in San Diego, California. Lau also experimented with using different frequencies of magnetic pulsation. He found that blood flow for 60 percent of his test subjects increased the most at magnetic-pulsation frequencies of about 15 cycles per second, just as Warnke had observed. It is interesting to note that 15 cycles per second might be a higher harmonic of the natural frequency of the earth's magnetic field of 7.8 cycles per second. If you multiply the earth's predominant geomagnetic frequency by two (2 x 7.8 = 15.6), you arrive at a figure of 15.6. This figure is very close to the 15-cycles-per-second pulsation of magnetic fields that were found to maximally enhance blood flow in Dr. Lau's research and tissue oxygenation in Dr. Warnke's studies. Another interesting coincidence about the 15 to 16 cycles per second pulsation relates to magnetic research by Dr. Ross Adey, also at Loma Linda University. Dr. Adey found that 16 cycles per second was the optimal frequency for increasing the migration of calcium ions in nerve cells. The ion migration was the kind that might alter a nerve's ability to transmit pain messages to the brain. All three studies suggest that stimulation of the body with fairly weak (about 5 gauss, or ten times the strength of the earth's field) pulsed magnetic fields (in a higher harmonic of the geomagnetic field) can enhance the magnetobiological effects of decreasing pain and enhancing tissue healing. The positive effects of magnetism are probably accomplished, in part, both by inhibiting pain-signal transmission and by increasing local circulation and oxygen delivery to the body's tissues. These studies also point to another important connection between the frequency of the earth's magnetic field and the biological responses of the human body. As we will see in a later chapter on the subject of psychic healing, the magnetic frequency window of 7.8 cycles per second has important implications for subtle magnetic healing phenomena as well.

While these three studies point to the importance of magnetobiological effects upon human physiology, mainstream physicians have been slow to incorporate magnetic therapies based on magnetic principles into treatment regimens. In certain specialized cases, however, novel magnetic adaptations of traditional drug treatments have been successfully explored. For example, cancer researchers have capitalized upon the phenomenon of the magnetic attraction for aggregates of iron and ferrite molecules (and the magnetic enhancement of local circulation) to selectively target chemotherapy agents to specific areas of the body. The resulting unique blend of drug and magnetic therapy consists of targeting cancer-fighting drugs into tumor tissues via a technique of binding the drugs to small aggregates of ferrite molecules. First a patient is given an intravenous infusion of the iron-bound chemotherapy agent. Then a powerful magnet is placed over the site occupied by the cancer. The strong magnetic field attracts the ferrite-bound cancer-fighting drugs out of the cancer patient's general circulation. The ferrite-bound cancer drug thus becomes concentrated mainly inside the cancerous tissue bathed by the magnetic field. By focusing the ferrite-bound chemotherapy drugs mainly in the tissues targeted for destruction, the oncologist hopes that the drug will produce fewer systemic side effects by spending less time in the general circulation. Also, in theory, since the magnetic field concentrates the drug in the cancer tissue, a lower drug dosage needs to be administered to the patient, and thus fewer side effects. While the technique still involves the use of toxic chemotherapy agents, the future use of such magnetic delivery systems might ultimately allow oncologists to decrease some of the terrible side effects of cancer-killing drugs.

Although we have talked mainly about the flow of ions and electrons altered by magnetic fields, there are other important biological effects produced by magnetotherapy. Earlier we discussed some of Dr. Albert Roy Davis's magnetic research. As early as 1936, Dr. Davis had discovered that the north and south poles of the magnet seemed to produce different biological effects on plants, animal, and even humans. But later, in the 1960s and 1970s, Davis began to conduct more extensive magnetic studies with Walter C. Rawls in order to elaborate on the varying effects of different magnetic polarities. The researchers were especially curious as to how magnetic polarity might be used for healing different medical disorders in human beings. Davis and Rawls found that the south pole of the magnet (the pole that orients toward earth's magnetic north) had a powerful stimulating effect on living systems. South-pole energy accelerated the growth rate and stimulated biological activity in both plant and animal cells. South-pole magnetic energy also changed the pH or acid/alkaline balance of tissues by causing them to be more acidic in nature. Conversely, the north pole of the magnet (the pole that would orient to geomagnetic south) was found to have an inhibitory or relaxing effect upon biological activity and would cause the tissues of the body to become less acidic and more alkaline in nature. Furthermore, north magnetic energy decreased general sensitivity to pain in areas directly exposed to the north pole of the magnet. In general, in their human studies, Davis and Rawls found both pain and localized tissue swelling to decrease after north-pole magnet applications for thirty to forty minutes once or twice a day.

Let's just stop here for a moment and see what happens if we combine Davis' and Rawls's research with some other studies highlighted so far. Many kinds of pain, especially muscle pain, are frequently associated with increased tissue acidity in the region around the pain. This acidity is sometimes caused by a buildup of lactic acid, a by-product of muscle metabolism. Application of the north pole of a magnet to an area of painful muscle spasm has been reported to bring about significant pain relief by a variety of magnetotherapy researchers. It is possible that north magnetic energy achieves this through a number of different energetic mechanisms. First of all, the north magnetic field may decrease localized muscle pain by decreasing the acidity of the muscle tissue (Davis/Rawls pole studies). Second, magnetic-field stimulation increases local circulation to the spasming muscle, thus helping improve local blood flow to carry away built-up lactic acid and other by-products of muscle metabolism (Lau/Warnke circulation studies). Third, the north-pole energy of the magnet has a relaxing effect on cells, causing contracted muscle cells to relax and release their stored-up muscle tension (Davis again). Finally, the magnetic field may alter the transmembrane flow of ions, which helps transmit pain signals along nerves that relay pain information to the brain (Becker's research). By blending the results from these different research studies, we begin to see some possible explanations for why magnets could be effective in relieving pain. With this in mind, let's return to Davis and Rawls's research for further insights into magnets' ability to promote healing.

During their research, Davis and Rawls claimed to have produced significant reductions in tumor growth in animals by exposing them to relatively intense (greater than 2,500 gauss) north magnetic fields. North-pole energy was found to inhibit the growth of tumors in test animals (causing either a reduction in size or an arrest in the growth of the tumor) consistent with the previously observed inhibitory effect of north magnetic energy. Conversely, Davis and Rawls found that south-pole magnetic fields actually caused tumors to grow faster and spread more quickly. Animal studies indicated that south magnetic energy also stimulated other aspects of cell growth, too, as south-pole-treated chicks produced roosters and hens that grew faster and stronger than those exposed to north-pole magnetic energies. In mice treated with south-pole magnetic fields, offspring were larger than those in the north-pole-treated groups.

Also, after birth, the offspring of south-pole-treated mice developed faster than the (non-exposed) control groups. As for the north-pole infants, they took longer to develop, were thinner, weaker, and did not feed as much as the mice in the control group. However, Davis and Rawls also discovered that north pole energy was very effective in decreasing pain from arthritis in (arthritis-prone) test animals. In fact, north pole-treated mice even showed a slowdown in the progression of their arthritis. Sometimes calcium deposits that had built up around joints of the north-pole-treated mice were even seen to gradually dissolve. Perhaps this was just another form of magnetic mobilization of calcium. Through animal studies, our diligent researchers also discovered that north magnetic energy attracted body fluids, as well as red and white blood cells. In animal experiments, the north side of a (3,000-gauss) magnet was placed against the body of a test animal in an area directly above or below a region of tissue swelling (for forty-five minutes). This particular treatment was found to draw the fluid out of the swollen region and into the adjacent region within a twenty-four hour period following placement of the magnet. Of great interest is some data showing (at least in animals) that the north-pole magnetic field also slowed down the aging process. In other words, test animals had their life spans extended by as much as 50 percent in some cases. Continuing with Davis and Rawls's research, test animals exposed to north magnetic energy were found to produce higher levels of circulating red blood cells. Furthermore, north magnetic energy decreased general sensitivity to pain in areas directly exposed to the north pole of the magnet.

In general, human studies by the aforementioned researchers found that both pain and localized tissue swelling decreased after north-pole magnet applications for thirty to forty minutes once or twice a day. With regard to south-pole magnetic fields and their effects on plants, Davis and Rawls noted that south-pole energy increased seed germination and plant growth in seeds and plants exposed to it. Even the length and size of the root systems in south-pole-treated plants were longer and more extensive. Sugar beets from south-pole-treated plants yielded a higher sugar content. South-pole-treated peanuts were found to have more peanut oil and higher protein content. Conversely, exposure of seeds to north-pole energy produced stunted growth patterns, an inhibitory effect consistent with other north-pole biological effects. Continuing with the pole studies, the two researchers' findings suggested pole-specific applications for specific medical problems. While most magnetic applications developed by Davis and Rawls involved north-pole energies, there were situtations where south-pole energies, correctly applied, were found to be of greater benefit. For instance, the south-pole energy was found to strengthen deteriorating muscles, joints, tendons, and ligaments; to improve the function of glands that had diminished in their normal functioning; and to increase local blood flow. In general, south-pole magnetism seemed to stimulate or enhance most forms of bio-

Possible Magnetic Mechanisms of Healing

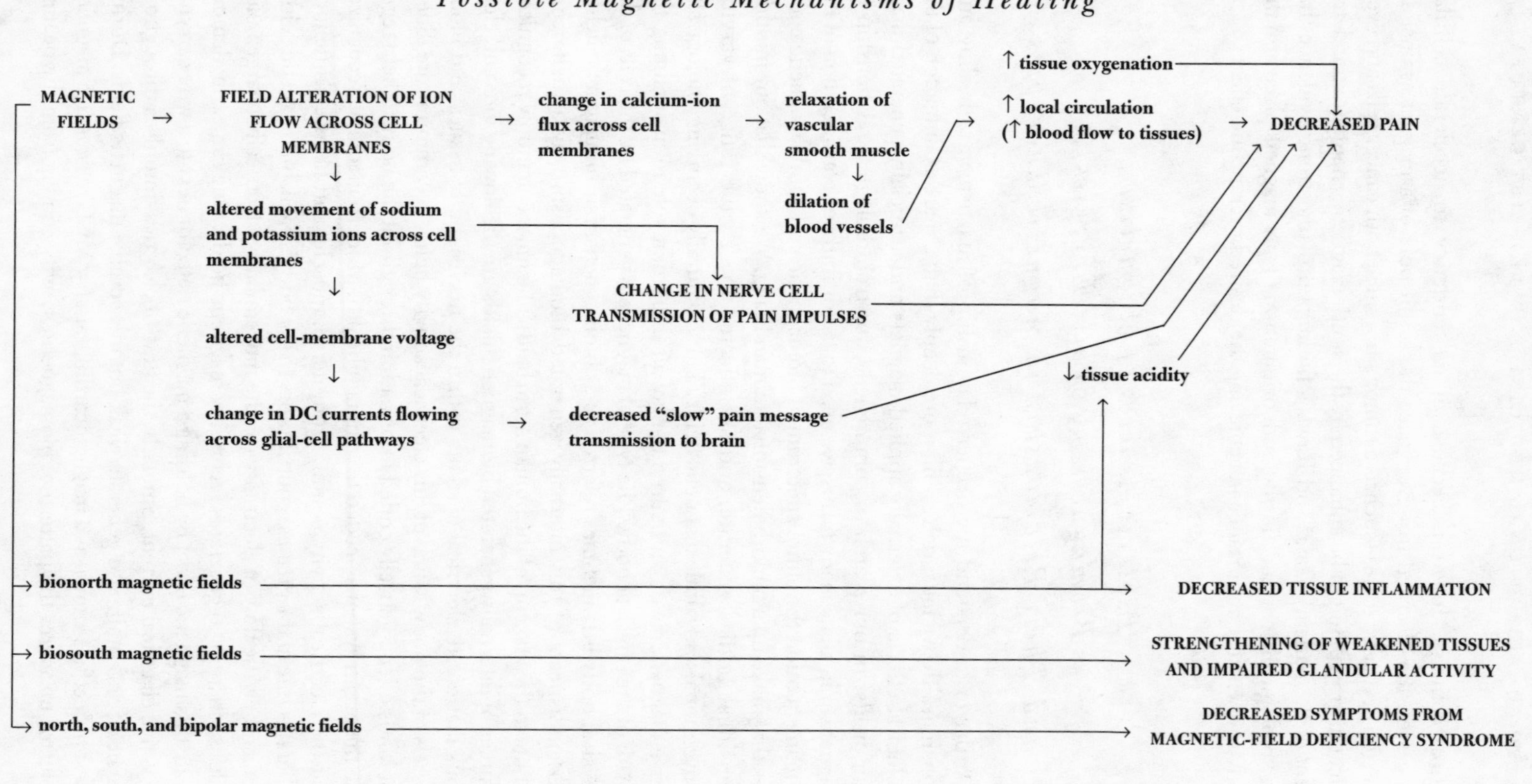

logical activity. However, because of its tendency to accelerate cellular processes, they learned to never place the south pole of a magnet against an infected or diseased area (such as a tumor site), since both cancer cells and even bacterial growth might be enhanced by the south magnetic energy (to the detriment of the patient). Davis and Rawls's findings provide strong evidence that the "biologically correct" north-south orientation of any magnets placed on the body is of critical importance in achieving the desired therapeutic effects.

Esoteric Aspects of Magnetism: Biomagnetism, Subtle Energies, and Their Relationship to Magnetic Therapies

In order to understand the deeper relationship of magnetism to healing and human health, we must delve a bit more deeply than a mere examination of the biological effects of externally applied magnets and artificially generated magnetic fields. In our search to comprehend our true biological relationship to magnetism in all its many forms, we need to factor into our consideration of the human equation the entire spectrum of biomagnetic and subtle magnetic energies that make up a multidimensional human being. We know that our multidimensional bodies generate, and are affected by, a wide range of varying conventional and subtle magnetic fields. These naturally occurring, biologically active magnetic fields and currents play an important role in maintaining the health of the physical body. The type of magnetism generated by the cells of the physical body is sometimes referred to as biomagnetism or biomagnetic fields. Certain aspects of this internally generated biomagnetism may actually contribute to the makeup of the human auric field. The most commonly recognized sources of internally generated biomagnetic fields are the flowing ions and electrons that create electrical currents within the nerves, body tissues, and blood vessels of the body. Much of this electrical activity helps to communicate different kinds of energetically coded information throughout the body to help regulate the organ systems needed to maintain health. Modern medicine recognizes that the electrical activity coming from nerve, muscle, and heart cells (among others) is one of the greatest contributors to the body's weak biomagnetic fields. Remember, the flow of electricity (or the movement of electrically charged particles such as ions) generates secondary magnetic fields. One can easily demonstrate this magnetic effect by holding a magnetic detector next to a wire carrying an active electrical current. Similarly, the flow of electrical impulses throughout the heart and brain also generate weak but measurable magnetic fields. During a test procedure known as a magnetocardiogram (or MCG) a sensitive magnetic detector measures the pattern of biomagnetic activity coming from the beating heart, caused by the electrical activity of pacemaker cells and the contracting chambers of cardiac muscle cells. Similarly magnetoencephalograms (or MEGs) measure weak magnetic fields produced by electrical brain activity. Researchers using these relatively new magnetic-measurement techniques are trying to study the magnetic-activity patterns of the heart and brain to better understand diseases that affect heart and brainwave rhythms. But aside from the biomagnetic fields generated by the heart and brain, there are still other, less obvious sources of biomagnetic activity that have yet to be acknowledged by modern medicine.

Though not widely familiar to American physicians, European research into the energetic aspects of cellular biology has shown that cellular division has both a biochemical as well as an energetic side, in that actively dividing cells seem to generate a certain amount of weak electromagnetic radiation. Cell division, the process of creating new cells, produces weak light photons in the ultraviolet frequency range. But in addition to this UV emission, dividing cells appear to radiate weak biomagnetic fields as well. French and Soviet research, dating back to George Lakhovsky in the 1930s, has provided experimental evidence that our DNA may function as an electromagnetic oscillator. Ukrainian nuclear physicists have measured the oscillation (vibrational movement) of DNA at a frequency in the range of 52 to 78 gigahertz—billions of cycles per second! The intensity of electromagnetic emission from DNA seems to be in the range of only a billionth of a watt. By the way, the Ukrainian physicists also believe that each human being has his or her own unique (optimal) frequency, say, 63.57737 gigahertz. Now, since all electromagnetic activity just happens to carry with it some type of magnetic-field component, the Ukrainian research would seem to indicate that vibrating human DNA should generate ultraweak biomagnetic fields. Indeed, our DNA does have the expected magnetic field, which oscillates at up to 70 billion times a second! Interestingly, this vibrational frequency just happens to be in the frequency range of energies that make up the spectrum of solar magnetic energies bathing our planet! Anyway, back to cell division's generating weak biomagnetic fields. Since human DNA is most active when it is being synthesized to make new gene copies for the process of cell division (mitosis), and since DNA activity generates oscillating electromagnetic fields, it makes sense that actively dividing cells would be both energetically and biochemically active.

The electromagnetic activity associated with mitosis was first noted by Soviet researcher Alexander Gurvitch in 1923 and later studied more intensively by German researcher Fritz Popp in the 1970s. Both researchers noted that actively dividing cells gave off certain types of electromagnetic radiation, and this radiation appeared to communicate information between cells. Gurvitch named this energy emission "mitogenetic radiation" because of its association with mitosis or cell division. Part of this radiation seemed to be in the form of UV-light emission. But in addition to ultraweak UV light, the fact that DNA molecules oscillate in the very fast gigahertz-frequency range suggests that mitogenetic radiation could also have a high-frequency biomagnetic component. What does this actually mean in terms of the biomagnetic activity of the body's cells? Well, we know that the human body is constantly renewing itself. As old cells die off, new cells are created through cell division to replace them. Such constant cellular division to replace aging cells throughout the body would also generate weak but significant biomagnetic energy, too weak to be measured by most conventional magnetic detectors. The magnetic-energy signals generated by the heart and brain are usually so strong they tend to drown out most of this weaker but significant biomagnetic activity produced by the DNA of our cells. But that cellular biomagnetic activity is there nonetheless. It may be that the energy field produced by weak, cell-generated biomagnetic activity throughout the body forms what clairvoyants perceive as the innermost layer of the human auric field.

Until very recently it was thought that the entire human biomagnetic field was created mainly by the electrical activity of the nervous system, the beating of the heart, and the weak fields generated by the body's metabolic processes. However, what many biomagnetic medical researchers fail to take into account is that in addition to the cell-generated biomagnetic fields, there are also a variety of subtle magnetic life energies that play a key role in the functioning of the multidimensional human's energy system. It is perhaps these subtle magnetic energies, which are currently ignored by mainstream science, that form the most important part of our total biomagnetic energy pattern. These subtle magnetic fields and currents are made up of the life energies flowing into our physical bodies from our acupuncture meridians, our chakras, our etheric bodies, and from our higher spiritual bodies. Vibrational-medicine researchers have long realized that these "nonphysical" biomagnetic currents and bioenergy fields are just as important as the cell-generated biomagnetic fields in determining whether the physical body will succumb to illness. Contrary to current medical thinking, clairvoyant medical research seems to suggest that the body's subtle-energy "field" actually generates the physical form and not the other way around. While the basic biomagnetic activity of the body may be a result of cellular processes, clairvoyants have observed that the subtle magnetic fields of the etheric and higher spiritual bodies appear to show abnormal changes long before any overt illness can be detected within the physical body. Clairvoyantly observed etheric biomagnetic disturbances may precede changes in the physical body by days, weeks, or even months. The subtle biomagnetic energies of the etheric body appear to play a key role in both providing structure and animating life energies that help to guide the growth, development, and repair of our bodies' cells and organs during states of illness, trauma, and stress, as well as during times of wellness. Current medical studies of conventional cell-generated biomagnetism observe only the tip of the iceberg. As the greater portion of icebergs remain hidden below the water line, the greater field of biomagnetic energies that needs to be studied remains hidden just below the surface, unless you happen to know where to look.

A major premise of subtle-energy work is that our higher spiritual bodies—the etheric, astral, and mental bodies—are composed of different varieties and grades of subtle magnetic matter. The subtle matter possesses unique magnetic properties with both similarities and dissimilarities to conventional magnetism. One illustration of the magnetic nature of astral and mental thoughtforms has already been mentioned in our discussion about the law of subtle magnetic attraction. We tend to "magnetically" attract to ourselves those people and things that occupy our waking thoughts and emotions. This is one aspect of subtle magnetic attraction that differs from conventional magnetic attraction. That is, with conventional magnets, similar poles repel while opposite poles attract each other. In the law of magnetic attraction, which guides the energetic behavior of astral and mental bodies and thoughtforms, like attracts like. The idea that our thoughts and emotions affect our physical bodies is not a new idea. It is the cornerstone of mind/body medicine. However, most mind/body researchers have looked only to biochemical correlates of mind, emotion, and immunity during illness and stressful states, while ignoring the possibility that our emotional states might also be accompanied by unique subtle biomagnetic energy patterns that indirectly influence our physiology through our multidimensional energy systems. Thus, to the student of vibrational medicine, it can be seen that the full human biomagnetic spectrum originates in the magnetic fields generated by cellular activity, as well as by the overshadowing and guiding spiritual and life energy currents and fields that help to animate and sustain the physical body.

Only a few creative individuals have gone down this road of subtle biomagnetic life-energy research. Going back to Mesmer, vital biomagnetic "life energy" currents might have been the ultimate source for the "animal magnetism" that Franz Anton Mesmer spoke about when he tried to explain the bioenergetic mechanism behind his "magnetic" healing experiments. It seems likely the energy which Mesmer referred to as animal magnetism was actually human etheric energy or the subtle magnetic life force itself. Since there were no recognized scientific theories in Mesmer's day supporting the existence of a human magnetic energy, critics of his magnetic-healing work attributed any healing successes to the effects of "suggestion." Thus, it seems that Mesmer became forever linked with the practice of hypnosis, suggestibility, and the term "mesmerism" that still bears his name. (We still describe someone who is totally "entranced" by a fascinating speaker as being "mesmerized.")

Let's continue with a few more thoughts about Mesmer's animal-magnetism energy. In addition to directly healing patients using both permanent magnets and the "magnetic passes" mentioned earlier, Mesmer also did demonstrations to show how animal magnetism could be stored in "magnetic batteries." The "stored-up" animal magnetism was then transmitted to patients via special magnetic-energy circuits. The magnetic storage batteries he developed were metal vats of water that Mesmer referred to as "baquets." He found he could magnetically "charge" water with his own hands and then later use the charged water to heal groups of patients. The biomagnetic circuit connecting Mesmer's patients to the magnetically charged water-filled baquets was made up mainly of silk cords that would be held in the hands of the patients. Unlike conventional magnetism, animal magnetism as Mesmer characterized it flowed not only through metals but also through certain organic materials, such as silk. As you may know, silk is normally considered to be an insulator to electrical-energy flow. More recent experiments with laying-on-of-hands healers have confirmed some of Mesmer's claims, specifically his claim that animal magnetism, in the form of "magnetically charged" water, could actually be used to heal sick patients. During the 1960s, Dr. Bernard Grad at McGill University in Montreal studied the ability of laying-on-of-hands healers to charge water with a subtle energy in order to affect plant growth. Grad confirmed, at statistically significant levels, that healer-charged water increased both the germination of seeds and the overall growth of plants. Grad also found that he could duplicate or even improve upon this growth-stimulating effect on plants if he used magnetically treated water "charged" using only a conventional permanent magnet.

Mesmer was not the only scientist of the time to be fascinated with the subtle-energy aspects of magnetism. In the mid-1800s, the German industrialist and chemist Karl von Reichenbach investigated the energetic properties associated with magnets. He used clairvoyantly sensitive observers to provide feedback for his work. Some of the observations made by Reichenbach's nineteenth-century clairvoyants about magnets and sunlight have been repeated by contemporary clairvoyants. Reichenbach's clairvoyant studies of the unique subtle-energetic aspects of conventional magnetism and even common sunlight now make certain researchers question whether mainstream science truly understands the energy phenomenona associated with magnets and sunlight. For instance, Reichenbach's sensitives could see luminous fields of color energy around the north and south poles of strong magnets. To the clairvoyants, the fields were indistinguishable from the auric fields of human beings. Reichenbach, therefore, came to believe that magnetic and auric fields were both composed of the same energy. He called this clairvoyantly perceived subtle energy "odyle" or the "odic force." The sensitives consistently saw blue energy around the north pole of the magnet and pink or reddish energy fields around the south pole. Their observations were quite reliable enough. In fact, the clairvoyants could correctly identify the north or south end of an unlabeled bar magnet merely using clairvoyant sight. They could also distinguish magnetized from unmagnetized water.

Based on feedback from the clairvoyant research team, Reichenbach came to believe that magnets, crystals, human beings, and even sunlight were a source of a subtle magnetic odic force. His sensitives also noted some unusual properties of the odic force associated with sunlight. Most dramatic was the clairvoyantly observed phenomenon of sunlight being conducted through metallic wires. This observation was later confirmed by twentieth-century radionics researcher T. Galen Hieronymous, who called the subtle-energy component of sunlight "eloptic energy" because it had both electrical (the ability to be conducted along metal wires) and optical properties. Reichenbach's sensitives could see sunlight being emitted as tiny beacons of luminous energy coming from the opposite end of a metallic wire held up to sunlight. Reichenbach also noted that the odic force from magnets could be conducted through silk, just as Mesmer had claimed. In addition, the odic force could be conducted through glass, resin, and water, and the force could also be used to "charge" certain substances. Reichenbach's discovery that magnets and sunlight radiate a subtle energy with properties similar to Mesmer's animal magnetism suggests that some of the biological effects attributed to solar magnetic fields and permanent magnets may not be strictly due to conventional magnetic effects. Their effects may be partly related to the unexplored subtle-energy component Reichenbach discovered over a century ago.

Nearly a hundred years later, Dr. Wilhelm Reich, a contemporary of Freud, studied subtle energies that he felt were an unexplored part of human physiology. Reich claimed to have discovered a subtle energy, which he named orgone energy, that was somehow related to the life force. Using special devices known as orgone accumulators or orgone boxes—constructed using alternating layers of metal, cellulose, and other organic substances—Reich was able to capture and store quantities of orgone energy, almost as the electrical pioneer Benjamin Franklin had used a Leyden jar to capture electricity. Having a readily available source of orgone energy, Reich experimented with treating a number of illnesses, ranging from various forms of neuralgia to cancer. Reich and many of his followers claimed many successful healings using their orgone therapy. One of Reich's theories about the nature of orgone was that it was a kind of life-force energy that circulated slowly through the human body. Partly from his psychiatric perspective on human emotional dynamics and partly from his experiments, Reich came to the conclusion that chronic emotional tensions could create blocks in the normal flow of this unique subtle life energy throughout the body. His theories sound like a discussion of emotionally related chakra blockages steeped in psychoanalytical terminology. Of interest is the parallel Reich noted between orgone energy and orgasmic energy. Since he was from the Freudian school of training, sexual psychological dynamics were of particular interest to Reich. He came to believe that the human sexual orgasm provided not only a release of emotional tension but also a discharge of orgone or life energy. Reich's thinking bore certain similarities to the writings of tantric-yoga masters who talk about the movement of life energy through the body during the sexual act. What is of particular interest to our discussion about subtle magnetic life energies comes from some of Reich's observations about the strange magnetic nature of orgone energy. Reich had noted that the energy from his orgone accumulators would deflect the north-south orientation of a nearby compass needle. In addition, he found that steel objects, such as a sewing needle or a scissors, actually became magnetized when they were accidentally charged with orgone energy. Based on this and further energy observations, Reich concluded that all forms of magnetism would eventually be seen as a function of cosmic orgone energy, a kind of primordial, pre-atomic life energy thought to pervade the entire universe.

After reexamining the pioneering research of Mesmer, Reich, and Reichenbach, it seems likely that there are subtle-energetic aspects to magnetism that as of yet have not been fully explored or even accepted by mainstream science. In time, I believe, scientific research will verify that these very subtle magnetic forces of life energy are the kinds of energy fields that are perhaps the most critical to achieving and maintaining optimal health.

Modern-day researchers have confirmed certain similarities between conventional magnetic fields and the subtle magnetic fields produced by the hands of healers. For example, Dr. Justa Smith of Rosary Hill College in Buffalo, New York, found that both forms of magnetism—healer's fields and conventional high-intensity magnetic fields—consistently accelerated the activity of enzymes, the biochemical workers of the cell. Where these two forms of magnetism differed was in their ability to be detected by sensitive magnetometers. Using a standard magnetometer, no significant conventional magnetic activity could be detected around the hands of the healers as they held test tubes of enzymes to be "healed." In earlier experiments, Dr. Smith had found that the strength of conventional magnetism needed to produce significant enzyme acceleration was about 10,000 gauss or nearly twenty thousand times that of the earth's magnetic field (about 0.4 to 0.5 gauss). If the enzyme-acceleration effect produced by healers was really due to conventional magnetic fields, the fields around their hands would need to be of enormous intensity. Such a field strength coming from a healer's hands would be something very difficult to miss, even with a conventional magnetometer. Interestingly enough, some form of magnetic-field emission from healers' hands was eventually confirmed by Dr. John Zimmerman of the Bioelectromagnetics Institute in Las Vegas, Nevada. Using ultrasensitive SQUID detectors (special magnetic-field sensors), Zimmerman demonstrated that healers did in fact emit weak, low-frequency, pulsed magnetic fields from their hands. But the intensities were well below those needed to produce the observed enzyme effects! The question became, What kind of magnetic fields were the SQUID detectors measuring? It is unlikely that they were being influenced only by conventional magnetic fields. Perhaps, instead, the SQUID detectors were being affected by a subtle magnetic life energy similar to Reich's orgone energy with its ability to deflect compass needles.

The point of all this discussion about biomagnetic phenomena is this: It seems that many different subtle and overt energies of life are magnetic in nature. It has even been said that the very energy of the soul is magnetic in nature. What we do know is that these forms of life energy that have been experimentally measured, especially the subtle energies associated with healing, are of a unique magnetic nature. According to Dr. William Tiller, a subtle-energy researcher and professor of material sciences at Stanford University, the magnetic energies of the etheric world should also have another unique property: negative entropy. This quality of negative entropy refers to the ability of etheric or life-force energy to push back and reverse the downward spiral of entropy. Entropy, the level of order in a system, usually moves in a positive direction, in which things naturally become more disordered and fall apart over time. So negative entropy is the ability of subtle magnetism to move systems in a reverse direction, toward states of higher order and organization. What etheric and other subtle magnetic life energies seem to do is to push biological systems toward physiological states in which the cells and even their enzymes behave in a more organized, highly structured, and energy-efficient pattern of cellular metabolism. The life force's negative-entropic drive for order has actually been verified in scientific experiments dealing with healing energy and its effect on human enzymes. When Dr. Justa Smith was carrying out her enzyme studies with healers and conventional high-intensity magnetic fields, she looked at the healer's ability to repair dysfunctional samples of the enzyme trypsin. The trypsin had previously been structurally damaged (made "sick," in a sense) by exposure to certain frequencies of ultraviolet light that destroy the molecule's active site. Such exposure disrupted the molecular structure of trypsin so that the enzyme's biochemical activity became impaired and nonfunctional. In Dr. Smith's studies, both the healer, and the intense magnetic-field source demonstrated the ability to "heal" the enzymes by causing the trypsin molecules to structurally refold and repair themselves so that they reassumed their former active shape. Following this instant self-repair, the trypsin's biochemical activity began to speed up the longer the healer held on to it (or the longer the high-intensity magnetic fields were bathing the trypsin). The ability to cause a structurally disorganized molecule with no biological activity to repair itself and become a repatterned molecule with full biological activity is certainly a demonstration of the negative-entropic drive of not only conventional magnetism but also of etheric, subtle magnetic life energy transmitted by healers' hands.

These unique observations add new dimensions to the question of how different magnetic therapies really accomplish their pain-relieving and healing effects. The magnetic nature of subtle energies—from ch'i to orgone to astral and etheric energies—plus the phenomenon of biomagnetism and the very magnetic nature of life itself can only make one wonder whether magnet therapies might also be providing a simulation of some of the natural, negatively entropic etheric life energies that normally support and maintain the health of the physical body during states of wellness. After all, in Dr. Smith's research, both conventional magnetic and subtle magnetic energy fields were able to produce a negatively entropic effect: the ability to increase molecular order within damaged enzymes. And aren't protein enzymes the main biochemical workers that carry out many of the most important functions critical to human health? If conventional or subtle magnetic fields can influence enzyme activity through negative-entropic effects, (including repairing damaged enzymes and other biological molecules as well as increasing their bioactivity), it would make sense that such energies could easily have healing properties, *if properly applied.*

This "negentropic" drive is a basic characteristic of the life force, the force largely responsible for the tendency of living systems to maintain order and organization at even the tiniest level of cellular structure. Without the life force, our lifeless physical form decomposes through decay and positive entropy, literally "ashes to ashes, dust to dust." Without the organizing influence of life energy pulsing through our cells, our bodies break down to their most basic chemical components, leaving us as compostlike food for scavenging bacteria and microbes. The soul and the spiritual bodies represent the system of magnetic, animating, creative life energies that inhabit and energize the physical form, providing invisible patterns that may lead to either illness or wellness, depending upon our state of spiritual, mental, emotional, and physical balance. The different forms of subtle and biological magnetic life energies can become distorted at various levels, from both inner disturbances and external electromagnetic-field disturbances. Magnet therapies might somehow provide a magnetic crutch or energetic stimulation that helps to support some of the natural bioenergetic processes of self-healing that can go awry when there is disease or distortion at different levels of the human multidimensional system. It is entirely possible that magnet therapy affects not only the physical body but the etheric body as well.

The subtle magnetic life force we've been discussing is also thought to influence our bloodstream, or more specifically, the hemoglobin that gives our blood its characteristic red color. Suppose that our blood carries not only oxygen but also a certain amount of life-force energy. This would mean that our blood, as it is pumped through the circulatory system, provides additional life force to the cells of the body above and beyond the pranic input of the chakras. There are certain aspects of blood's composition that lead to this speculation. For one thing, blood is composed largely of liquid plasma, which is mainly biological water that contains a variety of ions and protein molecules. As we have discussed, water is known for its ability to hold on to charges of various types of subtle energy, as suggested in homeopathic-remedy potentization as well as in the ability of "magnetized" water to accelerate plant growth. Besides water, blood also contains red blood cells filled with iron-rich hemoglobin. While the iron atoms in hemoglobin may not be directly attracted to magnets as iron filings are drawn to a bar magnet, iron is one of those unique elements whose atoms have the ability to hold on to magnetism. Is it possible that the iron in hemoglobin makes it not only a good transporter of oxygen but also an excellent carrier of prana or other subtle magnetic forms of life-force energy?

That there is a relationship between magnetic energy and hemoglobin cannot be denied. As mentioned earlier, Linus Pauling received the 1954 Nobel Prize in chemistry for his research into the magnetic nature of hemoglobin. But there is also a unique and unrecognized relationship among hemoglobin, the lungs, the kidneys, and the subtle magnetic energy of prana that has never been fully explored. Modern medicine has determined that both the lungs and kidneys work together to help regulate the body's pH, its acid/alkaline balance, so we can remain in the narrow pH range needed for the enzymes of the body to function optimally. When we breathe rapidly, as in the yogic deep-breathing exercises known as pranayama, we tend to hyperventilate, producing an alkaline condition of the blood called respiratory alkalosis. When we slow down our breathing, the opposite occurs, and our blood becomes more acidic. This produces a condition known as respiratory acidosis. The kidneys also regulate body pH by selectively excreting or reabsorbing positive (acidic) hydrogen ions versus negative (alkaline) bicarbonate ions as the kidney filters the blood to produce urine. If the blood's pH shifts even slightly and becomes either too acid or too alkaline, hemoglobin's attraction for oxygen is affected. If the acid/base balance of the body becomes too disturbed, the blood becomes less able to carry and deliver oxygen efficiently to where it is needed. Thus, the kidneys and the lungs, in their ability to influence blood pH and hemoglobin's ability to carry oxygen, help to regulate oxygen intake and distribution throughout the body.

The blood's capacity for carrying oxygen is also influenced by another aspect of kidney function. Our kidneys produce a hormone known as erythropoetin (EPO). EPO stimulates the body's bone-marrow cells, the site of active red-blood-cell production, to create more red cells. EPO levels are frequently low in patients with chronic kidney diseases, resulting in low hemoglobin levels. Conversely, synthetic EPO is sometimes administered to cancer patients to raise low hemoglobin levels caused by chemotherapy side effects. Because EPO affects hemoglobin levels, it indirectly influences the oxygen-carrying capacity of the blood.

But is oxygen the only life-promoting factor that our hemoglobin-filled red blood cells are carrying? I suspect that both hemoglobin and oxygen possess hidden, subtle biomagnetic functions that have never been fully researched. These two vital substances, oxygen and hemoglobin, may help to regulate the amount of magnetic-life force energy circulating through the bloodstream. Although the chakras assimilate prana directly by feeding life-force energy to the different organ systems in a specialized manner through the fine network of nadis, prana is also taken into the body via the lungs and may circulate through the bloodstream as well. Oxygen and hemoglobin may play a role in maintaining an optimal magnetic life-force charge within the blood. The reason for this speculation stems from certain statements in the yogic literature suggesting that oxygen may be a carrier of the subtle magnetic life energy known as prana. Furthermore, the iron-rich hemoglobin in our blood might make an ideal carrier for this magnetic energy. If these things are true, then the lungs and the kidneys, both able to influence hemoglobin and oxygen delivery, may form part of a life-force regulatory system that helps to maintain an adequate charge of magnetic life energy in the bloodstream.

As we breathe in air, oxygen molecules cross the membranes of our lungs and are taken up by hemoglobin molecules, which carry them throughout the bloodstream. The literature on yoga suggests that the oxygen we breathe becomes charged with prana by exposure to solar radiation before it is taken into our bodies through the lungs (and the chakras). A number of Russian physicists have confirmed that the energies traveling to the earth from the sun contain both visible light and a wide spectrum of magnetic and electromagnetic energies. These energies vary from extremely slow magnetic oscillations all the way up to electromagnetic energies vibrating in the gigahertz frequencies (frequencies that also happen to be in the same resonant frequency as our DNA). There are also energies contained within sunlight that seem to have unusual subtle-energetic properties. One such property of this subtle-energetic component of sunlight is the ability to be conducted through various organic substances (normally considered insulators) and metals. Radionics researcher T. Galen Hieronymous was able to demonstrate the conduction along a wire of some unrecognized subtle-energy quality within sunlight. This wire-traversing subtle energy could also be transmitted to plants kept in total darkness. Somehow, the subtle-energy component of sunlight that traversed the metal wire was able to stimulate plant growth and chlorophyll production, just as if the plants had been bathed in natural sunlight. I suspect that the subtle magnetic component within sunlight that Hieronymous was transmitting via wires was actually prana (or some similar form of subtle magnetic life energy).

If the preceding assumptions are true, then the life-giving qualities of the air we breathe could be due to both the biochemical properties of oxygen and to oxygen-bound pranic life energy. The fact that oxygen is transported throughout the bloodstream via hemoglobin-filled red blood cells means that a normal level of hemoglobin in the bloodstream is crucial to maintaining an optimal charge of both oxygen and life-force energy in the blood. Going back to the lungs and the kidneys, our lungs take in both oxygen and prana, which may be then transferred in part to hemoglobin for further transport to the tissues of the body. Because of our kidneys' ability to produce EPO, a substance that raises hemoglobin levels in our blood, our kidneys may be able to indirectly influence the prana-carrying capacity of our bloodstream by increasing the levels of this life-force-carrying molecule.

In addition to mentions in the yogic literature, the knowledge of a life-force-energy connection between the lungs and kidneys can also be found in texts of traditional Chinese medicine. In five-element theory, the diagnostic/treatment approach sometimes used to guide acupuncture treatments, the kidney and lungs are seen as energetically linked through their relationship with ch'i energy, a subtle magnetic life-force cousin to prana. In what is known as the "cycle of generation," ch'i energy is passed directly from the lungs to the kidneys via the acupuncture-meridian system. There is also another interesting correlation between the kidneys and life-force energy. According to traditional acupuncture theory, the kidneys are said to be storehouses of ancestral ch'i. Ancestral ch'i is the reserve of life force energy we each inherit from our parents. The preceding two observations provide additional supporting evidence for the notion that the lungs and the kidneys may be both physiologically and energetically linked. This means that the kidneys may function both as "biomagnetic batteries" for storing life energy, and as regulators of life energy carried through the bloodstream (via the EPO connection).

The idea that the kidneys function as bioenergetic batteries is also consistent with Dr. Bjorn Nordenstrom's modern-day theories on the biological energy circuits and batterylike electrical systems that exist throughout the tissues and organs of the body. These circuits form what Nordenstrom calls the body's "electrical circulatory system." While physicians have always assumed that the unique structure of the kidney was "designed" to promote optimal urine formation, the active pumping of ions across kidney tubules and the dense array of blood vessels in the kidney might also provide it with a unique bioelectrical function. Perhaps the specialized structure of the kidneys may somehow help to energetically contain or "store" the magnetic charge of ancestral ch'i energy.

That there may be a relationship between the kidneys and subtle magnetic life-force energy is also supported by certain observations on the physiological effects of Therapeutic Touch (TT). Experiments have shown that when a subtle magnetic life-energy transfer occurs between a TT practitioner and a client, this energy exchange (which often occurs without any actual physical contact) is usually accompanied by a significant rise in the client's blood-hemoglobin levels. Although this rise in hemoglobin was initially looked at as a biological marker of healer-induced physiological changes, this effect may be much more significant than previously thought. The elevated hemoglobin levels in TT-treated patients means that not only are they receiving more oxygen through the bloodstream, but they may also be carrying more life energy in the blood as well. One of the benefits of a life-force-energy transfusion may be a stimulation of the kidneys to produce more EPO, which can raise an individual's hemoglobin level. Although I am unaware of whether anyone has carried out this particular line of research, I think it would also be interesting to measure both hemoglobin and EPO levels in TT-treated patients before and after healing to see if increased EPO levels might be one of the reasons for this rise in hemoglobin. Another fascinating TT observation that ties in with this EPO theory is worth mentioning in this context. In the early days of teaching Therapeutic Touch to nursing students back in the 1970s, nurses were often instructed to perform an generally energizing TT treatment (to recharge someone with generalized, nonspecific complaints) by placing their hands over a person's midback, in the area directly overlying the kidneys. By sending the biomagnetic life-energy streams directly into the kidney region, the TT practitioner might just be directly stimulating EPO production while "recharging" the kidney's subtle magnetic storage batteries.

This discussion of the hidden energy relationship between the lungs and the kidneys, and their potential ability to influence blood levels of oxygen, hemoglobin, and prana, merely points out that there are biomagnetic aspects to human physiology that need to be more fully researched. Such research might eventually lead to new insights on the benefits of both life-force therapies, and magnetic therapies.

Besides having negative-entropic properties and influencing hemoglobin and oxygen, there are a variety of other subtle-energy mechanisms by which various kinds of magnetic fields may produce a healing effect on human beings. Some magnet researchers believe that the magnetic benefits of naturally occurring permanent magnets like lodestone have to do with the fact that these magnetized pieces of rock emanate the same magnetic energy as does the earth itself. Magnet researcher Don Lorimer, who has studied some of the more esoteric aspects of magnet therapy, considers the magnetic energy of the earth to be a kind of "loving" attractive force between the planet and all living things within the scope of its geomagnetic field. He believes that when the magnets are placed near the body, they help to reestablish the subtle supportive connections between human beings who may be energetically out of balance and the healing energy of the earth.

In this same vein, Lorimer suggests that magnets may also be helpful in balancing the energetic polarities of the body. Both Dr. Randolph Stone, developer of Polarity Therapy, and Drs. Davis and Rawls, the two noted magnet researchers, independently discovered that the human body possesses unique electrical polarities. Based on electrical measurements taken from different subjects, Davis and Rawls developed electrical body maps showing the voltages (along with their associated positive or negative charges) on the surface of various parts of the body. They found that the right side of the body was usually positively charged, while the left side was negatively charged. In addition, the front of the body was often positively charged, while the back was negative. Certain areas of neutral energy or polarity changeover were also found. This electrical duality between front and back, left and right, parallels the Chinese-medicine concept of yin and yang, with different aspects of the body electrically yin or yang, relatively speaking. The key to understanding and applying the principles of yin and yang, in this case, means that there needs to be a balance between these two different-polarity energies within the body.

The fact that the body possesses two different electrical polarities suggests that the north and south poles of magnets might actually be useful for polarity rebalancing. The polarity of the body apparently becomes disturbed when exposed to certain types of noxious environmental energies. According to Lorimer, different kinds of electromagnetic-field disturbances bathing the human body, including EM fields generated by electronic devices operating on 60-Hz electrical current as well as geopathic stress, can create imbalances in the electromagnetic polarity of the body, which, of course, would add to bioenergetic dysfunction within the body's various physiological systems. Lorimer found that placing two strong magnets on the balls of the feet for twenty to thirty minutes, with the north magnetic pole facing the right foot and the south magnetic pole facing the left foot, seemed to rebalance disturbances in electromagnetic polarity. When combined with other therapeutic approaches, this technique of magnetic-polarity rebalancing seems to aid the body in healing from a variety of different illnesses.

Magnet therapy applied to the feet in this fashion may not only rebalance the energy flow between the left and right sides of the body, it may also help to reestablish a more optimal magnetic-energy connection between the magnetically treated individual and the geomagnetic field of the earth. Perhaps someday a greater scientific verification of the more esoteric aspects of magnetism will help us to better understand our dependence upon the magnetic energy of the earth, the magnetic nature of the life force, and even the magnetic nature of our very souls. Future scholars and magnetic researchers will have to help fill in the details of our unique multidimensional relationships with the many forms of conventional and subtle magnetism.

Specific Magnetic Approaches to Healing

Probably the greatest application of magnetic therapy in healing is for the relief of various types of pain. There are a variety of reasons magnets may be effective in pain relief, including suppressed nerve conduction of pain impulses, decreased tissue acidity, improved local circulation, altered DC tissue currents, localized magnetic changes in the etheric fields of the body, and even increased levels of beta-endorphin (a naturally occurring painkiller, recently reported to be elevated after magnetic treatments). Probably one of the most common causes of pain is arthritis. Although many different pharmacological treatments for arthritis exist, each arthritis drug comes with its own potential side effects, ranging from stomach irritation and ulcer formation to liver-function changes and even suppression of bone-marrow activity. While newer arthritis drugs continue to hit the market, an energetic approach to relieving the painful joints of arthritis would offer a safer alternative, free of side effects when properly used.

Several researchers have discovered different types of magnetic-field therapies useful for relieving the pain and stiffness of arthritis. In Europe, doctors have found that treating arthritic joints with pulsed magnetic fields (delivered by specially constructed magnetotherapy devices) reduced pain, swelling, and stiffness after a number of treatment sessions. The magnetic treatments were noted to be effective for both rheumatoid arthritis (an inflammatory condition) and osteoarthritis (the more common form of wear-and-tear joint degeneration). In Italy, in 1992, the Italian medical journal *Panminerva Med* published an article on the effects of pulsed magnetic-field therapy in treating joint pain caused by inflammatory disorders. Pulsed magnetic fields of low intensity and very low frequencies were used to treat 3,014 patients with joint pains. Nearly 79 percent of all patients had a good response to pulsed magnetic-field therapy, while only 21.2 percent had a poor response. A good response was characterized by the disappearance of pain in the afflicted region, a 40 to 50 percent increase in joint mobility after one treatment rising to almost 100 percent improved mobility with repeated treatments, and a maintenance of the therapeutic benefits achieved after each magnetic treatment. In India, there has also been at least one large-scale study on the therapeutic benefits of pulsed magnetic field therapy for treating various forms of arthritis. Drs. Sarada Subrahmanyam and Sanker Narayan at the Madras Institute of Magnetobiology in India treated 1,000 people with both rheumatoid and osteoarthritis using pulsed magnetic fields (pulsing at 0.1 cycles per second with a field strength of 350 gauss) over the affected joints thirty minutes a day for thirty to forty days.

About 75 to 80 percent of the rheumatoid and osteoarthritis patients became free of pain and noticed an increase in their range of movement in joints that had previously been restricted by pain and swelling. With only two exceptions, no patients reported severe pain in the affected joints for up to two years after the study had been completed.

In the United States, a number of physicians have also reported good results in using magnetic therapy for treating arthritis. Dr. Benjamin Lau, mentioned earlier, treated nineteen patients with severe to moderate joint stiffness caused by arthritis. After three weeks of therapy, sixteen of the nineteen arthritis patients reported no more stiffness and near-complete relief of pain. Dr. Ron Lawrence, president of the North American Academy of Magnetic Therapy, has also had good success in treating osteoarthritis with static (nonpulsing) permanent magnets placed upon the afflicted joints of the body. Lawrence has found magnets to be effective in reducing pain from arthritis of the spine (including the neck and lower back), painful knees, elbows, wrists, fingers, and even TMJ (temporomandibular-joint problems that cause jaw pain and headaches). One typical case was a forty-year-old patient who suffered from a two-year history of low-back pain. The patient had played football in high school and had been injured a number of times while playing. He experienced typical back pain traveling down the backs of his legs to his knees. An MRI exam of his spine showed only degenerative changes in the lower lumbar vertebrae. No herniated intervertebral discs could be seen on the MRI. Dr. Lawrence gave the patient a back brace containing ceramic magnets that rested upon the painful lumbar region. He was advised to wear the magnetic brace whenever his pain really bothered him, which was usually about two to three hours a day. He also used a magnetic pad with neodymium magnets on the back of his office chair. Additional treatments, including massage and an adjustable footstool, were recommended to optimize pain relief. The patient noted significant improvement in the chronic back pain. He eventually found he had to use the magnetic back brace only when he had an acute pain flare-up.

Another typical arthritis patient treated by Dr. Lawrence was an eighty-four-year-old woman with osteoarthritis in both her knees. Her arthritis was so severe that her knees were swollen to four times their normal size. Because of the intense pain, she had become less active and had begun to gain weight, a common problem for arthritis sufferers who are afflicted with painful weight-bearing joints. Dr. Lawrence prescribed a special type of ceramic magnet to be held in place over the knees using a special neoprene knee brace. Almost overnight, the magnetic therapy reduced the size of her knee joints by nearly 200 percent. After a few days, she could move about more easily, and the knee swelling continued to go down. Within a week, her knees were almost back to their normal size, and she had a 90 percent decrease in knee pain. The patient found that when she was very active, she would get some increase in her knee pain the next day. However, wearing the magnetic knee brace overnight seemed to be of great help in providing further pain relief.

Dr. Lawrence also had as a patient a sixty-year-old part-time machinist who suffered from degenerative arthritis in the fingers of both hands. In his spare time, he had loved to sculpt and paint, but the pain in his hands was preventing him from pursuing this hobby. Dr. Lawrence gave him neodymium magnets, which he wore around both his wrists. In addition, he was given special cotton gloves peppered with strategically placed neodymium dot magnets (very small but potent magnets). He was instructed to wear the special gloves and wrist apparatus when he was at home during the daytime and while he slept at night. The patient found that the wrist magnets and magnetic gloves reduced the arthritic pain in his hands by nearly 80 percent from their previous levels. Four years later, the patient was still wearing the gloves intermittently (to accommodate the body's tendency to adapt to certain chronic forms of stimulation) and continued to get good pain relief. The effectiveness of magnets in relieving joint pains appears to be partly a function of weather conditions, so that joint-pain relief may be greater than 80 percent in warmer months but less during the colder winter months. Whether there might be a geomagnetic or solar magnetic correlation with such seasonal weather changes and the intensity of arthritic pain is an interesting point for speculation. But this relationship between geomagnetic effects and the severity of arthritis flare-ups has never been extensively studied.

Besides placing larger magnets directly over arthritic joints, a number of magnetic therapists, including Dr. Lawrence, use smaller neodymium and other potent tiny magnets placed over key acupuncture points to relieve arthritic pain. Depending upon which joint or joints are painful, dot magnets attached to an adhesive tape are placed directly over local acupoints near the joint (especially those points used in acupuncture for the relief of joint pain). In addition, one can also select certain distant acupoints, far from the affected joint, that are known to enhance overall pain relief in the body (so-called big points, like Large Intestine-4/LI-4 in the webbing of the thumb, also known colloquially as the "hoku point"). By combining magnets placed over the painful joint along with dot-magnet stimulation of key acupoints, many different forms of arthritis and pain syndromes can be effectively treated. In Japan, many cases of arthritis of the neck and shoulders have been effectively treated using magnetic necklaces. Research at the University of Tokyo in the mid-1970s tested the efficacy of magnetic necklaces (which delivered about a 1,500-gauss magnetic exposure) in relieving the pain caused by various forms of arthritis, including neck and shoulder pain, as well as arm, hand, and leg pains. In studies involving over 400 patients, 60 to 80 percent found significant pain relief for all the aforementioned forms of arthritic pain. It is interesting to note that Japanese magnetic researcher Dr. Nakagawa, discoverer of the "magnetic-field deficiency syndrome" (MFDS), found a commonality of symptoms between people suffering from MFDS and arthritis. Many MFDS patients often had no significant X-ray findings to pinpoint a cause for their joint pains, though. But the reported pains went away with the use of magnets during MFDS therapy. Perhaps some of the pain-relieving effects of the magnets, especially the magnetic necklaces (which are sometimes worn on an area of the body that is far from the site of pain), could be related to restoring a critical level of magnetic-field exposure to the body, something the MFDS sufferers might have been missing because of environmental factors such as steel shielding in their offices and apartments. By the way, magnet therapy has become so popular in Japan that magnets are used for some form of healing in nearly one out of every six Japanese households!

Pulsed magnetic fields have been found to be of benefit for treating neck pain as well. A small double-blind study on magnetic treatment of neck pain was carried out at Mater Misericordiae Hospital in Dublin, Ireland. Researchers studied twenty patients who complained of chronic neck pain that had been present for at least two months. In the first part of the study, ten patients were given electromagnetic collars to wear. The collars produced a low-energy but high-frequency pulsed magnetic field. The remaining ten patients formed the placebo group, who wore a similar-looking collar but without the magnetic coils turned on, so that there would be no actual magnetic therapy received. At the end of three weeks, nine of the ten patients in the "treatment" group found a significant decrease in neck pain, while in the placebo group, only two of the ten patients reported any pain relief. During the next three weeks, all twenty patients wore "active" pulsing magnetic collars. At the end of the second three-week period, seventeen of the twenty patients reported at least "moderate" improvement in their neck pain.

Magnetic therapy is often used for many nonarthritic forms of pain as well. For treatment of carpal tunnel syndrome, magnetic therapists will often tape permanent magnets (with the north pole placed against the body) directly over patients' painful wrists. Although vitamin B6 in doses of 200 to 600 milligrams is frequently helpful for relieving the pain of carpal tunnel syndrome, magnetic therapy can bring additional pain relief. Leg pains can also be treated with magnetic therapy, especially pain caused by nerve problems from diabetic neuropathy and by circulation problems as well. One orthopedic surgeon who was skeptical of magnet therapy was convinced to try it when he suffered increasing leg pain with even minimal exertion while walking from hole to hole on the golf course. The doctor suffered from vascular disease that had not only affected his heart but was causing leg pains due to decreased circulation to his lower extremities. After trying a pair of magnetic insoles placed in his shoes, he was amazed to find that his exercise tolerance had improved dramatically and that he could walk much farther without pain. One day on the golf course, he felt his legs becoming tired and painful. He was puzzled by the painful relapse.

It was only after he went to change his shoes that he discovered he had forgotten to place the magnetic insoles in his golf shoes. Since then, this orthopedic surgeon has become a magnet convert and now uses them on many of his patients, some of whom have had such dramatic pain improvement that they ended up canceling their joint-replacement surgeries! One patient who had magnet treatment for leg pain was a nurse. Following surgery for a bone fracture, she'd needed some hardware and screws to be placed into the ankle to hold together the healing bones because the fracture had been so severe. Even after healing from the surgery, she continued to experience ankle pain while walking any significant distance. She was given magnetic shoe inserts and soon found a dramatic improvement in her pain. Even though the pain never completely disappeared (because of the degree of injury she suffered), she was able to return to normal activities with only minimal discomfort.

A number of magnetic therapists have also had some success in treating fibromyalgia, a painful muscle and joint disorder, by having patients sleep on a magnetic mattress pad. Although sleeping on the magnetic pad may not necessarily relieve all the pain experienced by fibromyalgia sufferers, the magnetic energy often enables patients to significantly decrease their use of anti-inflammatory medicines and painkillers.

A number of magnetotherapy researchers have used certain types of magnetic fields to inhibit the growth of cancer cells. Dr. Benoytosh Bhattacharyya of West Bengal, India, discovered that certain types of magnetic fields affected cancer tissues that had been artificially implanted into mice and rabbits. After repeated exposure to magnetic-field treatments, the growth of the cancer cells was halted in its tracks. The doctor also found that magnetic field therapy increased the life span of the treated animals well beyond that of the control animals in his experiments. In one case, a dog, diagnosed with a brain tumor received magnetic treatment from Dr. Bhattacharyya. Prior to treatment, the dog could not walk with its hind legs. The animal's legs would drag behind it whenever the dog attempted to move around. The north pole of a small permanent magnet was taped to the dog's head for five minutes every morning and evening. Within a short time, the brain tumor started to shrink. As a result, the dog began to walk and even run within only a few weeks' time of initiating magnet therapy.

But what about the effect of magnetic-field therapies on human cancer cells? In 1996, Dr. Raymond Raylman at the University of Michigan reported in the journal *Bioelectromagnetics* on experiments measuring the inhibitory effect of magnetic fields on human cancer cells. In his research, Raylman exposed three different lines of human cancer cells to strong, static (nonoscillating) magnetic fields over a period of sixty-four hours. He found that all three lines of cancer-cell cultures showed evidence of suppressed growth. Other research into magnetotherapy has also suggested that there are certain types of magnetic fields that might be useful in enhancing the effectiveness of conventional cancer treatments. In one study conducted as early as 1980, Soviet researcher L. S. Ogorodnikova found that twenty to thirty sessions of magnetotherapy, given preoperatively to lung-cancer patients, reduced the size of their lung tumors prior to tumor removal by conventional surgery. In the late 1980s, Dr. Y. Omote, a Japanese cancer researcher, used pulsed magnetic-field stimulation to increase the incorporation of antitumor drugs into cancer cells, thus increasing their cancer-fighting ability. While Dr. Omote's study dealt more on a basic-science level with cancer cells, one must ask whether this finding might be relevant to the use of antitumor drugs in human cancer treatment? Support for this theory comes from Soviet magnetotherapy research conducted in the mid-1990s. In 1994, Dr. V. Smirnova experimented with whole-body exposure of cancer patients to certain types of magnetic fields (known as eddy fields), along with local magnetotherapy at the tumor sites, in order to improve patient responses to conventional cancer treatments for a variety of different malignancies. The experimental results proved quite promising.

The Soviets have also studied the use of magnetotherapy as a primary therapy, without the added pharmacologic effects of conventional anticancer drugs. In 1991, the magazine *Soviet Medicine* reported on the case of a woman with metastatic breast cancer who was successfully treated using only magnetotherapy. Local spread of the invasive breast cancer was markedly decreased after thirty magnetic treatment sessions consisting of whole-body exposure to eddy magnetic fields for sixty minutes using special Soviet magnetotherapy devices. Following 60 magnetotherapy sessions, one metastatic lymph node completely disappeared, while other nodes were reduced in size. After 110 treatment sessions, the breast-cancer patient was in complete remission, having had both her breast tumor and all metastatic sites in her body returned to a normal healthy state. While such Soviet reports of cancer regression by magnetotherapy may be tantalizing, these systems have never been evaluated by, or even been made available to, mainstream cancer researchers in the United States. Most U.S. researchers tend to focus mainly on surgery, chemotherapy, and radiation therapy. In addition, the FDA has never approved the use of magnetotherapy in treating cancer. Thus cancer patients seeking magnetotherapy have had to travel to some of the Mexican cancer clinics that use magnetic cancer treatments.

One anticancer application of magnetic therapy worth mentioning was researched by the late Swedish scientist Dr. Goesta Wollin. Back in 1987, Dr. Wollin used neomax supermagnets (a 4,000-gauss magnet as small as a stack of six quarters) to heal different forms of human cancer. Wollin placed a supermagnet with the south pole facing the body directly over the thymus gland, the master immune gland (located behind the upper portion of the breastbone). Wollin created a special "bib" containing the magnet, which patients would wear day and night. This would assure that the magnet would remain over the correct location to provide maximal south-pole magnetic stimulation to the thymus. As you will recall, the south pole of the magnet appears to have a stimulating effect upon all biological processes. In this case, Wollin theorized that south magnetic stimulation might increase the thymus gland's ability to provide a strong immune response against the cancer cells in the body. Later, Wollin experimented with applying an additional supermagnet directly over the tumor site with the north pole facing the body. The north magnetic pole has the opposite energetic effect of the south pole, providing an energy that relaxes or suppresses biological activity. Wollin's thought was that the north magnetic energy would work locally to directly decrease the growth of cancer cells, while the south-pole-stimulated thymus gland would trigger an aggressive immune response to destroy the remaining cancer cells. Wollin claimed to have had many successes with his magnetic cancer approach, both in Sweden and in the United States. In one notable case, Wollin worked with Dr. Erik Enby of Clyde, North Carolina, to magnetically treat a woman with metastatic lung cancer. The cancer had failed to respond adequately to several years of treatment with surgery, radiation, and chemotherapy. The patient was told by her physicians in June 1997 that she had only two months to live. She had a positive response to Wollin's magnetotherapy approach and a year later appeared to be in good health, having outlived her two-month death sentence.

While Wollin's research on using permanent magnets to heal cancer is certainly interesting, it should be approached with a certain degree of caution. One must remember that Wollin was using the south magnetic pole against the body in his treatments. Since the south magnetic pole stimulates biological activity, placing the south pole of a magnet directly over a collection of cancer cells could easily stimulate their growth, producing an enlargement of a preexisting tumor. Because of this potential untoward effect of the south-pole energies, magnetic treatments for cancer should be done only under the guidance of a health-care professional competent in the use of magnetic therapy.

There are other illnesses besides cancer for which magnetic therapies may prove beneficial. We have discussed how magnetic fields can affect nerve function, causing a decreased sensitivity to painful stimulation. A number of other aspects of neurological functioning may also be affected by magnetic therapies. While it is usually not a good idea to bring strong magnets near the head region, some researchers have found that weaker north-pole energies near the head are able to bring about a relaxing effect upon consciousness, whereas south-pole energies can have a stimulating effect, sometimes even enhancing extrasensory perception. For purposes of relaxation, however, it is safer to place the north pole of a magnet over the midchest region to produce the same calming effect. Certain neurological diseases, such as Parkinson's disease, have been shown to be profoundly influenced by magnetic-field treatments. In 1996, Dr. R. Sandyk reported in the *International Journal of Neuroscience* on experimental attempts to use pulsed magnetic fields to decrease the abnormal muscle movements caused by Parkinson's disease. Using extremely weak, pulsed magnetic fields to stimulate the brain via magnetic coils placed around the head, Sandyk was able to reverse tremors and other motor abnormalities in a number of different patients. Some of the magnetically treated Parkinson's patients improved dramatically even after only a single treatment session. In many Parkinson's patients, the decrease in symptoms would last for up to seventy-two hours before returning to the baseline state. Some patients with Parkinsons's disease also noted improvements in their short-term visual memory. In a similar vein, Dr. J. Bardasano, an Italian researcher investigating magnetic-field treatment for Parkinson's disease, developed a plastic helmet with internal magnetic coils that could be worn by patients receiving magnetotherapy. The helmet produced a weak, pulsed magnetic field (in the range of 8 cycles per second, the same as the earth's geomagnetic field) that after thirty minutes of treatment was found to decrease symptoms in patients with both Parkinson's and multiple sclerosis.

Magnet-field treatments have also been proven useful for relieving depression. One neuropsychiatric researcher, Dr. M. S. George, investigated the use of magnetic stimulation to the brain for this purpose. George used a slightly different method of applying pulsed magnetic fields to the head region, referred to as transcranial magnetic stimulation (TMS). The doctor and his colleagues reported in the Fall 1996 issue of the *Journal of Neuropsychiatry and Clinical Neuroscience* that they were successful in reversing depression with pulsed magnetic treatments (directed to the head) in a number of different patients. Some psychiatrists, such as Dr. T. Zyss, have speculated that TMS as a magnetic therapy for depression may ultimately prove better than electro-convulsive therapy (ECT), as it is painless, noninvasive, and even more effective than ECT in therapeutically influencing deep structures of the brain.

Personal Experimentation with Magnet Therapy: Some Guidelines to Follow

As we have already discussed, there are a wide variety of magnetic-therapy applications for healing and rebalancing the human body. Perhaps the safest is in the form of permanent magnets for healing. There are, however, certain do's and don'ts to follow when applying magnets to the body. Some key things to avoid:

1. Try to avoid using strong magnets upon the head. Using magnets as strong as 1,000 gauss or more against the head for more than ten to fifteen minutes (especially with the south pole facing the body) can produce headaches or throw the entire endocrine system out of balance. Magnetic researcher Dr. William Philpott has even noted seizures to occur (albeit rarely) in susceptible individuals, especially with south-pole cranial stimulation.
2. Do not treat cancer, infections, or inflammatory problems with the south pole of the magnet, as the stimulating effect of this type of energy could actually worsen the problem.
3. Avoid placing magnets anywhere near a woman's abdomen during pregnancy.
4. Do not bring strong magnets anywhere near the chest region of individuals who have pacemakers, as magnetic signals are frequently used for reprogramming pacemakers.
5. Do not use strong magnets around the head, neck, glands, or organs or the body for long periods on a daily basis. For one thing, the body needs to recuperate from the magnetic stimulation for there to be an optimal therapeutic effect. For another, continual exposure to a strong magnet may interfere with some of the normal biomagnetic and biological rhythms of the body.
6. Don't use a magnetic bed for more than eight to ten hours at a time.
7. Wait sixty to ninety minutes after meals before applying magnets to the abdomen (some magnetic fields may interfere with peristaltic movement of food through the intestinal tract).
8. Don't treat a condition with magnets if either south or north polarity seems to worsen the condition. Magnetism is not like homeopathy, in which you may experience a "healing crisis" before you get better. If north- or south-pole magnetic energy aggravates your symptoms, magnetic therapy (at least the type you are using) is not the best approach for your particular ailment.
9. It is generally unwise to use magnets to suppress recurrent pain without undergoing some type of medical evaluation first. For instance, recurrent abdominal pain might be indicative of a serious underlying medical condition that, while temporarily improved by magnet therapy, might suppress symptoms that could lead to an early and accurate diagnosis. Whenever possible, consultation with a health-care practitioner who is knowledgeable about, or at least open to, magnet therapy is highly recommended.

With the varying medical effects of north- and south-pole energies, how do you best determine which pole is most appropriate for treating a particular health problem? A number of magnet researchers, including Dr. Buryl Payne of

Psychophysics Labs in Santa Cruz, California, advocate the use of kinesiology or muscle-testing in determining which magnetic polarity to use on the body. As previously discussed, there are a variety of ways that muscle-testing can determine the correct homeopathic remedy to use, which vitamin to take, or in this case, the correct magnetic polarity to apply. Although two people are usually needed for muscle kinesiology (a tester and a patient), the so-called bidigital O-ring test advocated by Perelandra flower-essence developer Machaelle Small Wright, can also be used for self-testing (see Chapter 7 for further details). In general, one can try placing the magnet against the region of the body that is causing problems (e.g., pain) and then proceed with muscle-testing. If the arm muscle tests strong when the north side faces the painful area but tests weak with the south pole against the same region, then the north magnetic pole would be indicated for magnetic treatment. Even if the test is positive for magnetic north, kinesiological retesting at a later time is usually a good idea. For instance, in treating a painful knee afflicted by osteoarthritis with superimposed inflammation and joint swelling, the arm muscle may initially test stronger when the north pole of the magnet is placed against the knee, because the acute pain is being caused by an inflammatory process that can be calmed by north magnetic energy. After a period of time, when the swelling and inflammation subside, the muscle test may become positive for the south magnetic pole. This could happen because, after the inflammation is gone from the knee, the remaining source of knee pain might be bone rubbing against bone caused by degeneration of cartilage and ligaments in the joint. In such cases, the south magnetic energy can be helpful in stimulating repair of weakened joint structures.

It is this same physiological reasoning that may explain why some people find that sleeping on a magnetic bed actually aggravates some of their physical problems. Most magnetic beds have the north magnetic polarity facing the body, but they can be turned over to have the south energy for treatment as well. Some beds use arrays of special bipolar magnets that have both north- and south-pole characteristics in a single piece, while others use grid patterns of alternating north- and south-pole-facing magnets for generalized north-south stimulation of the entire body. Muscle-testing can be used to determine which polarity may be most compatible with the body's general energetic needs. On the other hand, there may be selective regions of the body that "prefer" the opposite polarity of magnetism. As an illustration, in some chronic illnesses, a weakened thymus gland might need stimulating with south-pole energy, while overstimulated adrenals could benefit from the calming influence of north magnetic energy to optimally improve symptoms. Also, a small percentage of people with whole-body degenerative arthritis can actually develop aggravations of joint pain with north-pole magnetic beds, because north magnetic energy can sometimes inhibit healing and repair of already weakened cartilage and connective tissue, producing more pain with movement. Conversely, a patient with leukemia could find that a south magnetic bed aggravates symptoms, because the south-pole energy might actually stimulate progression of the cancerous condition. While muscle-testing may seem silly to some medical professionals, it appears to be a valid technique for querying the inner wisdom of the body as to what type of energy stimulation it "knows" would be best for its current state. Therefore, I believe that kinesiological muscle-testing can be a valuable technique for selecting which type of polarity (and even strength) of magnet should be used for healing.

Muscle-testing with magnets can also provide a means for energetically testing the underlying metabolic activity of specific organ systems of the body. As an example, let's look at certain individuals who suffer from a disorder known as pernicious anemia. These individuals have a lower level of hemoglobin in their bloodstream, caused by vitamin-B12 deficiency due to a problem they have in assimilating vitamin B12 through the intestinal tract. Normally, our stomachs secrete a protein cofactor that assists in the process of vitamin-B12 absorption through the intestines. Many individuals with pernicious anemia produce less of this B12 cofactor because they suffer from a degenerative stomach disorder known as atrophic gastritis, a disease of decreased gastric-hormonal function. Since people with degenerative diseases of their organs, such as atrophic gastritis, are likely to benefit from south-pole magnetic stimulation of an already weakened cellular metabolism, they would be more likely to obtain a positive muscle test when the south pole of a magnet was placed over their stomach region. On the other hand, what kind of individual is more likely to muscle-test positive for the north magnetic pole? Take your typical stressful executive type, worried about a new corporate merger in the wings, who is nursing his latest ulcer by drinking from a nearby bottle of liquid antacid. Because an acute ulcer is an inflammatory disorder of the stomach, with active inflammation of gastric tissue, an individual with an ulcer would be more likely to muscle-test positive (muscle stronger) with the north pole of the magnet placed over the stomach (which would "cool down" the inflammation). By observing the kinesiologic reaction to north- versus south-pole energy, one can actually go through the entire endocrine system and other organ systems (via muscle-testing of key organ reflex points) to diagnose hyper- or hypoactivity of any area of the body. As an aside, people who are diabetic will usually test stronger when the south pole of a magnet is placed over their solar plexus center, the upper abdominal region beneath which sits the pancreas, the gland that secretes insulin. The decreased insulin-producing capablility of a diabetic's pancreas shows up as a positive response to the magnetic stimulation provided by south-pole energy. Muscle-testing positive with the magnet's south pole over the pancreas suggests that the weakened diabetic pancreas might actually benefit from the stimulation provided by south magnetic energy. In support of such kinesiological reasoning, some diabetic patients have actually found that wearing a south-pole-facing neodymium magnet directly over their solar plexus region results in better blood-sugar control with fewer requirements for diabetic medication.

The ability of south-pole energy to stimulate endocrine hypofunction is not limited to the pancreas. Other underactive glands may respond to south-pole magnetic stimulation as well. For instance, some magnetotherapists advocate treating sexual dysfunction with south-pole stimulation of the male and female endocrine glands located in the second-chakra region. While no large studies have been carried out to verify the effectiveness of this form of therapy, it is always important to bear in mind the potential hazards caused by misusing south-pole energy in that region of the body, such as the acceleration of preexisting cancer cell growth or increasing the proliferation of pathogenic bacteria and viruses.

There are a wide variety of magnets, available through hardware stores and mail order, that can be adapted for therapeutic use on the body. Magnets are available in numerous sizes, shapes, strengths, and polarity arrangements. One can obtain tiny beadlike magnets for use in acupoint stimulation, oval-shaped button-size magnets, dime-size neodymium magnets, cookie-size flat circular magnets, typical bar magnets, large (4 x 6 x 3/4 inch) flat block-shaped magnets, and even flexible magnetic strips of any shape and size. There are three main types of artificially produced magnets: Alnico, ferrites, and rare earth. The names come from the materials added to iron to increase its ability to hold on to magnetic fields. The Alnico magnets contain iron along with *al*uminum, *ni*ckel, and *co*balt (hence the name Al-ni-co) in addition to trace amounts of copper or titanium. Ceramic or ferrite magnets are made from barium-ferrite and strontium-ferrite powders pressed together in a die under many tons of pressure. After pressing, the mixture is fired at extremely high temperatures to form a ceramic. Then the ceramic is magnetized in special devices. Strontium-ferrite materials are also combined with a vinyl compound to make the typical flexible magnetic strips that seal refrigerator doors tightly. Some of the strongest magnets available are rare-earth magnets made from samarium or cobalt and mixtures of other elements. The very strongest commercially available magnets are the neodymium magnets, sometimes called neos, which combine small size with high magnetic-field strength. Small neodymium magnets with a strength of 4,000 gauss (8,000 times the strength of the earth's field) are perhaps the most potent permanent magnets used by magnetotherapists today.

Generally, the most useful magnets are those having only one direction of magnetic force at each end. Thus, the old horseshoe-type magnets would be unsuitable for therapy. The exception to this pole-at-each-end generality is seen in some of the Japanese magnets that are arranged with alternating north/south polarities in either concentric rings or checkerboard-type patterns in a single magnet. Some of these magnets were designed mainly to influence ionic currents in blood vessels crossing from one magnetic pole to another. While certain forms of pain are amenable to treatment using various types of mixed-polarity magnetic arrays, the differing biological effects of north and south magnetic energies on the body make me reluctant to recommend mixed-polarity magnets for other applications. The typical rubber-based refrigerator magnets and the larger, thin, flexible magnetic strips also tend to have alternating zones of north and south magnetic polarities along one side (giving them multipole magnet surfaces).

Not all countries use the same pole designation as we do in the United States. For instance, the British north pole of a magnet is what we usually refer to as the south pole. A number of magnet researchers also use the terms "positive" and "negative" polarities as opposed to south and north, respectively. For our purposes, and to keep confusion to a minimum, when I have referred to using the north pole of a magnet, I am referring to what magnetic researchers like Dr. Buryl Payne refer to as "bionorth." Bionorth is equivalent to negative magnetic energy, while biosouth equates with positive magnetic energy, at least according to the use of most magnetotherapy researchers.

For most uses, you will want to select magnets with larger, flat-surfaced magnetic poles. This way you can place the correct polar orientation against the body. In some instances, two flat magnets, one worn inside the clothing and one outside, can be used to hold each magnet in place over the appropriate body location. Alternatively, with larger flexible magnetic strips, adhesive tape or even elastic wraps may be more appropriate to hold the magnets in place. For treating a localized problem such as a sore tooth, the smaller neodymium magnets may be more useful for focusing strong magnetic fields on a relatively small area. Stronger magnets like neos are also helpful for penetrating thick casts surrounding a fractured bone. But the strength of the magnetic field drops off quickly the farther away from the magnet one goes. A painful bone fracture could also be treated with one of the larger, rectangular block magnets placed against the cast. The tiny disc magnets used for magnetic acupoint stimulation can be taped in place over skin acupoints and left on for several days. It is always prudent to check (and label) the polarity of such tiny magnets before placing them on acupoints. The north pole of the magnet placed against the acupoint is said to be relaxing or sedating (in acupuncture terms), while south-pole magnetic energy is said to be tonifying or energizing to acupoints.

Depending upon the type of problem you are trying to treat, you need to decide whether the north or south pole of the magnet should face the body. Generally speaking, most pain-related problems will usually require north-pole magnetic energy to decrease nerve sensitivity to pain. Sometimes, when pain is caused by a degenerative process, though, south-pole energy may actually be more appropriate. As said earlier, some experimentation may be needed in order to find an optimal pain-relief orientation for the magnet. If you decide to use magnet therapy in conjunction with body acupoints for pain treatment, keep this guideline in mind. In order to decide which acupoints on the body to magnetically stimulate for achieving optimal pain relief, simply try to stimulate the tenderest spots in the painful body region (sometimes referred to by acupuncturists as "ah-chi" or "ouch-y" points). Even if the tender spots are only "trigger points" and not acupoints, magnetic stimulation with a magnet's north pole can often relieve muscle spasms, which can be a contributing factor to certain kinds of body pains.

Perhaps the best method of selecting acupoints to stimulate in magnet therapy is to refer to a book that illustrates which acupoints to treat for specific types of pain. Many of these books have "recipes," with specific well-known acupoints to use for relieving knee pain, back pain, elbow pain, and such. There are many good acupuncture guidebooks around, such as Leon Chaitow's *Acupuncture Treatment of Pain* (an invaluable resource), which provides good visual diagrams to guide acu-magnet therapy. The index lists different types of pain according to the area of the body affected. Nearly every page in the book has a body diagram showing where the points are relative to well-recognized anatomical landmarks. Acupuncture books can be quite helpful in deciding which acupoints to stimulate, not only with magnets but also with other noninvasive energy technologies (including focused ultrasound, laser light, and non-needle electroacupuncture stimulation). When stimulating acupoints magnetically, we apply tiny dot magnets (north side facing the body) to the skin with a special adhesive patch or simply with half of a flexible fabric Band-Aid. By leaving the magnet in place against the key acupoints, a continuous magnetic stimulation of the selected acupuncture points and meridians is provided. Some therapists advocate leaving the magnets in place for three to five days to achieve the maximal beneficial effect. Acu-magnet therapists advise against using acupoint magnets on skin that looks irritated. Reactions to acu-magnetic stimulation may rarely result in a small red skin zone under the magnet, said to heal more easily if the skin is cleaned and left bare.

Dr. E. Holzapfel, a French acu-magnet practitioner, often disposes of the acupoint-stimulating magnets after one or two applications, since he feels that the magnets seem to pick up some of the "negative energy" of the condition they are treating. This is one of the few cases in which the magnets used for acupoint therapy may need to be considered "disposable" as opposed to "reusable" as in the case of other magnets used for therapy. While it may be ideal to dispose of such magnets, it is not always economically practical, since some of the more potent dot magnets are not necessarily inexpensive. This question of whether magnets can become contaminated with the imbalanced energies of their users is something, to my knowledge, that has not been fully studied. Perhaps clairvoyant investigation of this subject might prove fruitful. While it seems that the larger magnets may be less likely to pick up negative energies, the smaller dot magnets could be more susceptible to picking up undesirable subtle energies. The approaches used for "cleansing" crystals of negative energies might be valuable for cleansing magnets as well, since both magnets and crystals appear to work with different types of subtle magnetic energy. As an illustration, quartz crystals are able to hold on to subtle (etheric) magnetic-energy patterns, almost the same way a floppy disk can hold on to magnetic computer-memory patterns. Some of the methods used to cleanse quartz crystals of abnormal subtle magnetic energies include placing them in a running stream, placing them in sea salt, burying them in the ground for a period of time, leaving them in bright sunlight for several days, placing them in water mixed with a few drops of Pennyroyal flower essence, or running them through a tape-head demagnetizer. I personally prefer water and Pennyroyal, as this usually takes only fifteen to twenty minutes. It is possible that placing one's acupoint dot magnets in a bowl of water containing seven drops of Pennyroyal flower essence (for fifteen minutes) may be helpful for removing some of the negative energy patterns picked up by the tiny magnets after a two-day acupoint treatment. It is certainly cheaper than throwing them away after one or two uses.

While small acupoint magnets can safely be left on the body for days at a time, the slightly larger and more potent magnets, such as the neodymium supermagnets, should probably not be worn for more than a few days at a time (especially with the south pole facing the body) without a rest period. Such intense fields can potentially overstimulate the body, unless one is following the specific advice of a magnet-oriented health-care practitioner. Similarly, the larger four-by-six-inch plate magnets can be left against the body for an hour or more. Sheet magnetic material, which is lower in magnetic strength, can be wrapped around or taped to a limb or other body part and left in place for several days to get a maximal healing effect.

The larger rectangular plate magnets have another interesting therapeutic use. They can be used for magnetically "charging" large bottles of water. There are many magnet researchers who advocate drinking "magnetized" water for a variety of health reasons. As we have mentioned earlier, magnetized water has been shown by Dr. Grad and others to accelerate the germination and growth of plants. Interestingly enough, magnet researchers Davis and Rawls found that this effect depended upon the magnetic polarity of the water. Water energized by the south pole of a magnet stimulated seed germination and plant growth, while north-charged water seemed to have the opposite effect. There are reports of various health benefits from drinking magnetized water, although the magnetic polarity of the water is not always mentioned. Some, like the Soviets, use water charged with both magnetic poles, while Indian magnetotherapists use both mixed-polarity water as well as north- and south-pole-charged water for specific health problems.

In his book *Magnetotherapy*, Dr. R. S. Bansal, director of the Indian Institute of Magnetotherapy in New Delhi, India, reports that drinking magnetized water can produce a wide variety of health benefits, including boosting energy levels, aiding digestion, improving normal bowel function, helping to regulate abnormal menstrual cycles, increasing urinary flow while also dissolving kidney stones, and even reducing fevers. Also, north-pole-charged versus south-pole-charged water seems to produce different effects in the body. Depending upon your current state of health, one magnetic polarity of water might be better for you than another. But at other times, the opposite-polarity water could be indicated. For instance, if a person were suffering from chronic fatigue, drinking the more stimulating south-pole-magnetized water on a daily basis might provide an energizing boost to a chronically weakened system. On the other hand, someone who experiences the stress of daily life with anxiety and agitation may benefit more from the relaxing magnetic energies of north-pole-charged water. In general, the north-pole-charged water is the safer. South-pole-charged water should not be used when there is an infection or cancer present, since its stimulating energies can potentially aggravate such situations. North-pole-charged water is usually soothing and relaxing in its effects upon the body, helping to decrease pain and reduce fevers. The use of magnet kinesiology may ultimately be the best method of determining your own body's magnetic-water needs with regard to south- or north-pole-charged water.

The methods of magnetizing water vary a bit from country to country. In the former Soviet Union, they allow water to drip, drop by drop, between the poles of a powerful horseshoe magnet, after which it is collected in a large jar. This allows the water to be imprinted by both north and south magnetic energies at the same time. Some Indian magnetotherapists actually place a clean, large magnet within a glass jar filled with water, leaving the water to be charged by magnetic force over a period of twenty-four hours. The cleanest method is simply to place a bottle of water on top of a large, flat magnet. The pole you choose to charge with should be facing up toward the bottle. The bottle of water is left on top of the magnet for at least twenty-four hours, although magnetic effects can be seen after as little as thirty to sixty minutes of charging. Dr. Bansal recommends drinking two ounces of magnetized water before breakfast and after major meals of the day. For children, he suggests one ounce of magnetized water three times daily; for infants, one to two teaspoonfuls, three times a day appears most appropriate. In his book *The Art of Magnetic Healing,* Dr. M. T. Santwani of New Delhi recommends charging two bottles of water simultaneously. One bottle would be placed upon the north pole of a large magnet, and a second bottle would be placed on top of the south pole of another, similar magnet. After allowing the bottles of water to become charged by magnets for at least twelve to twenty-four hours, the two bottles can be used alone or combined to make a solution having mixed magnetic polarities (similar to the Soviet magnetized water). Santwani recommends using mixed-polarity water for treating cases in which an individual experiences difficult urination or is producing inadequate amounts of urine. However, if the urinary problem is caused by an infection, he suggests using only north-pole water, since the biostimulating effect of south magnetic energy can actually worsen an infection by stimulating bacterial growth. Santwani states that all infectious diseases should be treated only with north-pole-charged water. Regarding the "shelf life" of magnetized water, Santwani believes that magnetically charged water will retain its potency for five to six days if kept at room temperature and for as long as ten days if refrigerated. Perhaps the easiest way to produce and keep magnetized water is simply to keep a jug or bottle of water resting on top of a magnet in the refrigerator. More water is added to the bottle as the magnetized water is gradually consumed. North-pole-charged water is generally considered the safest to drink on a daily basis. There are even magnetic researchers who suggest that since north-pole energy slows down biological processes, drinking north-magnetized water may actually slow down the aging process. Whichever polarity water one decides to use, it appears that the drinking of either north- or south-magnetized water may subtly aid the body in reestablishing its inner energy balance.

One may ask, Does magnetization of the water do anything to change its basic physical properties? The answer is yes. The changes are very small but measurable. Magnetized water exhibits a change in the strength with which water molecules are able to bind with one another. Magnet researchers Davis and Rawls found that south-pole magnetic force appeared to make water molecules bind to each other more weakly than normal, thus giving it a lower surface tension than normal. Surface tension is the membranelike layer on top of water that water-strider bugs are able to walk upon. It is this same phenomenon of surface tension that is responsible for the ability of plants to suck water up through their roots from the surrounding soil. Alternatively, Davis and Rawls noted that north-pole-charged water had the opposite characteristic, demonstrating an increased surface tension (due to stronger water-molecule binding).

While the ability to alter the surface tension of water may not seem like much, a magnet's ability to change the way water molecules interact with each other within the body could theoretically produce major changes in enzyme activity. The shape an enzyme takes (and thus the shape and efficiency of the active enzyme site where it does its catalytic work) is strongly influenced by how the enzyme's protein backbone interacts with surrounding water molecules (a process that can be affected by magnetism). The fact that magnetized water possesses altered binding forces between water molecules and various proteins, salts, and mineral substances appears to endow it with the ability to dissolve buildups of various types of salts and minerals.

This phenomenon turns out to have industrial as well as biological applications. Magnetized water has been shown to be effective in breaking down and dissolving mineral and salt buildup within pipes and boiler systems. It is used this way in Russia. In a similar vein, Russian and Indian magnetotherapists have reported that magnetic water, ingested by patients with kidney stones and certain types of mineralized gallstones, allowed patients to slowly dissolve and excrete the abnormal buildups of salt and minerals from their gallbladders and urinary tracts. Following this same line of reasoning, some Indian magnetotherapy researchers assert that magnetized water has similar value in helping to remove calcium (and even cholesterol) deposits from critically clogged arteries in people suffering from arteriosclerosis and other types of vascular disease. Since such an approach is inexpensive and easy to do, it would be interesting to see how drinking "magnetized" water could affect the course of coronary artery disease over time. Some Indian and Russian magnetotherapists even claim that drinking magnetized water (of mixed polarity) may lower cholesterol levels within three weeks. A number of Russian magnetotherapy researchers believe that many of the positive effects magnets have upon the body may be related to their ability to magnetize the water content of the human body itself. Keep in mind, we are more than 90 percent water.

It's also possible to use a slight variation of magnetized water to achieve subtle-energetic healing effects that extend beyond the physical body. As mentioned earlier, one of the ways to produce magnetic water is to place a strong magnet in the water itself, as opposed to the more common method of energizing water via the magnetic forces which penetrate the glass walls of the water bottle. A slight variation of this method is used to make a form of magnetic vibrational remedy known as a gem elixir. Gem elixirs are similar to flower essences in the way they are prepared. Both are made by placing the flower or gem into a clear bowl of water in bright morning sunlight for several hours. The magnetic influence of solar prana energetically imprints some of the subtle-energy patterns of the gem or flower into the water.

There are two unique gem elixirs prepared from magnetic materials. They are lodestone elixir and magnetite elixir. The actions of these vibrational remedies are known primarily through intuitive research methods. Lodestone elixir is said to align our physical body's biomagnetic field with the earth's magnetic field. It may also enhance transmission of information throughout the body in its many forms. It seems to accomplish this in part by enhancing nerve-to-nerve communication throughout the entire nervous system. Lodestone elixir is reported to be a general tonic for the endocrine system and may also be valuable for stimulating tissue regeneration. It is said to balance the acupuncture-meridian system and bring balance to the opposing forces of yin and yang within the body. In addition, the etheric body may be strengthened by lodestone elixir. Lodestone elixir can also be helpful in detoxifying from any kind of radiation exposure. The other magnetic gem elixir (with slightly different properties) is magnetite elixir. This elixir is said to enhance blood circulation throughout the entire endocrine system. Magnetite elixir may be helpful for people who have had radiation exposure caused by working in a mine rich in radioactive minerals or from general overexposure to radiation. It is also believed to enhance meditation, allowing a greater inner focus. Magnetite elixir is said to energize and align all the chakras, meridians, and subtle or spiritual bodies.

Indian magnetotherapist Dr. M. T. Santwani recommends magnetically charging certain types of oils for use in healing. Although he uses mainly coconut and sesame oil, I suspect almond oil could be substituted. To make magnetized oil, place a bottle (preferably cylindrical) of natural vegetable oil on top of the pole of a strong magnet for fifteen days. Santwani recommends making a bottle of north-pole oil and another of south-pole oil. After being left to "charge" for fifteen days, the bottle is shaken vigorously and labeled appropriately. North-pole oil is said to be useful for treating painful conditions caused by infections like boils or abscesses. South-pole oil is recommended for painful degenerative problems such as sciatica, low-back pain, and cervical disc disease, but should be avoided whenever coexisting infection is suspected. When using either north- or soul-pole oils, it may also be advantageous to add a few drops of healing essential oils to the bottle. For instance, essential oils like helichrysum, peppermint, and birch added to a bottle of magnetized almond oil could provide additional medicinal and subtle-energy healing benefits for relieving arthritic pains.

Magnet researcher Buryl Payne suggests creating a "magnetic first-aid kit" for use around the house in health emergencies and acute pain problems. The ideal magnetic first-aid kit would include magnets of different shapes, sizes, and strengths for different applications. After purchasing a new magnet, it is a good idea to check and then label the north and south polarities. To do this, you can either suspend the magnet from a string and watch how it orients in the earth's magnetic field (the north end of the magnet will point in a southerly direction) or use a labeled bar magnet to identify north and south poles. (Remember, opposite poles attract.) After identifying the poles, you can put a drop of red nail polish on the south side of the magnet so you can quickly identify the north and south poles. For most common uses, such as in treating aches and pains, you will want to place the north side of the magnet against the body. If treatment with the north side of the magnet is ineffective, you can always flip the magnet over to try the opposite-polarity magnetic energy on the problem. When in doubt, try magnet kinesiology. Most minor aches, muscle pains, and sprains can be effectively treated with north-pole energy from medium-strength, flat ceramic magnets taped to the painful body region. Alternatively, flexible (multipolar) magnetic strips, wrapped around the painful joint or muscle, have also been found to provide effective pain relief.

There are a variety of magnetic bracelets that can be worn for pain relief or for a general energizing effect. Many of these are available through medical-supply stores, television ads, and a variety of specialty mail-order catalogs. Dr. Payne finds the Stress Bracelet to be one of the more useful magnetic bracelets.

This twenty-four-carat gold-plated bracelet contains three or four magnets and is made of copper, zinc, and a combination of other alloys. In his book *Magnetic Healing: Advanced Techniques for the Application of Magnetic Forces,* Payne states that wearing the Stress Bracelet makes him more energetic when he is fatigued. He will also put on the bracelet when he feels a sniffle coming on. He says this usually cures him before he develops a full-blown cold.

Magnetic necklaces can also be included in a magnetic first-aid kit. Payne reports that the Longevity Necklace, a gold-plated, lightweight necklace made with samarium-cobalt magnets, kept his neck feeling loose when he spent hours at a time working in front of a computer screen. Considering the number of secretaries, office workers, students, and other individuals who work at computer screens all day long, such necklaces could prove very useful for preventing or relieving the neck and shoulder problems that frequently plague computer users. The only warning I would offer to those who would use magnetic necklaces (or any therapeutic magnet) while working around computers is to keep them away from floppy disks, which could be accidentally erased by coming into contact with a strong magnet. The types of magnets used in making magnet necklaces can vary greatly. Besides using conventional spherical iron-based magnets on a string, some magnetic-necklace manufacturers use magnetized balls made from hematite (a polished, dark, iron-rich mineral) or from samarium-cobalt-iron mixtures. Payne suggests wrapping a magnetic necklace around the forehead to relieve a minor headache, or around the wrists, elbows, knees, feet, or hands to relieve pains in those body regions. Payne notes that if the necklace is formed into two small circles resembling the number eight on its side, these circles can be placed over the eyes to deliver a soothing magnetic-rejuvenation effect. While magnetic necklaces may actually improve the condition of the entire body for some individuals, other people with more serious forms of illness seem to obtain less profound health improvements from such necklaces, unless magnetic therapy is synergistically combined with other healing modalities. The magnetic necklaces appear to work the best for relatively healthy individuals whose only problem is minor aches and pains.

Magnets can also be sewn into special "magnetic garments" that can be worn for pain relief. The magnetic back braces or neck braces are the simplest of these and can be worn during the day for back-pain relief. You might also consider using one of the available variety of magnetic insoles in your shoes. Magnetic insoles have recently been proven effective in relieving foot pains caused by diabetic neuropathy, but other forms of foot pain have been known to respond as well. The magnetic insoles vary somewhat in construction and magnet placement. Some insoles are made of a flat, pressed, rubber-type magnet with multipolar north and south zones throughout the same side of the insole. The insoles often have tiny raised beadlike surfaces, which are thought to provide a kind of acupressure massage to certain acupoint regions or reflex-

The Many Forms of Magnetic Therapy

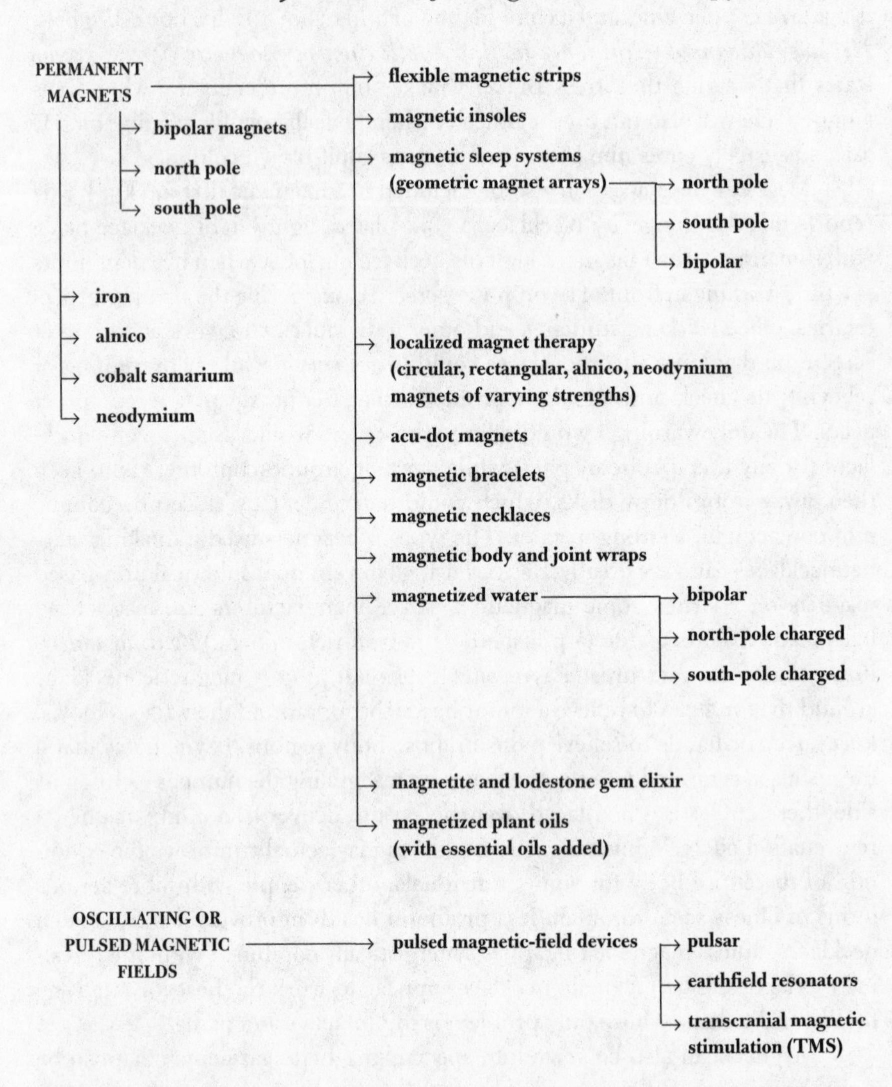

ology zones of the feet as well. Other magnetic insoles have small flat magnets incorporated into a flexible material. The small magnets are positioned at key locations to stimulate certain zones of the feet.

We have already mentioned magnetic beds, sometimes referred to as magnetic sleep systems. Different manufacturers from the United States, Canada, and Japan use magnets of different sizes, different polarity arrangements, and even incorporate wire-grid patterns into the system. Metaforms, a company in Boulder, Colorado, produces a unique fabric sleeping pad that incorporates magnets, crystals, and a geometric wire gridwork array into its design. Not only does this unique assemblage appear capable of helping to magnetically balance the body's energies, but the wire grid within it can also be accessed through metal alligator clips. This access allows the pad to be connected to an external vibrational remedy so the pad can function as a "substance biocircuit." A substance biocircuit allows subtle energy from a particular remedy to pass from a bottle directly into the bioenergy field of the individual lying on the circuit. The assumption is that the vibrational energy of a flower essence or homeopathic remedy in solution, connected via a wire passing through the remedy, could be transmitted to the body of the person lying on the Metaforms sleep pad. Magnet researchers note that magnetic beds are useful mainly for people whose bodies are already in balance but need a little energizing or tonifying.

There are also a variety of pulsed-magnetic-field devices available for self-treatment. One interesting magnetic-treatment system, the Pulsar, was developed by Dr. Payne. It consists of a small box (the size of a Walkman tape player) connected to a wire that runs to a small, flat, disk-shaped pad containing an electromagnetic coil. The device sends a pulsed magnetic field through the pad. The magnetic field pulses about sixteen times per second (the frequency of magnetic pulsation found to maximally stimulate blood flow). The three-inch-diameter treatment pad can be taped or held against the body, and is said to produce bionorth energy on one face and biosouth energy on the other side. A stronger version, known as the Power Pulsar, comes with a nine-inch-diameter flexible treatment coil that can be bent around the shoulders, elbows, legs, etc. The Power Pulsar produces a much stronger field than the smaller Pulsar and is said to penetrate the entire body with magnetic energy. The larger system comes with a box (containing the electronics and power supply) about the size of a small laptop computer. There are also a number of pulsed-magnetic therapy devices in use by physicians in Germany and England. These European devices are obviously less accessible to patients living in the United States. But you can always travel to special clinics or research centers if you're in search of a particular, unique form of magnetic therapy.

Regardless of which magnet is eventually chosen, you must follow the basic magnet-therapy guidelines. Increasing numbers of people are beginning to experiment with using magnets for pain relief. More and more sports-medicine doctors are using magnets for treating a wide range of musculoskeletal disorders. And, as mentioned, a variety of other medical problems, ranging from depression to Parkinson's disease, are being helped with experimental magnetotherapies. Magnetic therapy, once thought to be in the realm of medical quackery, is gradually gaining acceptance and support from both old and new scientific studies. It is perhaps the cheapest, and probably one of the safest (when properly used), forms of vibrational therapy available today. Yet its benefits remain unknown to the majority of Western-trained physicians. The emerging vibrational-medical model points to new ways of understanding how magnetism may actually be one of the key energetic phenomena behind the process of life itself. Perhaps in the future, all physicians will come to better appreciate the many health benefits of this ancient yet powerful form of healing.

FOOTNOTE

1. Guy Playfair and Scott Hill, *The Cycles of Heaven* (New York: Avon Books, 1978), p. 116.

N i n e

RADIONICS AND DISTANT HEALING: THE BLACK SHEEP OF VIBRATIONAL MEDICINE

RADIONICS IS A vibrational therapy of the twentieth century. Currently, England is the greatest center for radionics research. But radionics is not a recognized form of medical therapy in America. Thus, most radionics devices are confiscated by the FDA when brought to the United States. FDA officials are quite sure radionics machines have no medical value whatsoever. In other words, radionics systems are essentially considered illegal technologies, at least with regard to their being transported and marketed across state lines. America provides a rather unwelcome atmosphere for radionics practitioners attempting to practice their skill. Despite the controversy around radionics, though, time must be taken to explore the theories and history of such a complicated, controversial, yet fascinating therapy.

What Is Radionics? *The History of a Magical Science*

Radionics is a science of device technologies. All radionics systems are built to interface human consciousness with a patient's "person-specific" subtle energies in an effort to diagnose and treat illness from a purely vibrational perspective. Radionics is one of the few areas in vibrational medicine capable of providing tools that allow a trained radionics operator to measure and quantify the energy characteristics of the physical body, its organs, the chakras, the etheric body, and the higher spiritual bodies. So bizarre and irrational to the left-brain, analytical, scientific mind are some of the phenomena observed in radionics that its detractors have claimed that radionics is not truly science but something more akin to magic. To the uninformed, it actually appears as if something magical is taking place during radionics therapy. But to those who are better informed about the nature of radionics and what it is capable of, this developing science provides a whole new understanding of the far-reaching potentials of human consciousness and the hidden healing capabilities of the multidimensional human being. The unique discoveries made by pioneers of radionics research, taken as a whole, have begun to provide us with a new window into a world of vibrational healing that until recently did really seem to be almost magical. Only in the age of quantum physics and laser holography have scientific models begun to evolve that may help explain radionics. It is easy to see how many early radionics practitioners were persecuted for what must have sounded like insane, heretical notions of invisible life-energy waves.

The science of radionics employs a variety of specially contructed devices. Under carefully controlled conditions, the device can both diagnose illness and provide a purely vibrational form of therapy for healing various diseases. Radionic "machines" sometimes seem to act as a kind of psychic amplifier for the unconscious energy-sensing capabilities of their operators. The amplifying capability of radionic devices allows a radionics practitioner to make an accurate diagnosis of a patient's illness without ever having to physically examine the patient. In addition, radionic devices may also amplify innate individual psychic-healing skills. Radionic systems are capable of broadcasting highly specific vibrational healing energies to patients even at great distances. But the challenge of radionic technology is explaining how it works.

The story of radionics is strange and convoluted, filled with weird discoveries by some rather brilliant but unorthodox thinkers and inventors, most of whom were misunderstood by their scientific peers. It is a story of scientists who were just a little too far ahead of their time and of the incredible potential of human consciousness. One of the earliest radionics pioneers was Dr. Albert Abrams, an early-twentieth-century San Francisco–based neurologist. Like most physicians in 1910, Dr. Abrams typically performed routine examinations of his patients' abdomens using the physical-examination technique known as percussion. During percussion, a physician places the finger of one hand on the abdomen and then taps firmly on the resting finger using the middle finger of the opposite hand to create a specific sound. Percussion is sort of like thumping a melon to test its ripeness. The percussion technique gets its name from its similarity to the process of creating sound by hitting a stretched membrane covering a drum or other percussion instrument. The sound produced by the tapping finger, whether hollow or dull, provides the physician with information about the acoustical density of the underlying tissue structures just below the exploring finger. During a typical physical examination, the nature of the sound heard while tapping on different areas of the abdomen helps to provide feedback about the size of a patient's liver, the occurrence of abnormal abdominal masses, and even the presence of excessive gas or fluid in the abdomen.

One day Dr. Abrams was routinely performing abdominal percussion on a patient with a prominent lip cancer. The doctor noticed a dull-sounding note in one area of the abdomen that he could not explain. Oddly enough, when he had the patient move from an east-west orientation to a north-south position, the dull note immediately changed to a hollow note. This change in the percussion note's character would happen repeatedly when he had the patient change positions back and forth between north-south and east-west. The dull note would occur only when the patient was facing west, a strange phenomenon Abrams had never before encountered. Dr. Abrams was intrigued. It seemed to him that a change in a patient's orientation to the earth's magnetic field was producing physiological changes in the body as detected by the percussion technique. Furthermore, Abrams speculated that the dull note he was getting from percussing the patient's abdomen had something to do with energetic emanations from the cancer on the patient's lip. Another odd observation Abrams made also seemed to confirm the actions of a "disease radiation." When he had a healthy patient lie on an exam table pointing due west while placing a bottle containing a cancer-tissue specimen on the table next to the patient, the healthy patient would immediately exhibit the same abdominal-percussion reflex as seen in the cancer patient. If the bottle of cancer-specimen tissue was removed from the room, the cancer-associated abdominal reflex would disappear entirely when the healthy patient was reexamined using percussion. Abrams believed that something entirely energetic, emanating from the cancer-tissue specimen in the glass bottle, must be passing through the glass and into the energy field of the nearby healthy patient. Somehow the cancer tissue's energy produced a reflex change in the healthy patient's abdominal muscles, as demonstrated by a change of sound from hollow to dull on abdominal percussion.

Abrams began to do experiments using percussion of the abdominal region as a form of feedback to energetically detect different diseases. The human abdomen literally became his diagnostic "sounding board." As part of his standardized test, Abrams would first carefully percuss the abdomen of a healthy young male volunteer. The doctor made sure there were no abnormal areas of dullness in the volunteer's abdomen no matter what direction the volunteer faced. Abrams would then hook up the young surrogate tester with a wire that led to the patient who needed to be tested for illness or to a wire linked with a surgical-tissue specimen. When using cancer tissue in circuit with the healthy tester and the metallic wire, a healthy young tester would produce the same dull percussion note in the area of his abdomen that Abrams had detected in the first lip-cancer patient. For optimal results, Abrams determined that the wire from the tissue specimen needed to be connected to the surrogate tester's forehead (via a headband mounted electrode). Furthermore, the abdominal-reflex phenomenon was reproducible only if the surrogate tester was facing west. The necessity for maintaining a particular geomagnetic orientation was never fully explained by Abrams, but it was later discovered to be a necessity for reproducing other radionic phenomena as well. Abrams began to experiment with percussion of the abdomen on patients suffering from different diseases and found that each disease repeatedly produced a dull note in a very specific area of the abdomen. He felt he had stumbled onto a new diagnostic method that might be of great benefit in precise disease detection. It appeared to Abrams that various diseases produced some kind of unique body reflex involving a change in the character of the abdominal muscles. This reflex could be detected only by using the percussion method of physical examination (at least in Abrams's early experiments). At first Abrams was convinced that each disease produced its own unique abdominal-reflex change. However, one day he used the abdominal-percussion procedure on a patient with syphilis and found that the syphilitic patient had dullness to percussion in the same region he had come to associate with cancer. This meant there were now two diseases that produced the very same identifying abdominal-reflex muscle change. It eventually occurred to Abrams that the stomach reflex was not so disease-specific after all.

While initially dismayed, Abrams soon began to study the various muscle-reflex changes in greater depth. After much thought and experimentation, he became convinced that what he was detecting in patients with various types of illnesses was a disease reflex produced by some type of "electronic energy." The disease energy seemed electronic in nature to Abrams because it could flow through metallic wires from sick patients (or abnormal tissue specimens) to the surrogate testers, whose abdominal muscles were temporarily "altered" by its influence. According to Abrams's theory, this electronic disease energy, emanating from the abnormal body tissues, energetically influenced the abdominal muscles of both sick patients and surrogate testers to produce reflex changes that could be detected using the percussion technique. He speculated that the unusual disease radiation was due to an abnormal arrangement of electronic

Basic Principles of Radionics

1. **Each disease is associated with a specific energy with unique "frequency" characteristics.**
2. **Disease energies are radiated from a patient's body, a specimen of diseased biopsy tissue, and even from the blood.**
3. **Human beings react to this disease frequency, especially when they are oriented to geomagnetic west.**
4. **When facing geomagnetic west, human beings will react to disease frequencies in the body by producing a unique abdominal reflex—a temporary change in the abdominal muscle detectable by percussion.**
5. **Healthy humans (facing west) will produce the characteristic abdominal reflex if they are close to or electrically wired to disease specimens or patient blood spots from sick individuals.**
6. **Disease-specimen energy frequencies are easily conducted along metallic wires (passive conduction).**
7. **Variable-resistance devices (radionic instruments) act as specialized tunable electronic filters that only allow one disease frequency to be conducted through the device circuitry at a time, depending upon the variable tuning setting.**
8. **The radionics operator becomes psychically linked to the patient at a distance via the blood spot.**

energy in the diseased tissues of the body. Abrams came to this conclusion after observing many healthy surrogate testers manifesting the same diagnostic abdominal reflexes as the patients who had a particular illness, as long as they were connected via a wire to the sick patient (while facing west). In addition, the phenomenon of activating the disease reflex merely by placing a bottle containing a diseased tissue specimen close to, but not touching, the surrogate tester suggested that this peculiar disease radiation behaved like a kind of "wireless energy" or energy field. It was clear that some energetic aspect of the disease was being transmitted over the wire, as well as through glass and air, and from one individual to another, as demonstrated by the temporary "transfer" of the disease-reflex reaction to the healthy surrogate tester's abdominal muscles.

Abrams soon began experimenting with this "electronic" energy. He began by having a special variable-resistance box built. The device could vary the resistance to "electronic energy" flowing through the wire of the three-part radionic surrogate/wire/patient circuit. Abrams wanted to use the resistance box as a discriminator for filtering out one disease frequency from another. The variable-resistance box was introduced into the circuit that carried the disease energy along the wire between the patient in question and the surrogate tester. Abrams tuned the resistance-meter dials to different settings (in ohms) and listened for feedback from the so-called stomach reflex on surrogate testers. He determined which settings would allow the disease frequency of, say, syphilis to pass through (while simultaneously blocking the "cancer frequency"). Using this bizarre method, he eventually discovered the frequency-specific numerical codes that would allow him to detect the energy produced by different diseases using his resistance box. Each "disease code" was based upon how many ohms of resistance it took to filter out all bioenergies except for the particular disease frequency. By tuning the dials of his resistance-meter box to one setting versus another, Abrams had come upon a way of using the human abdominal reflex to effectively distinguish cancer from syphilis, as well as from other diseases. (For instance, when the dials of the resistance box were tuned to, say, 1,427 (ohms), a positive abdominal reflex in the surrogate tester would only be detected only if the patient being tested had syphilis.) Eventually, in this manner, Abrams and his students learned to successfully diagnose a wide variety of ailments. His discovery became known as the "Electronic Reaction of Abrams" (ERA). It could be said that Abrams's ERA device was the very first radionic instrument ever built.

Based on initial successes, Dr. Abrams went on to create a guide containing the appropriate diagnostic areas of the abdomen and the resistance (in ohms) on his device necessary to detect all major diseases. At the same time, Abrams made another great discovery about his three-part diagnostic circuit. He found that a drop of the patient's blood placed on a piece of filter paper could be connected in circuit by a wire to the radionics device and to the surrogate tester in order to obtain a diagnosis. Even with this alternative hookup, using only a drop of blood, it was still possible for his equipment to diagnose an illness. In other words, the unusual electronic disease vibrations of illness seemed to emanate from a single drop of the patient's blood! This discovery meant that a patient did not need to be physically present in order for a diagnosis to be made. Only a drop of the patient's blood, Abram's radionic device, and a healthy surrogate tester were required for a patient's diagnostic workup. Abram's discovery eventually led to the modern-day practice of using blood spots on filter paper as a means of "distantly" diagnosing a patient's illness using radionic technologies.

Abrams successfully diagnosed hundreds of patients with his radionics device. But he received complaints from some of his student doctors. It seemed that his specialized percussion method of disease diagnosis was very difficult to learn. Therefore, Abrams set about devising a different form of feedback to be used as an alternative to percussion. His students were given a glass rod that could be stroked across the patient's abdomen. When the student doctors encountered an abnormal stomach reflex (SR), the glass rod would encounter resistance against the skin of the abdomen, making the skin pucker. This reaction of the glass rod's "sticking" to the skin of the abdominal wall seemed to work just as well as the percussion method and was easier for student doctors to learn. However, there was yet another technical problem with Dr. Abrams's radionic diagnostics system. The challenge involved his dependence on having human testers as a key part of his disease-detection system, young, healthy surrogate testers who could consistently produce the requisite "stomach reflexes" that Abrams's equipment relied on for feedback. The young men who volunteered found that they could tolerate being used as "guinea pigs" only for several hours before becoming fatigued. This fatigue factor resulted in unreliable reflex readings after a period of time. Another odd phenomenon sometimes occurred during the radionic testing period. In addition to the necessity for facing west, Abrams eventually discovered that testers needed to avoid the color yellow during the test procedure. This meant not wearing any yellow clothing or even eating any yellow food before or during the radionic testing. The color yellow, as we discussed in relation to color therapy, is said to resonate strongly with the solar plexus chakra. Since the solar plexus chakra is probably the dominant chakra involved in producing Abrams's stomach reflexes, perhaps I might offer an explanation for the need to avoid yellow in any form. As yellow is said to energize and influence the third or solar plexus chakra, its frequency-specific energy input might have interfered with the sensitive process of disease diagnosis using Abrams's radionics device. It is possible that his diagnostic system somehow relied upon the solar plexus chakra and its form of subtle-energetic "gut intution" to produce the necessary abdominal-reflex effects. Alternatively, the color yellow might have energetically "overloaded" the solar plexus center, making the tester's abdominal reflex an unreliable disease indicator in the presence of yellow.

Although Abrams sought to replace the percussed abdominal reflex and the abdominal glass-sticking response with a variety of devices and detectors, he remained unsuccessful in his attempts. In order for his radionic system to work, a human-feedback instrument had to serve as part of the circuit. At one point Abrams even offered ten thousand dollars to any electronics expert who could provide him with an objective diagnostic-feedback system for his radionics instrument, but to no avail. This dependency of early radionic systems upon the human mind, body, and the chakra system to achieve successful disease detection suggests that the human element is a key factor in radionics. However, to the linear, dogmatic thinking of Abrams's medical colleagues, such a subjective system of diagnosis was too unbelievable and strange for many to accept. The concept of an interdependent human/machine system was too difficult for many of the doctors of the early 1900s to grasp. But Abrams's research paved the way for an entirely new science of subtle-energy diagnosis that would be refined and even computerized in the years to come. Dr. Albert Abrams was certainly ahead of his time in many respects. The results and implications of his research are only now being used to understand the subtle- and spiritual-energy dimensions of human beings.

What Abrams appeared to have stumbled onto was the field known as medical radiesthesia. Radiesthesia is a term that literally means "sensitivity to radiation." Medical radiesthesia is frequently used to describe the ability of certain individuals to use various forms of dowsing in order to intuitively diagnose medical ailments. Medical diagnosis by dowsing is theorized to work in the following way: Medical dowsers are believed to psychically tune in to a distant patient based upon some type of information-carrying subtle energy or radiation emanating from the patient. Medical dowsers seem to be sensing some aspect of the human energy field made up of biomagnetic energies radiating from the cells of the physical body, as well as the vibrational fields of the etheric and higher spiritual bodies. Recent dowsing research has suggested that there may also be areas in the physical brain (such as the hypothalamus) that are actually sensitive to weak magnetic fields. However, a number of radiesthesia researchers lean toward viewing both radionics and dowsing as operating primarily from psychic input via the chakra system. The psychic input (coming through the chakras) is thought to be expressed through the nervous system of the radionics practitioner or medical dowser. As mentioned earlier, the chakra system is closely linked with the nervous system in the physical body. Many vibrational-medicine researchers believe that the chakras are not just subtle-energy processors involved in supporting the tissues of the physical body but psychic-energy processors as well. And different chakras seem to be linked to different forms of psychic sensing. It has been suggested that we take in unconscious psychic information through the chakras via their innate psychic sensing capacities. Often, the psychic information received through the chakras is outwardly expressed through the nervous system, but at a totally unconscious level. The postulated unconscious link between the chakras and the nervous system would explain how unconscious neuromuscular actions (in the form of unconscious muscular arm motion in a pendulum dowser or in the form of involuntary abdominal-muscle reflexes in an Abrams radionics tester) could provide feedback about the nature of pathological disease energies coming from a particular patient or a specific biological specimen (such as biopsied tissue or a drop of blood). Some medical dowsers, studied by medical researchers in France and Europe, have had amazing reputations for the accuracy of their intuitive medical diagnoses, arrived at using only specially constructed pendulums.

In general, the broad categories of dowsing and radiesthesia are based upon the idea that all things radiate an invisible subtle energy that unconsciously affects human beings. Some professional dowsers have studied the nature of the subtle-energy emanations coming from different physical substances (such as minerals) and have put their research to practical use in successfully locating valuable underground deposits of minerals and semiprecious metals. What Abrams and his early radionics colleagues were probably doing was merely exploring a different way of detecting subtle-energy radiation coming from the tissues and blood of sick patients, almost in the same way a professional dowser tries to detect various types of underground mineral deposits. Just as the energy sensed by dowsers allows them to locate underground deposits of gold, uranium, and other precious minerals, so did Abrams's radionics process help a radionics practitioner locate and quantify disease. In a way, both procedures might be considered a form of "psychic prospecting." Abrams's discovery of the strange, unconscious abdominal-muscle reflex offered a simple way of using the human energy system, and its unconscious neuromuscular reflexes, to provide a sophisticated, subtle-energy-based disease detector. But Abrams went far beyond the dowsing rods and pendulums used by dowsers, incorporating electronic equipment into the human-detection circuit to aid in quantifying and diagnosing specific illnesses accurately through their unique disease energy signatures.

Abrams was intrigued by the fact that his surrogate testers (the human detectors in his radionic feedback system) would produce the requisite stomach reflex only if they were facing due west. If a disease's radiation passing via a wire to the tester could somehow be neutralized by the magnetic fields of the earth (through changing the tester's direction), Abrams wondered, could other forms of electromagnetic radiation counteract a patient's abnormal disease radiations in order to actually heal the disease? Abrams began to do radionic experiments looking at the energy qualities of different known cures for specific illnesses. He eventually experimented with quinine and malaria. In Abrams's day, quinine was prescribed as a cure for malaria in both pharmacologic doses and homeopathic potencies. In a strange discovery, which may lend credence to the resonance hypothesis of homeopathy, Abrams found that when quinine was introduced into his radionic circuit, it produced an abdominal reflex in the test subject identical to the reflex produced in a tester radionically wired to a malaria patient. In other words, the abdominal reflexes Abrams observed for both the disease and its cure were identical.

This close match suggested some type of energy resonance between the illness (malaria) and the remedy (quinine) used to treat it. In classical homeopathy, healing is thought to take place only when the vibrational frequency of the remedy matches (and somehow neutralizes) the "disease frequency" of the patient. The matching of homeopathic and disease frequencies seems to produce a canceling-out process that ends up neutralizing the disease frequency and thus cures the illness energetically. Dr. Abrams's hypothesis was given further weight by another observation made during his malaria research. While testing his surrogate testers using a drop of a malaria patient's blood together with a few grains of quinine in the radionics circuit, Abrams noted that the abdominal-reflex reaction for malaria had completely disappeared. As long as the tester remained radionically connected to both malaria-tainted blood and quinine, the malaria abdominal reflex could not be reproduced. The quinine seemed to be energetically canceling out the disease radiation produced by the malaria-infected blood. Abrams tried a variation on the experiment by radionically testing a second infectious disease, syphilis, comparing it with its then-

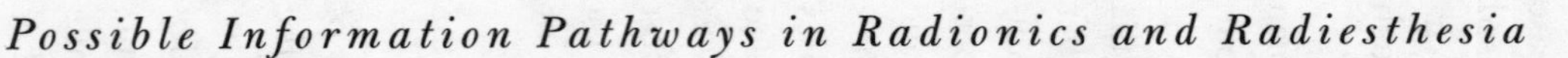

Possible Information Pathways in Radionics and Radiesthesia

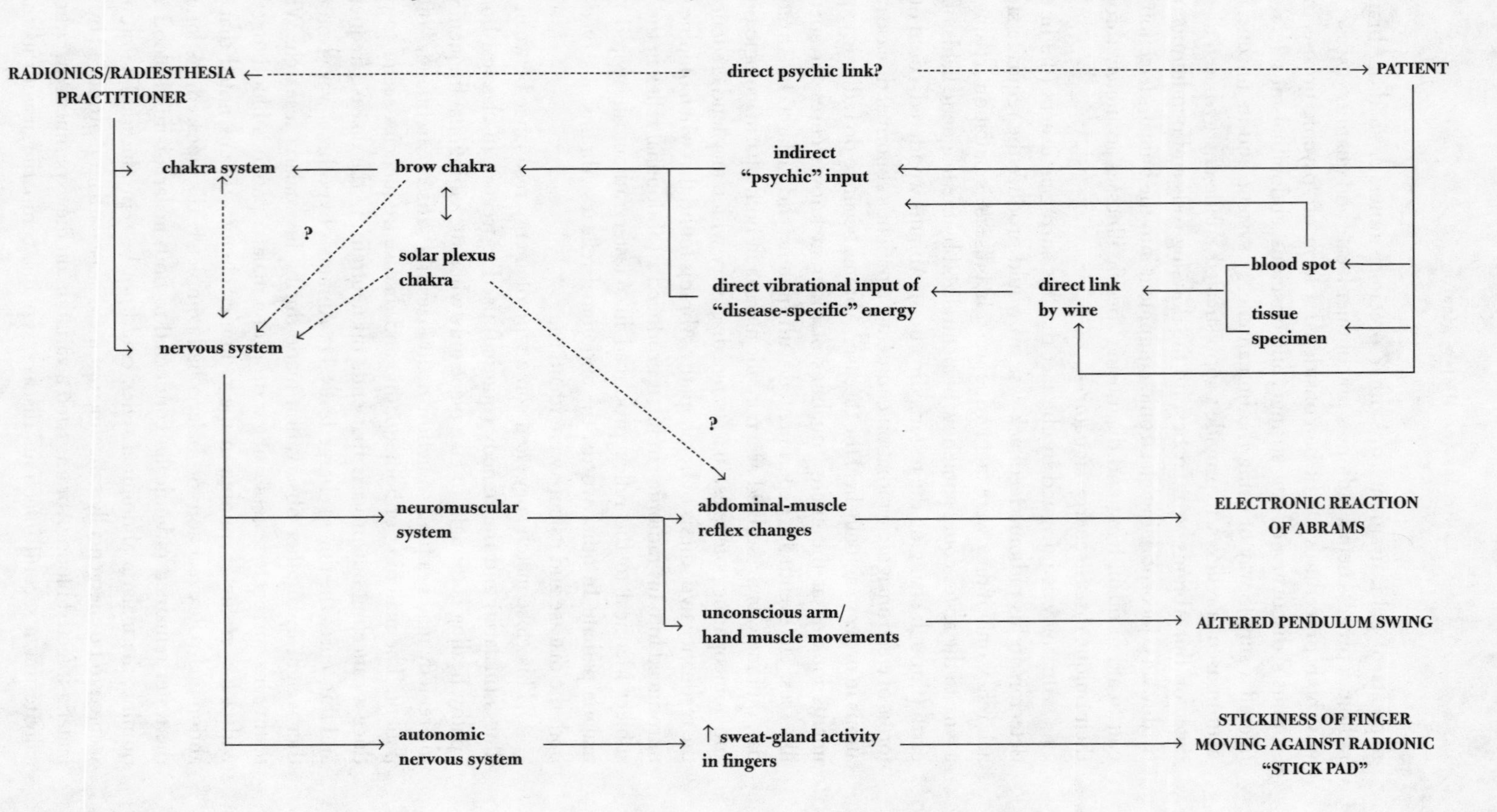

popular cure, mercury. He introduced a drop of blood from a patient with active syphilis into the radionic circuit and found that his testers showed the indicative abdominal-reflex characteristic of syphilis. If, however, the toxic metal mercury (which in Abrams's day was a recognized cure for syphilis) was added to a container holding a drop of blood from a syphilitic patient, there was a disappearance of the syphilitic abdominal-reflex reaction in the surrogate tester. This provided further evidence to Abrams that aside from their chemical effects in the body, there was another important reason certain medicines cured specific illnesses. He believed that healing substances had certain energy qualities that neutralized the abnormal disease radiation produced by the diseases for which they had been prescribed. Abrams wondered if there might not be an entirely energetic method of healing illness that fit his theory of electronic vibrations of diseases.

Working with one of the foremost radio and electronics experts of the time, Abrams developed a new radionics treatment device that became known as the Oscilloclast. During a typical session, a patient would be connected to the Oscilloclast by a wire and an electrode. After initially making a medical diagnosis of the patient's illness using his basic radionics test equipment, Abrams then tuned the dials of the Oscilloclast's radionic circuit until it was set at the correct one of ten special energy-treatment settings. Each of the ten treatment settings produced a different "treatment rate" or energy frequency which would be delivered by the device to the patient. Unlike his previous "passive" radionic systems, the Oscilloclast contained electronic equipment that was actually powered by an electrical current. The different settings on the instrument consisted of specific frequencies of electrical and radiofrequency energy that Abrams and his colleague had empirically discovered to be helpful for treating particular illnesses. After being tuned to the right setting, the Oscilloclast would broadcast electrical and radiofrequency energy to the patient via a wire connected to an electrode placed over the area of the patient's spleen. From a subtle-energy perspective, this is interesting, since the spleen chakra (one of the important minor chakras) is a major processor and distributor of prana and subtle energy to the other chakras and energy systems of the body. Although repeated treatments by the Oscilloclast were often required for long-lasting healing effects, Abrams and his students had great success in using his new radionics device to energetically heal a wide variety of medical disorders.

Many health professionals studied Dr. Abrams's work in an attempt to carry it further. The most successful of his initial successors was Dr. Ruth Drown, a chiropractor who studied Abrams's instuments and methods. Dr. Drown attempted to refine the apparatus to an even greater degree. She was among the first to introduce into radionics instruments a new feedback system, a rectangular rubber feedback plate that became known as the "stick plate." This name came from the process by which the radionics instrument was "tuned" to the appropriate diagnostic and treatment setting. After connecting her instrument to either a patient or a spot of the patient's blood placed in a special area of the device, the Drown radionics-device operator would stroke his or her finger against the rubber membrane of the stick plate while simultaneously tuning the resistance-meter dials of the device to different settings until a "sticking reaction" was achieved. When a dial became tuned to the appropriate resistance setting, the rubber membrane became noticeably "sticky" to the moving finger of the radionics operator. This sticking sensation acted as feedback to the operator that the correct dial setting had been reached. One might compare a stick-plate reaction to the sound of a tumbler in a lock mechanism clicking into place as a burglar listens attentively to the turning dial of a safe he is attempting to break into. It was also reminiscent of the "sticking reaction" of Abrams's glass rod as it was stroked against the abdomen of a surrogate tester.

After tuning the various dials to the appropriate settings based upon feedback from the stick plate, Drown would read off the numbers from the system to arrive at the "radionic rate" of the patient. The multidigit rate number would be compared to a reference table listing the numerical rates or frequencies Drown had determined were associated with different diseases. It is interesting to note that a number of dowsing researchers used Drown's instrument to compare the radionic rates for certain inorganic mineral specimens with the numerical frequencies dowsed with pendulums or dowsing rods. They found that the dial readings of Drown's equipment were almost perfectly correlated with the dowsing values for certain minerals, and the dowsing values matched the minerals' atomic weights precisely. This peculiar finding seems to further support the hypothesis that dowsing or medical radiesthesia is involved in radionics. It also provides additional support for the idea that Drown's equipment really did allow the operator to acquire accurate information related to specific atomic structures and energy characteristics of the various mineral or biological specimens being analyzed!

Dr. Drown even refined Dr. Abrams's technique of using blood spots in both diagnosis and treatment of illness. She began using the blood spots to radionically "broadcast" subtle-energy treatments to distant patients. In the case of distant radionic treatments, each blood spot appeared to somehow resonantly link the radionic instrument with the distant patient, no matter how far away the patient might be. Drown was among the first radionics practitioners able to use a blood spot to diagnose and treat an individual even when the patient was miles away from the location of the radionics procedure. The ability of Drown's equipment (known as Radio-Vision) to function successfully in this manner supported her claim that she was able to accomplish healing at a distance with radionic technology. But the rationale behind the operation of Radio-Vision made little sense to classically trained scientists of the time, or to modern-day critics, for that matter. You see, according to the so-called inverse-

The Strange Case of the "Electronic Reaction of Abrams" and the Evolution of Radionic Instruments

THE "BASIC" ERA OR ELECTRONIC ABDOMINAL REFLEX

patient with cancer facing west → abdominal percussion → positive abdominal reflex = ERA (Electronic Reaction of Abrams) → confirms presence of cancer in body

patient with cancer facing any other direction than west → abdominal percussion → negative (absent) abdominal reflex → falsely suggests no cancer present

SECOND-LEVEL ERA (SURROGATE TESTINGS)

healthy individual (surrogate tester) facing west → surrogate tester linked via wire to cancer specimen or cancer patient → positive abdominal reflex → confirms malignant nature of tissue specimen or cancerous growth in patient

surrogate tester facing any direction other than west → surrogate tester linked via wire to cancer specimen or cancer patient → negative (absent) abdominal reflex → with surrogate nonaligned with magnetic west, results of ERA surrogate test is invalid

VARIATION NUMBER ONE ON SECOND-LEVEL/SURROGATE RADIONIC TESTING— THE INTRODUCTION OF THE BLOOD SPOT

surrogate tester facing west → tester becomes linked via wire to blood spot → variable-resistance box introduced into circuit between tester and blood spot → resistance box acts as variable bandpass filter, selectively allowing only certain disease frequencies to pass through the circuit, according to the tuning setting → abdominal reflex of surrogate tester becomes feedback for disease presence or absence

resistance box acts as variable bandpass filter … ↓ radionics device operator tunes the resistance (ohms) dials on the instrument (based on feedback from abdominal reflex) ← (feedback from abdominal reflex)

radionics device operator tunes the resistance (ohms) dials on the instrument (based on feedback from abdominal reflex) → limited radionic disease diagnosis

NEXT GENERATION IN RADIONIC TESTING AND DIAGNOSIS OF DIFFERENT DISEASES THE "RADIONIC STICK PAD" REPLACES THE ABDOMINAL REFLEX (ERA)

blood spot on filter paper placed in radionics device "witness well" (permanent magnet in internal radionic circuitry does away with need for geomagnetic alignment to west) → radionics operator intuitively tunes dials of resistance device (direct psychic link replaces wire linkup with patient) → positive "sticky" sensation on stick pad = equivalent ERA (positive reaction) = correct dial tuning on radionic device

correct dial tuning on radionic device ↓ procedure is repeated for sequence of three or more resistance dials → multidigit numeral frequency code read off the dials of the "tuned" radionic instrument (the so-called radionics rate) → radionics operator looks up frequency (radionics rate) in disease table that lists specific frequencies for specific diseases → based on analysis of radionic rate and disease-rate table, diagnosis of illness is made

square law of physics, any kind of electromagnetic energy decreases in intensity according to the square of the distance from the source. In other words, the farther from the energy source, the more rapidly the energy intensity drops off. That radionics treatments could work even when patients were many miles away suggested that something other than conventional electromagnetic fields was operating. However, scientists of Drown's era (as well as scientists of today) did not accept the possibility that radionics operated by means of subtle energies with properties quite different from conventional electromagnetic fields and currents. As we are now finding out, though, the subtle energies of the human multidimensional system and of the life force itself may have the unique property of being able to act even at great distances!

An even stranger phenomenon reported by Drown involved unexposed photographic film that was placed into a specially made cameralike radionic instrument. If Drown also added a patient's blood sample to the circuit connected to the camera loaded with film, she was able to obtain photographs of a given patient's disease-affected internal organs. Oddly enough, if the patient was directly wired to Drown's radionic camera device, a picture of a particular diseased organ's external surface would be produced that was quite similar to what might be seen by a surgeon during exploratory surgery. Conversely, if a blood spot was used to obtain the picture, the developed radionic photograph would display an image of the inside of the same diseased organ very similar to the image produced by modern flexible fiberoptic endoscopes. The ability to produce this strange photographic phenomenon seemed to be a skill demonstrable only by Drown and a few others, though. You see, it seems Dr. Drown was a practitioner with unique psychic abilities. It is believed that Drown might have developed many of her radionic innovations based primarily upon her marvelous psychic insight. Some later radionics researchers were also able to produce radionic photographs of patients' internal organs. The most notable of the group was George DeLaWarr. He developed the famous Mark VI DeLaWarr camera. Like Drown, DeLaWarr apparently had the gift of psychic sight. And again, it does look as if practitioners with psychic insight do achieve the best results in radionics diagnosis.

The potential for distant radionic diagnosis and healing of patients via blood spots, along with other strange claims attributed to Drown, undoubtedly made radionics seem more magic than science in the eyes of its harshest medical critics. Both Abrams and Drown had many physician followers and supporters who successfully used their radionics systems to heal patients. But radionic healing was still considered extremely controversial and was attacked by many who held positions of power in the dominant medical establishment of the day. Drown was eventually persecuted by the medical profession and the court system. Both mainstream bodies were interested in conventional scientific explanations for how radionics devices worked. They were not moved by the testimonies of many patients who had been cured by radionic treatments. To the court system, the term "radionics" seemed to imply some sort of relationship to radio waves, the only form of energy then known to transmit or receive information at a distance. Therefore, the legal system sought an explanation of radionics that involved "radio waves." During Drown's trial for medical fraud in California, experts on radio equipment and electronics stated that there was no way Drown's radionic equipment could have any basis in radio technology, especially after closer examination of the equipment. The testimony of the radio experts carried great weight with the jury. Ultimately, Drown was convicted of medical fraud. The conviction happened despite a continuous stream of successfully treated patients who testified that they had been cured by Drown of a variety of diseases, some of which conventional doctors had deemed incurable. Following the long court battle, Dr. Drown ended up serving a short jail sentence. She died of a stroke in 1966. While she was in jail, all of Dr. Drown's radionic instruments and technical notes were destroyed by zealous authorities on a crusade resembling the persecution of Dr. Wilhelm Reich in the 1950s. Reich, a student of the famous psychoanalyst Dr. Sigmund Freud, professed to be able to heal illness using devices known as "orgone accumulators." Like Drown, Reich claimed to use a subtle unseen, unmeasurable life energy (which he called "orgone") to heal individuals. He also asserted that some of his orgone devices (known as "cloudbusters") were capable of altering local weather patterns. Like Drown's, Reich's heretical notions were met with his imprisonment and the destruction of his research, books, and equipment by government authorities. Reich died soon after going to jail. In comparison, Drown's persecution was so extreme that certain states made mere possession of a Drown radionics instrument punishable by imprisonment. This was the McCarthy-like atmosphere of fear and persecution created around radionics in the United States during the 1950s (and up until the present) that has led most radionic research in this country to go "underground."

One other radionics researcher who further refined Abrams's and Drown's instrumentation was inventor T. Galen Hieronymous, a pioneer in the field of radio. Hieronymous has the distinction of being awarded one of the most unusual patents ever granted by the United States Patent Office. His patent, number 2,482,773 granted in 1949, describes an instrument for detecting emanations from materials (both organic and inorganic) and for measuring the characteristics of these energy emanations. His device (which was not actually referred to as a radionics instrument) had a classic rubbing "stick plate" (à la Drown) as well as a variety of electronic devices, including a radiofrequency amplifier, variable condensers, and noninductive resistance devices. At the heart of Hieronymous's device was a glass prism component that he said refracted, focused, diffracted, and otherwise manipulated the energy emanations picked up from different substances. The fact that Hieronymous's device had a prism to refract subtle energy made it clear that whatever the energy being dealt with was, it certainly wasn't electrical in nature, as Dr. Abrams had once surmised many years before. Hieronymous named this strange force "eloptic energy" because it seemed to have both electrical and optical properties.

To provide evidence for this naturally occurring subtle energy possessing both electrical and optical properties, and to be able to show its relationship to living systems, Hieronymous devised a unique and clever experiment. He planted identical seeds in two groups of potted soil in a darkened basement. To the soil in the experimental group of potted seeds, Hieronymous ran a wire connected to an outdoor metallic plate. The plate was exposed to sunlight. The control group of potted seeds remained unconnected to these plates. They merely sprouted and grew in the dark, after being watered like the other seeds. In several weeks, Hieronymous examined and compared the two groups of plants that had sprouted in total darkness. He found that the experimental group, whose pots were wired to the outdoor sun-exposed plates, grew normally and were bright green in color. The control-group plants were pale and deficient in chlorophyll, just as plants deprived of light would be. Hieronymous believed that eloptic energy, a subtle-energy component of sunlight with growth-enhancing properties, had been transmitted through the wires from the outdoor plates to the experimental group of potted plants in the basement. This experiment was successfully repeated by a number of people, including some skeptical fellow researchers. Hieronymous seemed to prove his theory that an energy existed in nature that could demonstrate both the characteristics of electricity, in its ability to be transmitted through metalllic wires, and the characteristics of light, since the energy needed by plants to photosynthesize chlorophyll was light in the visible spectrum.

One of the unique innovations made by Hieronymous was later incorporated into many other radionics devices. He devised a new method of using radionics instruments to energetically diagnose and treat a patient's disease. Hieronymous knew that the strange bioenergetic radiation emitted by people (or plants) could be measured directly by using radionic testing of blood spots or by hooking the patient to the device via a wire and electrode. But he showed that this same bioenergetic radiation could also be captured and measured in a patient's photograph. The idea of some human-energy essence being captured on film is reminiscent of "primitive" African and South American tribes' superstitions about not being photographed for fear of capturing and taking away a part of their spirit. Perhaps radionic systems are just indirectly demonstrating the hidden truth behind this so-called primitive superstition in a new way. Radionic systems use photographs of subjects to be tested or treated as a means of diagnosing the "disease frequency" of a patient. But the photos are also used to broadcast subtle energy in the appropriate healing frequency back to the patient at a distance. In other words, it is possible to energetically manipulate people at a distance using only their photographs. This strange idea of sympathetic magic, or using someone's likeness to distantly manipulate that person, does sound a bit like voodoo, but it has been proven to be a real phenomenon by different radionic practitioners using a wide variety of systems. Distance healing also raises certain ethical questions about "treating" patients without their consent, even if the intervention is of a healing nature. In fact, the opposite idea, the concept of "harming" individuals at a distance using merely a photograph, was an area said to have been of great interest to both the Soviets and to certain groups within intelligence agencies during the peak of the Cold War. The secret interest in "psychic warfare" by military and intelligence groups kept many radionics researchers such as Hieronymous from sharing "too much" of their proprietary technology.

In any case, Hieronymous demonstrated that with the right circumstances, the proper use of photographs in radionic devices could have amazing healing benefits. But as stated earlier, the practice of using radionics for treating human illness was essentially suppressed after Dr. Drown's court battles. So Hieronymous decided to maintain a lower profile with respect to his radionics research. He began adapting radionics technology for use in a less "politically charged" arena, namely the field of agriculture. He started with plant diseases and crop infestations. His work was similar to that of another radionics contemporary, Curtis Upton. Upton had adapted one of Abrams's original Oscilloclasts for use in the detection and treatment of plant diseases. Upton found that he could diagnose and treat crop disorders either by placing a leaf from the afflicted plants into his radionics device or by merely placing a photograph of the trees or crops in question into the device. Like Upton, Hieronymous experimented with a number of other agricultural applications of radionics. In one fascinating experiment carried out in the 1960s, Hieronymous was asked by Ed Hermann, an engineer at McGraw-Hill Publishing, to treat a caterpillar-infested tree in his front yard. The infestation had resisted treatment by conventional pesticides. Hieronymous requested from Hermann a photograph of the tree, the photographic negative of the tree, a few leaves from the tree, and several of the annoying caterpillars. It should be noted that Hieronymous was living in a city three hundred miles away from Hermann's home at the time of the experiment. Several days after sending the requested photo, leaf, and caterpillar samples, Hermann drove up to his home to find a great surprise. He was astonished to see a circular "carpet" around the previously infested tree, a carpet made up of thousands of dead caterpillars!

Hieronymous, Upton, and a number of other researchers were very successful in radionic pest eradication. Several radionics businesses, most notably the Homeotronic Research Foundation of Harrisburg, Pennsylvania, actually contracted with farmers to rid their fields of insect infestations. A client would take aerial photographs of the crops to be treated and insert them into a radionic instrument. The farmers were also supplied with radionic devices that were to be turned on for certain time intervals during the planting and growing seasons. Then the farmers would radionically broadcast the energetic frequency of a specific chemical known to be toxic to the invading insects via the radionic instruments and the aerial crop photos. Interestingly, radionic pest control was so effective that the Homeotronic group took money from their clients only if their efforts at pest eradication were successful. This work did not go unnoticed by the chemical-pesticide industry. There were strong attempts to suppress this effective nonchemical method of pest control. The attempts worked. Homeotronic Research Foundation could not compete with the huge chemical-pest-control industry. Although Hieronymous and his radionics colleagues were far ahead of their time, they were never able to get established scientists of the day to examine the success of radionics technology. Hieronymous lived well into the 1970s. And when the Apollo space missions were still exploring the moon, he sent NASA radionically derived physiological measurments he'd made of the Apollo 8 and Apollo 11 astronauts during their trips in space. Although Hieronymous's measurements actually agreed closely with NASA's data, the space agency showed little interest in learning more about his remote-sensing technologies.

Even though radionics had been such a difficult area for legitimate study in the United States, research continued in England. Two of the most respected English radionics pioneers were George DeLaWarr and his wife, Marjorie. George DeLaWarr has already been mentioned earlier in connection with the Mark VI DeLaWarr camera. DeLaWarr's interest in radionics began during World War II, when the importation of any devices, let alone radionic instruments, was almost impossible. DeLaWarr, a civil engineer in Oxford, England, was asked to build a copy of Ruth Drown's original radionic instruments. He was so intrigued by her device that he further refined them and developed new radionic systems. He did numerous experiments with radionic therapies, too. DeLaWarr's use of colored light and specific sound patterns in radionic devices brought about additional healing benefits. DeLaWarr Laboratories in Oxford eventually became one of the world's leading producers of standardized radionic equipment. While other practitioners had success in using DeLaWarr's radionic instruments, only certain individuals were able to use his radionic camera to produce the amazing pictures of internal organs described earlier. Nonetheless, some remarkable pictures were produced using DeLaWarr's technologies. One of particular interest produced by the DeLaWarr Mark VI camera was taken using only a blood spot of a pregnant patient who was nearly a hundred miles away at the time of the radionic photography. In the developed radionic photo was an image in the shape of a human fetus! Unfortunately, even in England, DeLaWarr had his detractors and medical critics. He had to spend a considerable amount of time defending his work in the English court system.

Another English researcher of radionics was Malcolm Rae. Rae developed radionic systems that went far beyond the simple rates and dials of the classic radionic instruments. He created one device, known as the Magnetogeometric Potency Simulator, a unique radionic system used to "deliver" subtle energy of the appropriate frequency and character to a patient in need of healing. The device works in several ways. In Rae's early version of the potency simulator, the correct radionically determined energy pattern for healing could be broadcast to the patient at a distance using only a blood spot of the patient. The blood spot, referred to as the "radionic witness," helped to make a subtle-energy link between the patient and the radionic instrument in both diagnosis and treatment. The subtle-energy treatment was broadcast through the blood spot back to the patient, wherever the patient might be. But the other method of using the Potency Simulator was mind-boggling. A vial of water and alcohol or a bottle of milk-sugar tablets (such as the kind used to make homeopathic remedies) was placed in the Potency Simulator to be "radionically imprinted" with the appropriate vibrational-healing pattern. The patient would then physically ingest drops of the liquid or the milk-sugar tablets that had been imprinted with the healing-energy pattern in order to "receive" the energy.

The concept of radionic imprinting has since been adapted to many other supposedly "nonradionic" instruments used in vibrational medicine, especially the computerized systems used in electroacupuncture diagnostic testing. The ability of such systems to imprint the vibrational pattern of a homeopathic remedy into a neutral vial of alcohol and water is a direct outgrowth of Malcolm Rae's radionic imprinting discoveries. Another version of the Rae Potency Simulator, still in use by health professionals, looks like a small wooden box with two metallic wells in it separated by a tunable dial (connected to a typical resistance meter inside the box). Next to one of the metallic wells is a thin rectangular slot into which special Rae Remedy Cards may be placed. Typically, the device is used to imprint "blank" vials of liquid or bottles of milk-sugar tablets with the vibrational patterns of specific homeopathic remedies or flower essences. To make a vibrational "copy" of a remedy, the practitioner places a bottle of the homeopathic remedy into the "in" well and a dropper bottle of alcohol and water into the "out" well (where "in" and "out" signify the direction of energy flow from the "original" to the "vibrational copy"). Prior to placing the remedy and the bottle of alcohol-water (the medium onto which the vibrational copy is to be imprinted) into the in and out wells, a dial is tuned to either "1"(for a standard one-to-one copy), or to a three-digit number. The number selected for setting the potency simulator is taken from a Rae reference chart indicating which numerical setting to use for creating a particular homeopathic potency within the vibrational duplicate.

A practitioner can dial the device so as to produce homeopathic potencies as low as 10X all the way up to potencies as high as 50MM. By using the

Rae Potency Simulator, a practitioner can take a simple herbal mother tincture or even a premade homeopathic remedy and produce "radionic potentizations" of those same medicinal substances in varying homeopathic potencies. For instance, a practitioner could place a 10X homeopathic potency of Arnica in the "in" well of the Rae simulator and use it to produce the equivalent of a 1M (1,000X) potency of Arnica in the "out" well simply by tuning the dial on the device to the appropriate numerical setting for a 1M potency listed in the reference chart. The Potency Simulator can create a unique homeopathic antidote to a toxic medication, such as a chemotherapy drug, by placing a vial with a few drops of the chemotherapy agent in one side of the device and a "blank" vial of alcohol-water in the other side. In this case, the meter would be tuned to the setting indicated for producing a 1M potency. This approach has actually been used to decrease the toxic side effects of chemotherapy agents by making a 1M radionically produced homeopathic remedy from the chemotherapy drug itself. Isopathy, a subcategory of homeopathy, uses this detoxifying principle to help patients release certain harmful chemicals that may have built up in the body. Isopathy accomplishes this through prescribing homeopathic potencies of the toxic chemical one wishes to detoxify from. The homeopathic remedy induces the body's bioenergetic systems to somehow neutralize or discharge the toxic substance from the body. This form of homeopathic detoxification is actually rather common in electroacupuncture-based homeopathy and is used for ridding the body of accumulations of heavy metals, pesticides, and retained petrochemicals.

According to radionics practitioners using the Rae Potency Simulator, one can almost magically create instant vibrational or homeopathic remedies without having to physically dilute and repeatedly succuss a medicinal substance, as in the classical method. There are sometimes instances when the homeopathic remedy needed for treatment doesn't even exist (as in the case of a 1M-potency homeopathic chemotherapy-detox remedy). In such a case, the Rae Potency Simulator provides an avenue for creating a remedy from nearly any substance that can be captured in a bottle, be it animal, vegetable, or mineral. To understand this radionic method of creating duplicate energetic copies of certain substances (either in a one-to-one or homeopathic potency), we will talk in terms of a useful analogy. Think about the way you could go about making a copy of a friend's musical recital that has been recorded on a tape cassette. You need to use a tape recorder made for copying tapes. A typical music cassette consists of a magnetic recording of particular vibrational patterns (musical harmonies and rhythms created by certain audible vibrational patterns of sound). When a musical selection is first recorded, a microphone picks up the unique vibrational patterns (notes of music) of the performer's voice combined with the vibrations produced by the musical instruments being played. The microphone translates the acoustic vibratory patterns into electrical rhythms and patterns that travel through the wires to the tape recorder's magnetic recording head. As the magnetic recording tape moves across the recording head, the electrical patterns of the music become translated into and stored on the tape as magnetic-energy patterns. Placing the tape into a tape recorder and pressing "play" reverses this encoding process, and the magnetic signals are again translated into sound patterns the human ear can appreciate. If you wish to copy your friend's tape for personal use, you can duplicate the cassette using a special tape recorder that does this automatically. In such tape recorders, the magnetically encoded vibrational patterns on the original tape are transferred by wire to the second, blank tape. The process makes a near-perfect copy of the musical recording. All this is accomplished by a form of "magnetic imprinting" on blank cassette tape, a readily available form of magnetic recording media. Now, compare this magnetic imprinting of music to the operation of a Rae Potency Simulator. The simulator is a somewhat similar, albeit radionic, device able to use "magnetic imprinting" to duplicate the vibrational patterns of one recording (the recording of a homeopathic vibrational frequency on potentized water) onto a second, "blank" recording medium (a vial of water, the universal subtle-energy storage medium, or, alternatively, a bottle of milk-sugar tablets). In other words, water or milk sugar is used to store the vibrational pattern of the "duplicate" in a fashion similar to the way the second, "blank" tape stores sound.

We know from experiments by Bernard Grad and Dr. Robert Miller that water is a perfect storage medium for "holding" subtle energies, especially magnetism. Remember, magnetically imprinted water has even been shown to accelerate the growth rate of plants. One of the important "ingredients" in a radionic system like the Rae Potency Simulator is the use of permanent magnets to impart certain energy characteristics to the circuitry component of the simulator. The energy of a permanent magnet seems to function as a kind of driving force that transfers a portion of the subtle magnetic energy of the remedy to be copied onto the "vibrational blank" of the alcohol-water mixture. The alcohol and water then take on the vibrational characteristics of the remedy that was radionically copied, just as a duplicate musical cassette will faithfully reproduce the same music recorded on the original cassette. The energy field of a permanent magnet seems to aid in imprinting certain energetic or vibrational patterns onto water in much the same way that bright sunlight is able to imprint the life-energy patterns of flowers onto the water during the process of manufacturing flower essences. Even as far back as Dr. Albert Abrams's original radionics work, the importance of magnetic fields to radionic phenomena was theorized. Recall that Abrams's famous "disease-specific" stomach reflexes could be elicited in healthy testers only when they were facing due west. Abrams knew that a proper (west-directed) magnetic orientation of test subjects in the earth's magnetic field was a key ingredient to successful radionic disease measurement.

Unfortunately, he never fully understood why this was so. But the ability of magnetic fields to energetically influence living systems and to produce certain types of subtle-energy flow in radionic instruments was something Malcolm Rae truly understood. Magnetism was a phenomenon Rae capitalized upon as he added a wide variety of innovations to his radionic devices. Many of these innovations employed permanent magnets in key radionic circuits.

Undoubtedly, Rae's most controversial discovery was that the energy pattern of a particular homeopathic remedy or flower essence could be energetically stored or encoded in the form of a geometric pattern composed of mere ink on paper. Furthermore, this same geometric pattern, in the form of something called Rae Remedy Cards, could be inserted into the Rae Potency Simulator to create instant radionically imprinted remedies with the same healing properties as a homeopathic remedy. This unorthodox process was really quite simple. First, the appropriate remedy card for, say, Arnica was placed into the card slot. Then a dropper bottle containing a neutral alcohol-water mixture (the vibrational "blank" used to create the "instant" remedy) was placed into the simulator's "out" well. The vibrational-healing pattern of Arnica is somehow transferred from the geometric pattern on the card into the water. Thus, a vibrational remedy with the same healing applications as traditionally produced Arnica was created. Theoretically, one could use the Rae Potency Simulator and the Rae Remedy Cards to create an acutely needed homeopathic remedy that might be unavailable or out of stock.

Rae had employed dowsing techniques, a form of radiesthesia, to develop the geometric patterns used to make his Rae Remedy Cards. In order to find the pattern for a particular homeopathic remedy, he would use pendulums and other dowsing tools to measure the two-dimensional geometric shape of the remedy's subtle-energy field with regard to the way it interacted with the earth's magnetic field. Rae was using the technique of dowsing to feel his way around the borders of different remedy fields so as to more clearly define the highly specific and unique field shapes of each remedy. What Rae might have been measuring was a kind of auric-field boundary zone. Clairvoyant healers have often referred to a uniquely shaped subtle magnetic field around particular vibrational remedies. They claim that the field is a product of the vibrational energy being radiated outward by a homeopathic remedy against the background of earth's magnetic field. In fact, Rosalyn Bruyere, a prominent clairvoyant and healer, has commented about her clairvoyant observations of a standard vibrational remedy, castor oil, that she sees a symmetrical six-inch field of white light around every bottle of the stuff! What Rae eventually discovered was that the magnetic energy of the earth's geomagnetic field could be used to produce instant homeopathic remedies using only the dowsing-derived diagrams of the remedies' energy-field shapes.

Once Rae had derived the shape of a remedy's field, he used the geometric pattern in the following way: He would place a small vial of water in the center of the geometric remedy pattern with the diagram oriented toward magnetic north. After waiting ten to fifteen minutes, Rae found that the vial of water in the center of the energy-field diagram had been charged with the same healing properties as those of the original homeopathic remedy used to create the initial geometric pattern. In other words, the shape of a remedy's field, even though it was only symbolically represented as ink on paper, could somehow be used to imprint a remedy's vibrational-energy pattern upon a vial of water. This form of imprinting worked only if the water was placed in the center of the remedy's field pattern after it was oriented by compass to magnetic north. It was the combination of earth magnetism with geometric patterns that led Rae to dub his new method of subtle-energetic therapy "magnetogeometric applications" of radionics. There appears to be a strange similarity between Rae's technique and some of Dr. Abrams's early radionic phenomena (the stomach reflexes) with regard to the dependence of both processes upon a precise geomagnetic orientation for each system to work. Yet another unusual correlation with geomagnetic orientation and subtle-energy phenomena can be found in something called "pyramid power." Proponents of pyramid power claim to achieve various subtle-energy effects from scale models of pyramids built to resemble the Great Pyramid of Egypt. In order for these model pyramids to reliably produce their subtle-energy effects, including food preservation and razor sharpening, they must also be oriented to magnetic north. The pyramid, of course, is a unique geometric symbol, bearing a shape strongly resembling the structure of the carbon atom upon which all earth life is based. The three separate yet connected phenomena—Rae's magnetically imprinted remedy patterns, Abrams's west-facing reflex, and pyramid power's magnetic-north orientation—all suggest an important and mysterious relationship between radionics, geometry, magnetic fields, and subtle-energy effects. This puzzling relationship is one that Malcolm Rae appears to have understood better than most. Based upon his revolutionary discoveries and intuitive understanding of geometric patterns, magnetic fields, and radionic principles, he created an additional variety of unique radionic instruments besides his Rae Potency Simulator.

The geometrically imprinted remedies created by the Rae's remedy diagrams appeared to possess the same healing properties as classically produced homeopathics. Rae's research led him to believe that the energy of the earth's geomagnetic field was actually imprinting the remedy field patterns onto the vials of water. In his early work with this magnetogeometric imprinting, Rae found it necessary to orient his remedy diagram in a very precise north-south compass orientation. But this procedure was tedious and time-consuming. So Rae expeditiously discovered how to do away with the orientation procedure entirely. He learned he could dispense with the remedy diagram's north-south orientation if a small flat magnet was placed under the center of the diagram, just below the vial of water. The small permanent magnet provided a similar magnetic driving

Creation of the Rae Remedy Patterns and Rae Remedy Cards

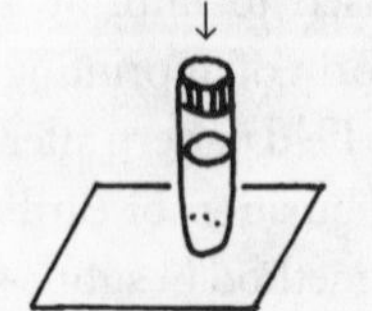

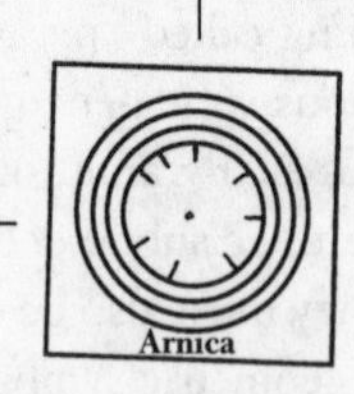

force for imprinting the water, just as the earth's geomagnetic field had provided the driving force in the first magnetogeometric experiments. During the late 1980s, publisher and subtle-energy researcher Don Gerard used a large number of the original Rae geometric remedy patterns in a privately published book called *The Paper Doctor: A Vibrational Medicine Chest.* The book's diagrams incorporated Rae's original magnetogeometric-imprint method for making instant vibrational remedies. *The Paper Doctor* contained the remedy patterns for a wide variety of homeopathic remedies used to treat common health ailments and came with a small plastic vial for water. The vial was to be placed upon the center of a remedy diagram. A board composed of a small flat magnet taped to the back of a square of cardboard was also included. The board was to be placed under the remedy diagram in order to provide the magnetic force for imprinting. I personally used some of the radionically imprinted remedies and found them effective for treating minor health complaints.

Rae went several steps further with his discovery of magnetically imprinted remedy patterns based on geometric shapes and magnets. He took his original diagrams and converted the angular and pointed shapes of the dowsing-derived remedy fields into purely symbolic forms that looked like thin outer concentric circles with a series of angular lines drawn facing toward the center of the circles. The forms looked something like pies with oddly cut slices. Rae believed that even these purely symbolic geometric patterns worked just as well as the original remedy-field diagrams in their ability to magnetically imprint water with the properties of homeopathics and other vibrational remedies. Rae, thus, had two different geometric patterns that could be used to imprint a particular vibrational remedy, one directly measured from the remedies and the other merely a symbolically encoded remedy pattern. Rae preferred using the second, more symbolic geometric patterns with his Potency Simulator. Remember that this device could use the energy of an internal permanent magnet to transfer and imprint a classically made vibrational remedy into a blank vial of water. It could also produce the same remedy without an "original remedy" to copy by the mere insertion of the appropriate Rae Remedy Card for that particular vibrational remedy into the device along with a "blank" vial of water or milk-sugar tablets. By changing the tunable dial, Rae was able to vary the homeopathic strength or potency of the final vibrational remedy produced by the Potency Simulator. The wide variety of applications for which the Potency Simulator may be used has already been discussed. A number of Rae's Potency Simulators are still in active use by many alternative-medicine practitioners.

Having now had a somewhat lengthy description of the history of radionic instruments, let us next examine exactly what is going on with such devices. How do they work? We have already alluded to the theory that radionic instruments are psychic-energy amplifiers. But such instruments may be much more than this. Dr. Abrams's first radionic instruments (from which all later radionic systems have been derived) were essentially amplified dowsing systems. The Abrams instruments used the so-called stomach reflexes produced on the abdomens of healthy young male volunteers as a means for providing dowsing feedback on different illness states. But what, exactly, is dowsing?

In a traditional dowsing arrangement, a dowser uses a forked branch, an L-shaped rod, or a pendulum to find a specific commodity or resource. For example, a dowser might try to locate a stream of flowing underground water in order to determine the best place for a farmer or a new business to dig a well. A number of European dowsing studies have theorized that water dowsers actually sense weak magnetic fields produced by flowing underground water. It is not entirely clear if the magnetic-field sense comes entirely from the brain, the adrenal glands, or even another area of the body like the chakras. But many good dowsers can use the visual feedback provided by the movement of their dowsing instruments to precisely predict the depth of flowing underground water as well as the flow rate. Unconscious muscle movements in the hands and arms of the dowsers are translated into larger movements of a pendulum or dowsing rod to provide observable, conscious feedback to the dowser about what he or she is looking for, whether it be the water in our example, oil, or gas. Dowsers have also used the technique of pendulum dowsing to diagnose illness. The method behind medical dowsing (medical radiesthesia) is largely the same as that in dowsing for water. In both cases, however, the key to what makes the dowsing instruments move in a particular way is the intention of the dowsers with regard to what questions they are mentally asking in looking for water, a disease, or something else. In other words, consciousness is an integral part of the dowsing process. The pendulums or dowsing rods act as amplifers of the unconscious psychic-sensing process in a way that helps to provide feedback to the conscious mind about the questions dowsers mentally pose to themselves about their target. The actual psychic sensing of water or illness is largely unconscious and is presumed to be mediated through the chakra system, our innate psychic-sensing network that is integrally linked with the nervous system.

In Abrams's initial system, the unconscious muscle movements used as feedback in medical dowsing were the abdominal reflexes elicited in healthy volunteers. Somehow, a reflex change in the contraction of a volunteer's abdominal muscles would produce either the dull or the hollow note on percussion, which Abrams used as a positive indicator in his radionic disease detection system. However, in order for his system to work consistently—meaning that Abrams would consistently get the correct "dull" abdominal percussion note with the appropriate disease-frequency setting on the radionic instrument—the volunteer tester had to be hooked up to Abrams's radionic apparatus via a wire going to a forehead electrode while the patient and tester were facing due west. Abrams's volunteer needed to have a direct electrical connection between the patient or the patient' blood spot via a forehead electrode. Radionic researchers believe this phenomenon suggested that the radiesthetic sensing capabilities of volunteers might have somehow involved either an unknown radiation-sensing capability of the physical brain or the psychic perceptual ability of the third-eye chakra (over which the electrode was placed). There is also indirect evidence that the solar plexus chakra is involved in the process as well. If you recall, the solar plexus chakra is believed to feed subtle energy to the nerves and muscles of the abdominal region, the area of the tester's body used to indicate a positive radionic reflex. Support for the role of the solar plexus chakra in radiesthesia also comes from observations by color therapists who have observed that the color yellow strongly influences the solar plexus chakra. Remember, Abrams discovered that anything with the color yellow—whether it was yellow clothing worn by the volunteers or yellow food they had recently eaten—could interfere with the reliability of the radionic abdominal-reflex test. Perhaps, then, yellow "anything" in the vicinity of the tester could theoretically upset the delicate balance of radiesthetic sensing or processing, if indeed the solar plexus chakra was involved in the radionic abdominal reflex. Other intuitive accounts, clairvoyant observations, and even Far Eastern doctrines also link the solar plexus chakra with "gut intuition." This would tie in with the dowsing effect in radionics, in that dowsing appears to be some type of unconscious psychic sense that can reliably detect the unique energy frequencies or vibratory rates associated with specific diseases.

This brings us back to Dr. Abrams's original hypothesis, upon which all his later discoveries were based. He speculated that all matter, especially biological matter from the tissues and organs of the body, gave off a distinctive energy or radiation with highly specific characteristics. It is interesting to note that Abrams's "biological radiation or electronic vibrations" are very similar to psychic Edgar Cayce's description of bodily electronic vibrations, which were said to produce either health or illness, depending upon their state of balance. The idea that all matter emits a distinct and identifying subtle radiation is also supported by dowsers who say that they can sense the subtle-energy radiation of specific substances. There are many good dowsers employed by industry and private mining firms who possess an amazing track record of successes in locating underground water, oil, gold, etc. Perhaps their success provides evidence that supports their claims of radiation sensing. Furthermore, it tends to suggest that dowsing could be an important diagnostic tool in many different sciences. Another observation reported by dowsers that lends further support to the concept of a substance-specific radiation relates to what happens when dowsers attempt to tune in to a specific numerical pattern for an underground mineral while using the pendulum method. The figures they come up with often strongly correlate with the atomic weight of the mineral or element they are seeking to locate. In addition, health practitioners working with Ruth Drown's radionic instruments also found that the radionic rates her devices associated with specific elements also matched their atomic weights. Again, this finding is quite similar to what the dowsers had found.

These observations suggest that dowsing does play an important part in radionics. But the dowsing reflex may be an unconscious nervous-system response to frequency-specific subtle energy emitted by different types of matter, both organic and inorganic. All these observations, taken as a whole, tend to lend support to Abrams's belief that all biological tissues radiate specific frequencies of "electronic energy," and that such energy could be accurately detected using radionic instruments connected to individuals functioning as "living radiation detectors." The idea of frequency-specific energy associated with all forms of matter is not entirely surprising if you think about the human body in terms of what we now know from vibrational medicine and quantum physics. We know that all matter, biological or inorganic, is actually a form of frozen energy with both wave and particle characteristics. The fact that these energies can be detected and quantified using the multidimensional human as a living detector actually makes sense in terms of the human capacity to psychically sense many types of information through both the chakras and the nervous system. Radionics makes use of our unique sensing ability and expands it even further, helping to quantify the highly specific energies emitted by different organs of the body in states of health or illness. Developers of radionic instruments such as Abrams, Drown, and others found certain rules of operation that were necessary for radionic devices to provide reliable information for the diagnosis of illness. The protocols of radionics would appear to be a function of both the human capability to sense subtle energies and the energetic characteristics of matter itself. Let us examine some of the methods used by radionic diagnosticians to better understand radionics' protocols and their implications.

We mentioned earlier that many radionic practitioners use a biological specimen from patients (such as a drop of blood on a piece of filter paper or even a lock of hair) to assist in radionically diagnosing their health status from a distance. In Abrams's day, he could diagnose only patients who were, at most, a few miles away. Eventually, Ruth's Drown's more advanced radionic technologies could diagnose patients from hundreds of miles away. In either case, the blood spot was often the focal point for tuning in to the patient at a distance. But how exactly does one use a spot of dried blood to tune in to a patient who may be many miles away? The blood spot, as mentioned, is referred to in radionics terms as a radionics "witness." It is usually inserted into a small "witness well" or chamber in most radionic devices while the operator attempts to fine-tune the dials of the device to determine the correct energy frequency of the patient. In some ways, this process is similar to tuning the dial of a radio to find just the right frequency setting for your favorite music station. You will hear the music only if your radio dial is tuned to the exact frequency of radio energy put out by the station's transmitter. But the radio analogy is just that, an analogy, because no actual radio waves are really involved in radionics. And yet it would seem that some form of energy (or information) is being transmitted over the great distances between the patient and the radionics instrument in order for the practitioner to be able to accurately diagnose the patient's health problem.

There are a number of different theories put forward to explain how a blood spot can provide accurate disease diagnosis without being submitted to any chemical testing. The use of blood specimens in disease diagnosis is quite common in conventional medicine. But the mechanism involved is strictly physico-chemical in nature. That is, clinical laboratories perform chemical analysis of blood components looking for elevated levels of proteins, enzymes, and other biochemicals specific to certain organs of the body and which may change in certain directions in various disease states. During the radionic analysis of a blood spot, only the energetic characteristics of the specimen are sampled, since no chemical tests are performed. One theory as to why the blood spot may reflect a patient's health has to do with the concept of holograms. A hologram is a three-dimensional picture of an object taken using beams of laser light. In a hologram, a single laser beam is divided into two beams by a device known as an optical beam splitter. The first of the two beams is used to illuminate the object to be photographed, while the second beam is bounced off mirrors until it arrives at the position of a photographic plate next to the object being holographically photographed. When the laser light reflected off the object combines with the other beam of laser light, the two beams mix to form an energy-interference pattern that is captured on the photographic plate. When the holographic film plate is developed and illuminated using a beam of the same type of laser light, the image reveals a three-dimensional view of the object photographed.

The objects in holograms are so three-dimensional that they even have physical properties resembling those of the original object. For instance, I have a hologram of a microscope with a computer chip on the microscope's slide. By changing my viewing angle of the hologram, I can either look at the microscope from the side or actually peer through the microscope lens to see the magnified object on the slide. In other words, the hologram actually possesses the optical magnifying properties of the microscope. This holographic image is very different from a conventional photograph. Even more amazing is the fact that you can slice off a piece of the holographic film, hold it up to laser light, and see an entire, intact object, the same as the one appearing in the uncut holographic film. The basic holographic principle of these unique energy-interference patterns is that "every piece contains the information of the whole." Every piece of the hologram's captured energy-interference pattern contains all the information of the original photographed object. A number of scientists have studied the nature of matter at the subatomic level and have concluded that all subatomic particles are essentially miniature energy fields of frozen light, as well as tiny energy-interference patterns. Since all matter at a quantum level exists as an energy-interference pattern, some scientists have suggested that all matter might also possess holographic properties at some level. In fact, these same scientists have theorized that the universe itself may actually possess the characteristics of one gigantic, real-time hologram. The key, of course, is in learning how to tap into the holographic properties of matter and space itself. It is possible that radionics is able to accomplish this to a certain degree with the aid of the consciousness of the human operator.

As for the radionics connection with this holographic concept, a drop of blood captured on a piece of filter paper might be considered a "fragment" of the biological hologram of a human being. According to this line of holographic reasoning, if you could read the holographic fragment contained within the blood spot, you could tap in to all the biological information of the individual who supplied the drop of blood. A radionic operator could tune into the holographic pattern of an individual to extract information about the energetic frequencies emitted by the patient's tissues and organs. Within those tissue frequencies would be the keys to radionic diagnosis of disease. As an aside, something unusual sometimes happens when a blood spot is used as the main radionic witness. What is very bizarre about the blood spot as a radionic witness is that the minuscule specimen often continues to show vibrational changes happening in the patient's body hours to days after the blood sample has been taken. George DeLaWarr's radionic blood-spot photos frequently demonstrated this strange change over time. An interesting example of what DeLaWarr often managed to capture on film is as follows: DeLaWarr performed radionics photography on the blood spot of a patient who was about to undergo surgery. When DeLaWarr developed the images produced by his radionic camera, taken during the actual surgery, they showed the unmistakable image of a hemostat (a surgical instrument) in the patient's abdomen. But the hemostat was clearly absent from the patient's body at the time the blood specimen had originally been taken. The vibrational-energy pattern generated by the blood spot (as revealed in the radionic photo) seemed to change according to what was happening to the patient at a distance, in spite of the fact that practitioner and patient were many miles apart. Such evidence suggests that the blood spot as hologram fragment maintains a dynamic energetic connection, a kind of holographic connectivity, between itself and the whole (the hologram of the entire patient).

Interestingly, sometimes DeLaWarr's radionic photography produced images that looked similar to, but not exactly like, a patient's diseased organs. As an illustration, a radionic image of a patient's liver filled with hydatid cysts showed many bubbles in the image of the liver. The bubbles looked very similar to (but not exactly like) the physical appearance of the actual liver cysts. Radionics photographers, including George DeLaWarr, assumed that their radionic imaging process was actually capturing an image of a patient's etheric body and not the actual physicial body. This claim of etheric imaging through radionics just may have some validity. If this is true, it is possible that this strange behavior of holographic connectivity between blood spot and patient may somehow take place at the etheric level of matter.

There has been some research supporting the idea that the etheric body has holographic properties. Dr. Ioan Dumitrescu was the Romanian developer of the electronography process. This electrical photographic technique of body imaging is similar to Kirlian photography. Dumitrescu did some investigational work using electronography to study the so-called phantom-leaf effect. As discussed earlier, the Kirlian phantom-leaf effect captures an image of a leaf tip that has already been cut off and destroyed prior to the leaf's being photographed. The phantom leaf tip is most likely an ephemeral image of the leaf's etheric

A Typical Radionic Diagnostic and Healing Procedure

RADIONIC DEVICE

banks of tunable resistance meters (with dials)

witness "well"

radionic "stick plate"

radionic "witness" (blood spot)

possible mechanisms by which blood spot functions as radionic witness

- may function as hologram fragment of "whole-patient" hologram (which is radionically decoded or "read")
- may function as waveguide to tune in to patient's coordinates in the larger "universal hologram"
- may emit unique life-energy radiation pattern specific for each patient (read by radionic device/practitioner)
- may provide a focus for psychically attuning to patient at a distance

PATIENT → ? direct psychic connection → PRACTITIONER

PATIENT
- direct wire-electrode connection
- direct tissue or blood specimen connection (via wire-electrode setup)
- RADIONIC WITNESS
  - blood spot on filter paper
  - hair sample
  - tissue or bodily-fluid sample
  - photograph of patient (especially photographic "negative")

→ RADIONIC DEVICE → practitioner tunes dials of radionic instrument while stroking finger against "stick pad"

↓

final dial settings give "radionic rate" or "disease frequency" of patient

↓

practitioner reads radionic rate/disease frequency from standard radionic "rates chart," looking for match with a specific disease or physiological imbalance

RADIONIC DIAGNOSIS

practitioner uses radionic device to tune to appropriate frequency of subtle energy for radionically healing patient's illness

← practitioner "broadcasts" appropriate frequency of subtle energy to patient via blood spot or other radionic witness → PATIENT

RADIONIC TREATMENT

body. Dumitrescu performed a variation on this experiment. He cut away a circular hole in the middle of a leaf prior to photographing it electronographically. In the final electronography image of the leaf, there is an electrical image of a leaf with a hole in it. What is unusual about this image is that, an observer can clearly see, within the circular hole in the leaf image, the outline of another, smaller leaf with a hole in its center! In similar fashion, when a fragment taken from the middle of a conventional hologram of a leaf is held up to laser light, you can see, that the smaller holographic fragment contains an image of the entire leaf. This is exactly what Dumitrescu saw in his electronography variation of the phantom-leaf effect! His experiment does seem to contribute interesting data supporting the possible holographic nature of the etheric body.

Now, if we accept the possibility that a blood spot might possess energy qualities similar to those of a hologram fragment in which the originating hologram is the energy pattern of the individual from whom the blood was taken, we can see how such a holographic fragment could provide meaningful diagnostic information. The key, of course, is in learning how to read or decode the energy pattern contained within the hologram blood spot in order to derive meaningful information about the health status of a particular patient. Maybe radionics instruments are merely tools for decoding holographically embedded information. As mentioned earlier, when using the blood spot for radionic diagnosis, radionics practitioners found that a strict protocol needed to be followed in order to obtain an accurate diagnostic profile. Part of the radionic protocol concerns blood transfusions. Radionics practitioners have long known that blood transfusions render a blood spot useless as a radionic witness for measuring changes in a patient's health. In one early case, a radionics practitioner had been doing daily health evaluations of a distant hospitalized patient. Suddenly the patient's blood spot seemed to indicate a total absence of life-energy readings. It was as if the patient had died. What had actually happened was not so alarming. You see, the patient had received a transfusion of multiple units of blood while in the hospital. You might ask why the administering of immunologically compatible blood products would invalidate a patient's radionic blood-spot reading. Well, it seems that the injection of blood from another individual does not supply merely fresh plasma and red blood cells loaded with hemoglobin. A blood transfusion also injects highly specific subtle-energy patterns that have come from the blood donor or donors. When foreign energy frequencies are injected into a person's bloodstream, the energetic pattern carried by the first (pre-transfusion) blood spot now no longer energetically matches the new (post-transfusion) subtle-energy state. The patient's post-transfusion blood spot now carries not only the energetic frequencies from the patient but also those of the different donors who supplied the units of blood used for the transfusion. It's almost as if one were mixing the energy-frequency patterns of different blood holograms into one confusing mishmash.

Up to now we have assumed that the radionics device (and its operator) may be using the blood spot to decode a fragment of the patient's etheric-body hologram. But just suppose that a blood spot is not acting as a hologram fragment to be radionically "read." After all, research results can sometimes support more than one explanation for a particular phenomenon. Maybe the radionics practitioner is merely using the blood-spot witness to make a psychic connection between his or her own unconscious mind and that of the distant patient. While tuning in to a distant patient, the practitioner could be tuning directly in to the patient's own personal hologram, which is actually a fragment of the greater universal hologram. In such a case, the mixing of blood energy frequencies may somehow interfere with the psychic connection made between the radionics practitioner and the patient, a link normally provided by the frequency-specific blood spot. In circumstances in which a patient has received multiple blood transfusions, radionics practitioners will use a lock of the patient's hair or even a photograph of the patient as a substitute witness for reestablishing the radionic connection.

Since radionics practitioners have successfully used a patient's hair or even a photo of the patient to access health information about a patient, then it seems the holographic fragment hypothesis of the radionics process might be a little too simplistic. Consider, if you will, that the entire universe may be a gigantic, real-time hologram, constantly changing as its components (people, animals, plants, climate, geology) continue to change. The holographic-universe concept implies there is a strange energetic connectivity between all parts of the hologram, no matter how small. Affect one small corner of the hologram and all the rest of the parts are simultaneously affected, albeit at a nearly undetectable subtle-energetic level. Presuming the universe really is a giant hologram, it might also be possible to read information energetically encoded within a specific region of the hologram, if only we knew how to tap in to and decode specific areas of the universal hologram. Since each living being is a unique entity with highly specific energetic-frequency patterns, a small sample of an individual's frequency pattern might allow someone with a universal hologram reader (a radionic instrument) to tune in to that person's portion of the universal hologram. Think of the blood spot in a radionic device as something akin to the shoe of a missing convict given to a bloodhound to sniff in order to aid the dog in hunting and tracking down the prisoner by zeroing in on his specific scent pattern. The radionic blood spot may provide an energy scent, a patient-specific frequency pattern that helps radionics practitioners use their instruments to psychically tune in to the location of the patient on the grid of the universal hologram. Some have likened the use of the blood spot in radionics to a kind of "witness waveguide," allowing practitioners to psychically zero in on the frequency and wave patterns specific to the one person in a billion about whom they are seeking to gain information.

The holographic-fragment concept of radionic witnesses and the witness-waveguide hypothesis both provide interesting explanations for how witnesses may function in radionic diagnosis. The most unusual thing about radionics witnesses, however, is that the information and energy flow through them appears to be a two-way street. For example, a photograph of a patient used as a witness may capture some small representation of a person's visual and subtle-energy patterns on film, but it is probably not capturing the entire holographic gestalt of the patient. More likely, the photograph provides a psychic-energy link between the patient and the radionics practitioner through which information, and perhaps subtle energy, may flow. Many radionics practitioners treat patients by radionically broadcasting specific radionically derived subtle-energy frequency patterns back to distant patients via their witness. Again, the radionic witness might be in the form of a blood spot, a hair sample, or even a photograph. Remember the case of T. Galen Hieronymous and the engineer's caterpillar-infested tree? Hieronymous was able to rid the tree of its caterpillars from hundreds of miles away using only a photograph of the tree as a radionic witness. The dying caterpillars might just as easily have been dying bacteria left over from a successfully treated human infection (treated, of course, by radionic broadcast). The point is that a simple photograph provides a unique energetic and informational link between the radionics practitioner and the distant patient that can be used for either diagnosis or treatment on a purely vibrational level. You can see why radionics is frequently considered the strangest of all vibrational-medicine healing modalities, even among alternative-medicine practitioners, because in radionics everything is literally based on diagnosing and treating human illness using only varying frequencies of subtle energy.

Radionics is also dependent in some way upon the consciousness of the radionics operator. The conscious and unconscious psychic abilities of the radionics operator remain a key part of successful radionics diagnosis and treatment. Also, radionic detectors usually have some kind of dowsing feedback system, such as the stick plate or the abdominal reflex. The inclusion of such human-machine links suggests that the human nervous system, with its theorized connection to the chakra system, is an important part of the total radionics system. But the dependence of radionics upon the clear and focused consciousness of the operator means that different operators may be capable of varying levels of success with the same instrument, depending on their level of training and their natural psychic sensitivity. This human factor makes radionics a very subjective science with varying results from practitioner to practitioner, even when using the same model instrument. Remember the famous DeLaWarr camera? It was was used to make over five thousand successful and high-quality radionic images of distant patients via their blood spots using a strictly defined protocol. But, alas, the majority of researchers after DeLaWarr failed to get the device to reproduce his results. It seems that a key ingredient eventually found to be important to the success of DeLaWarr's radionic photography was the unconscious participation of one worker, Leonard Corte. Some practitioners could get radionics pictures only if Corte touched the unexposed photographic plates prior to their insertion into the camera. It seemed as though Corte emanated a type of subtle energy that sensitized the film and equipment in such a way as to produce successful radionic pictures. As long as Corte had contact with the photographic plates either before or after photography and film development, the DeLaWarr camera produced high-quality etheric images of hundreds of patients' internal anatomies using only a blood spot as a radionic witness. Confirming this unusual "experimenter effect," radionics practitioners Elizabeth Baerlein and Lavender Dower found they could only get their DeLaWarr camera to work if Corte had handled the light-shielded photographic plates used in the camera at least two hours before they were exposed. After two hours, the film-sensitizing effect that Corte somehow provided seemed to fade away, and only blank images would be produced.

The Corte anecdote seems to indicate the presence of some unique psychic-energy interface that exists between multidimensional human consciousness and the life-energy frequencies used in certain aspects of a fully functional radionics device. If the phenomenon of etheric photography at a distance based only on a blood spot seems totally bizarre, then what are we to make of the ability of Rae Remedy Cards to heal? The behavior of the Rae Potency Simulator makes little sense to the linear, rational, analytical minds of most scientifically trained physicians, who are used to thinking in terms of purely physically based healing modalities. It is beyond the scope of many doctors to accept the possibility of transferring the healing electronic or magnetic vibrational patterns of a homeopathic remedy (already considered to be little more than a placebo) to a neutral bottle of alcohol and water. But even if anyone can believe it is possibile to transfer the subtle magnetic healing patterns of a homeopathic remedy from one liquid solution to another (via the Rae Potency Simulator), it becomes much harder to rationalize how a geometric pattern consisting of only ink on paper (the Rae Remedy Cards) could in any way produce a remedy with therapeutic properties. And yet the "instant" homeopathic remedies made using the Rae Remedy Cards and the Potency Simulator (or even made using the original Rae remedy patterns as seen in *The Paper Doctor*) actually do produce therapeutic results time and time again. Although their effects do not seem to be as "potent" as those of classically produced homeopathic medicines, the ink lines shouldn't produce any healing effects at all—or should they?

This is the paradox of many of the discoveries of radionics practitioners. Technologies have been developed that to the rational, analytical thinking of our left-brain culture produce effects that seem to be closer to magic than to science. They appear to operate more on a right-brain system of symbolic languages, holographic gestalts, and higher intuition. In a sense, the radionics device acts as a kind of a bridge between the left-brain scientist and the (higher-self-directed) right-brain intuitive in each radionics operator. If we look at the successful track records of certain individual radionics practitioners, subjective or anecdotal as such reporting might be, we must come to the conclusion that radionics is certainly capable of producing significant healing responses using a wide variety of radionic modalities, not the least of which are the Rae Remedy Cards. Since the cards do work and produce repeatable healing results (under appropriately controlled conditions), such a healing phenomenon must be based on some kind of vibrational science we are only now beginning to grasp.

Considering Radionics Today as Part of a Healing Approach

The long and winding history of radionics produced a wide variety of tools and devices that have evolved since the early days of Abrams and Drown. However, the times when practitioners needed to tap on the bellies of healthy young volunteers (as Abrams did) in order to perform radionic testing have long since passed, although some modern radionic devices are still used in conjunction with pendulums for a kind of marriage between pendulum dowsing and radionics. One contemporary practitioner using this synthesis of old and new radiesthetic techniques in the 1970s and '80s was the late David Tansley of England. Tansley developed radionic systems capable of diagnosing physical disorders within the body, and of quantitatively analyzing a person's chakra-energy levels. He also was able to determine radionically the energetic status of a patient's etheric, astral, emotional, and spiritual bodies. Tansley was one of the first radionics experts to introduce the concept that illness may begin first at the chakric, etheric, and spiritual levels. His discoveries have since been incorporated into a variety of newer radionic systems. Because of Tansley's work, today's radionic technologies provide one of the few tools that can attempt to "quantify" the functional status of the esoteric components of human multidimensional anatomy.

While some systems of radionics use the pendulum in conjunction with classical radionic boxes that have many tunable "rate" dials, other systems have made a tremendous leap into the computer era. An example of new computerized electronic radionic systems is the SE5, a small handheld device running off a Sharp handheld computer. Some of the newer computerized systems coming out of England are entirely automated, with only minimal participation of a radionics practitioner. With the automated systems, the patient is merely hooked up directly to the radionic computer unit via a series of wires and skin electrodes attached to the patient's wrists and ankles. First, an automated testing sequence is initiated by the patient, who presses a "start" button. The device first "scans" a patient, checking his or her unique energetic emissions patterns. After determining which frequencies of energy are required for healing and rebalancing, the computer device then automatically broadcasts healing frequencies of energy to the patient through the electrode connection during a twenty-minute automated sequence. Although preliminary results with these automated radionic systems have been promising, no wide-scale clinical trials have yet been published.

Some radionic technologies have actually become embedded in the inner workings of nonradionic computer technologies. The ability to "magnetically" imprint homeopathic-remedy patterns onto neutral solutions, as is done in the Rae Potency Simulator, is a radionic capability of many EAV-type computer systems. And thus, homeopathic practitioners who prescribe remedies based on computerized EAV-type acupoint-energy analysis will often give patients "imprinted homeopathic remedies" made by their computers when the key remedies a patient requires are not readily available.

Even the Internet has gotten into the act as a vehicle for transmitting different forms of radionic healing. In one instance, the Internet was used as a medium for "electronically transmitting" a radionically encoded homeopathic remedy from one computer location to another. French homeopathic researcher Dr. Jacques Benveniste has recently reported success in both "electronically" encoding and transmitting (via the Internet) homeopathic-remedy patterns used to "reconstitute" the imprinted homeopathic remedy at a second location. This is like an EAV-computerized system's homeopathic-imprinting capabilities. The encoded vibrational packet that carried the remedy patterns in electronic language (as opposed to the geometric language of, say, Malcolm Rae's Remedy-Card patterns) was actually sent along the Internet from one computer to another as an e-mail attachment! In yet a different example of using the Internet to promote radionic healing technologies, an India-based group working with geometric remedy patterns very similar to Malcolm Rae's Remedy-Card patterns is making their patterns available for free from their Internet site. The Sat Sanjeevini Foundation performs this free service in the hope of providing cheap, easy healing methods to anyone with computer access. The foundation approaches the concept of radionic imprinting via geometric patterns slightly differently from classical radionics. While the Rae radionics technologies are based largely on using permanent magnets or magnetic fields to accomplish the homeopathic imprinting, the Sat Sanjeevini group believes they can accomplish the same type of vibrational-remedy imprinting using the focused energy of prayer as the driving force for imprinting.

The idea of using focused prayer as an energy source for radionics is quite interesting, especially in light of recent speculations about the energetic, or nonenergetic, mechanisms behind prayer in distant healing. In his book *Healing Words: The Power of Prayer,* physician and author Dr. Larry Dossey examines a variety of medical studies that strongly suggest that prayer can indeed produce significant and measurable therapeutic healing benefits for patients who are "prayed over," even at great distances. In one famous study conducted by a San Francisco cardiologist, two groups of hospitalized cardiac patients were randomly assigned to either a control group or a group that would be prayed over by prayer groups. At the end of the study, those who had been prayed for had significantly fewer cardiac and respiratory complications. To provide a possible explanation of how an "act of focused consciousness" might achieve healing effects, Dossey invokes a little-known area of physics dealing with what are called "nonlocal forces" or "nonlocality." The quintessential example of nonlocal forces can be seen in the strange behavior of certain types of energetically paired particles or photons. The odd but experimentally verified fact is that when you "disturb" one of a set of two "spin-paired" particles, the sister particle always changes its spin in such a way as to suggest that instantaneous communication was received concerning what has happened to its companion particle, no matter how far apart the two paired particles may be. This is an example of action at a distance that defies explanations involving conventional electromagnetic (EM) fields. You see, it seems physics has determined that such EM fields rapidly decrease in intensity as the distance from the energy source increases. The idea of that particle seeming to simultaneously "know" what is happening to its paired companion particle has led to the theory in physics that the two particles must somehow be interconnected by so-called nonlocal forces. These nonlocal forces are invoked to provide an explanation for this strange communication and action at a distance. Don't these nonlocal forces seem oddly reminiscent of radionic distant-healing phenomena? Since prayer is really an act of consciousness, Dossey argues, it's likely that the mind itself is capable of healing action at a distance through nonlocal forces. Perhaps physicists will eventually prove that nonlocal forces are actually the magnetic conducting medium of the all-pervasive ether that belongs to the etheric plane of the earth (and of the universe itself). In a slightly different view of the relationship between nonlocal forces and radionics, David Tansley put forth an interesting idea. Tansley believed that certain aspects of the etheric plane possessed a triangular gridwork pattern he referred to as a psi field. Tansley postulated that this etheric psi field somehow allowed the etheric field of the earth itself to act as a subtle connecting medium between the consciousness of a radionics practitioner and a patient's blood spot for distant radionic diagnosis and treatment.

Consciousness, especially focused consciousness held in prayerful thought, has been shown to trigger healings at great distances. This focusing of consciousness also appears to be an important prerequisite to successful radionic diagnosis and treatment. When viewed from this perspective, the attunement a radionics practitioner goes through when working with each patient can be considered a kind of sacred healing prayer. The idea that prayers produce potent subtle-energy effects brings us back to the Sat Sanjeevini Foundation's use of focused prayer as a force for radionic imprinting. If indeed the Sanjeevini remedy-card patterns are capable of radionically imprinting healing remedies onto small vials of simple tap water, they demonstrate that prayer, combined with radionically derived technologies, can offer a source of cheap and easily distributable healing tools to people throughout the planet. The marriage of focused prayers with radionic geometric remedy patterns distributed free over the global information superhighway of the Internet is a revolutionary concept even in vibrational medicine. But it is a concept that shows how the principles of radionics may help to redefine the very nature of healing on a purely vibrational level.

Radionics research continues throughout the world. But radionics in the United States has been practiced largely in secret. All the bad press that radionics has gotten over the years in the United States—with devices confiscated by governmental regulatory agencies and professional reputations ruined—has forced many practitioners to remain underground. Thus, it seems unlikely you would be able to find a radionics practitioner through the Yellow Pages or via a magazine advertisement. There are, however, radionics and dowsing organizations that promote education and research in the field of medical radiesthesia and radionics. Most of these groups frequently host some fascinating yearly conferences at which radionics and radiesthesia practitioners and researchers share discoveries, ideas, and information about new device technologies. The conferences can often be a resource of information for those seeking to find out more about radionics' unique intuitive approach to diagnosis and healing. Some radionics groups also provide information as to how radionics can be used to help determine which other, nonradionic therapies may be of health benefit to people. For instance, in Europe, many radionics practitioners frequently use their device technologies to do more than diagnose illness and to broadcast healing radionic energies to patients. They also use them to determine what is the best complementary program of nutrition and dietary change, herbs, supplements, or even medical and surgical treatments that might be of greatest value in helping a patient to recover from a serious illness. This radionic approach to "dowsing the body's needs" is somewhat similar to using muscle kinesiology testing for determining exactly what a person's body needs to heal itself.

For those practitioners who do work with radionics, their typical patients can be those who are combating life-threatening diseases, as well as individuals who are merely looking for an energetic approach to healing day-to-day illnesses. There are no published studies on what radionics can and cannot treat, but a number of radionics practitioners claim to have had successes healing both simple and complex illnesses, some of which were unresponsive to conventional medical treatment. For instance, there have been case reports of patients with inoperable brain tumors going into remission after a multidisciplinary healing program that included radionic broadcast therapy. Radionics is a unique healing modality in that it is a system of diagnosis and treatment geared to more than just the physical body. Because of Tansley's contribution, radionics now focuses on analyzing and healing both the physical and the subtle-energetic and spiritual components of multidimensional human anatomy. Radionic devices attempt to heal and rebalance a patient's physical and subtle anatomy using classical radionic technologies, along with healing sound frequencies and colored-light therapy broadcast to the patient. And, of course, radionics also provides a unique way of working with homeopathy.

Radionics is certainly the strangest of all vibrational-medicine approaches, but it is one that may ultimately teach us the most about the vibrational nature of healing and human consciousness itself. To the uninitiated, radionics will forever seem like magic. Yet to those with an open mind, radionic technologies may provide a healing answer when all other treatment modalities have failed. It is certainly an approach to diagnosis and treatment worthy of further study and research, both to verify the claims of its proponents and to discover how this technology may aid us in releasing our own inner capacities for self-healing and healing at a distance.

Ten

THE VARIETIES OF HANDS-ON HEALING

KNOWN BY A variety of names—including bioenergy healing, psychic healing, paranormal healing, spiritual healing, Therapeutic Touch, Reiki, Johrei, Mari-el, SHEN therapy, pranic healing, and a bevy of other terms—laying-on-of-hands healing is perhaps the oldest form of vibrational therapy used today. In the past, people have tended to associate the laying on of hands with religious ceremonies or church gatherings. Indeed, Jesus was said to have performed many healings using laying on of hands, but it was also practiced by many spiritual leaders, healers, and teachers through the ages. In recent years, the hands-on-healing approach is finding greater acceptance among a wide variety of health-care providers, from nurses and doctors to chiropractors and massage therapists. Most forms of hands-on healing involve some use of the hands upon the patient (or in the general vicinity of the patient's body) combined with the channeling of healing energy from healer to healee. Rather than relying on a vibrational remedy, a healing essence, or some healing-device technology, this form of vibrational therapeutics uses the human capacity to consciously direct the flow of healing, multidimensional energies into the human body and its associated physical- and spiritual-energy systems to bring about healing changes.

The Scientific Study of Bioenergy Healing: Is It Real or Is It Just the Placebo Effect?

The increased acceptance of Therapeutic Touch–type therapies is due partly to a growing body of research that tends to validate some of the physiological changes and healing benefits claimed by TT practitioners. Scientific research has shown Therapeutic Touch healing to be effective in reducing anxiety, producing states of deep relaxation, and accelerating the healing of surgical wounds. For many, the idea of laying on of hands conjures up an image of evangelical religious leaders ministering to their congregations. Back in the 1950s, there were Sunday morning television programs with charismatic healers such as the late Kathryn Kuhlman. This woman would place her hands on the foreheads of congregation members, whereupon the faithful would fall backward in a dead faint and then be caught and carried away by Kuhlman's waiting aides. Many of Kuhlman's congregation claimed to have been healed of a variety of health problems. They believed that the power of the Lord had flowed through Kathryn and into their bodies, healing them of their ills. Such performances, while dramatic and intriguing to religious worshippers, brought ridicule and disbelief from most scientists of the day, who quickly dismissed any healing effects produced by contact with Kuhlman as the result of religious hysteria or the ever-popular "placebo effect." Still, there were some intrigued scientists who suspected that something other than the power of belief was at work in bringing about the many cures attributed to the laying on of hands.

One of the earliest pioneering scientists to study the physiological effects of hands-on healing was Dr. Bernard Grad, a gerontologist at McGill University in Montreal, Canada. During the early 1960s, Grad became intrigued by the phenomenon of laying-on-of-hands healing, so much so that he set out to find a way to establish whether such healings were the result of psychological effects (as most critics of the practice believed) or true energy effects produced by healers. He sought a way of proving or disproving the popular notion that healers' effects were simply due to the power of the mind triggered by a person's belief system (the so-called placebo effect). Grad chose plants and animals for his experiments, because physiological changes produced by laying on of hands in a sick plant or an ailing animal would have to be attributed to something other than the placebo effect. Grad initially chose sick plants as his first nonhuman test subjects. Obviously, plants or seeds cannot "believe in" the healing powers of a particular healer or of a higher spiritual power. Since laying-on-of-hands healing requires a "sick patient" to work on in order to demonstrate its healing effects, Grad had to devise a way to create a sick plant.

To do so, he watered barley seeds with a saline solution, since salt is a known inhibitor of plant growth. If a healer could positively affect the germination rate of saline-treated seeds, as well as their subsequent plant growth, the healer's influence would have to come from something other than the power of belief. Grad persuaded Oskar Estebany, a local healer, to "treat" one of two sealed flasks of salt water (simply labeled bottle X and bottle Y) using his healing touch. Only Grad would know whether bottle X or Y was the healer-treated salt water, thus excluding any influence on the experiment that might be unconsciously attributed to experimenter expectation. His lab assistants watered two groups of barley seeds with saline from either bottle X or bottle Y and then allowed time for the seeds to germinate and sprout. After the seeds had sprouted, the seedlings were planted in identical pots filled with soil and watered with normal, untreated water. At the end of several weeks, Grad compared the two groups of plants. Grad discovered that the plants watered with healer-treated saline were markedly different from his control plants. When compared with the control group, barley seeds that had been watered with healer-treated saline sprouted more frequently and produced taller, leafier plants with higher levels of chlorophyll, at statistically significant levels. This meant that the differences between treated and untreated plants could not have been due to chance alone.

Grad's results in the plant experiment pointed toward a number of interesting conclusions. Specifically, Grad had shown that laying-on-of-hands healing produced real, measurable effects in living systems above and beyond any possible psychological effects that might have been due to the power of belief. In addition, Grad had proven that healer-charged water could positively influence plant growth. Based on these findings, Grad suspected that the healer had actually imparted "something" to the water that was able to positively influence the growth of living organisms. Exactly what is exchanged between healer and patient during such therapeutic interactions remains something of a debate among alternative-health-care professionals. While many have suggested that laying-on-of-hands healing involves a transfer of life-force energy from healer to patient, there are others who propose that such "healings" are related to a resonant-field interaction between healer and patient. Still other theorists claim that all the observed effects result from the consciousness of the healer and the patient interacting at some "higher" or "nonlocal" level of quantum reality. But no matter which model of healing one favors, for the model to be valid it must explain how a healer can alter the nature of a simple saline solution so it boosts the growth rate of plants. In other words, does the healer merely transfer life energy to the water or, alternatively, might he or she actually transmit some kind of "healing bioinformation" into the water that triggers healing and accelerated growth in plants and other living systems?

Perhaps healers are able to charge water with subtle magnetic life-force energy that carries with it some sort of healing bioinformational pattern, not unlike the pattern one might find in flower essences, homeopathic medicines, and other aqueous-based vibrational remedies. On the other hand, the subtle life energy transferred to the water might have been all that was needed to stimulate the salt-soaked barley seeds to germinate and grow. Dr. Grad and other researchers have noted a striking similarity between magnetic fields and healer's energy fields. As an additional variation on his plant experiment, Grad watered barley seeds with magnet-treated salt water using the same protocol he had established earlier. He discovered that magnetized saline not only accelerated plant growth, it produced more vigorous growth-stimulating effects than even healer-treated water. There seemed to be a kind of "qualitative" similarity between a healer's energies and the energy field of a permanent magnet. Both healers' energy fields and magnetic fields could easily penetrate the glass walls of Dr. Grad's sealed bottles of saline, endowing the salt water with growth-promoting qualities. Grad's findings support the idea that the life-force energy emitted by a healer's hands is actually a kind of "animal magnetism," a distant cousin to the "iron-filings-attracting" form of ferromagnetism.

After his initial success, Grad wondered whether he might not be able to use his plant-based research protocol to verify another hunch. If healers were able to positively influence the energy field of living systems, was it possible to use this same protocol to verify a "negative" energy influence that could actually inhibit the growth of plants? Grad speculated that the subtle bioenergy fields of some sick individuals might possess "life-energy-sapping" properties that could cause other individuals (and perhaps even plants) to become sick. Using his same saline-treatment protocol, Grad asked neurotically and psychotically depressed patients from a psychiatric hospital to hold on to and "treat" bottles of saline that would later be used to water barley seeds in his laboratory. Subsequent analysis of the data showed that psychotically depressed patients did indeed impart an "energy" to water that had the reverse effect of magnets and healers. While magnets' and healers' energies caused plants to flourish, water energized by psychotically depressed individuals actually depressed the growth of plants. Grad's intriguing results with water treated by psychiatric patients hints at the possibility that people's bioenergy fields have the potential either to heal or to unconsciously harm people in their immediate environment. Most of us have had the experience of being in the presence of people who make us feel "drained" after being with them for only a few minutes. Perhaps the real reason such "energy sappers" make people feel drained lies in a bioenergetic effect as opposed to a psychological influence. One of the important implications of this aspect of Dr. Grad's research is that there may be hidden "bioenergetic hazards" associated with working around sick and emotionally disturbed indidviduals. If such hazards are real, they will need to be adequately addressed in future training of health-care professionals in order to prevent them from becoming ill due to such unseen and unsuspected bioenergy effects. One of the easiest precautionary measures to avoid picking up sick patients' bioenergies may be the simple act of hand-washing. Since Grad's plant experiments demonstrated that water is a "universal energy sink" for subtle energies, merely placing one's hands under running water could drain away some of the unwanted "bioenergy contamination" acquired from contact with "disturbed" or severly ill individuals (in addition to removing possible bacterial and viral contaminants).

Another implication of Grad's work with psychiatric patients has to do with the concept of "intentionality" and healing. In the experiment mentioned above, Grad had both neurotically depressed and psychotically depressed patients hold on to bottles of 1-percent saline (sealed in a brown paper bag) before using the "charged" saline soution to water barley seeds. One of the neurotically depressed patients that participated in the study was a young woman who, unlike the other patients, asked why Grad wanted her to hold the sealed bottle of salt water. When told about the purpose of the experiment, the woman became elated. When she brought the bag containing the sealed bottle of saline back to Dr. Grad, she cradled it in her arms as if it were a newborn baby. Interestingly enough, the water treated by this particular patient produced accelerated seed germination and plant growth above and beyond that of the control group of untreated water (which was the opposite effect from what Grad had expected). This unusual outcome suggested that the consciousness and intentionality of the individual participating in Grad's healing experiments was a key factor. The idea that a healer's intentionality and consciousness play an important part in healing outcomes was further supported by certain observations made by Grad of Mr. Estebany, the healer involved in his research study. Grad and other researchers had noted that Mr. Estebany seemed to have "off days" when he did not "perform" as well. For instance, if Estebany was distracted or had just had an argument with his wife, he did not produce as significant a healing effect as on other, more "focused" days. As mentioned earlier, Grad was of the opinion that laying-on-of-hands healing worked via a transfer of "vital energy" from healer to patient. Based on his observations of Mr. E's "off days," he concluded that if a healer was either emotionally upset or physically tired, the vital energy that flowed from healer to patient would not occur as easily. Thus, it seemed that not only bioenergy but also consciousness itself appeared to play a critical role in the process of hands-on healing.

Grad performed a number of other experiments in an attempt to measure Mr. Estebany's healing influence on animals as well as plants. For example, he found that Mr. E could greatly accelerate the healing of surgically created skin wounds on the backs of mice. Another experiment involved giving drugs to mice to produce thyroid goiters in the animals. Healing treatments by Estebany resulted in fewer drug-treated mice developing goiters than those in the control group, which Mr. E did not treat. All of Grad's findings pointed to some real energy transfer between healer and patient that could not be due solely to the power of belief or faith. One of the most unusual (and ironic) outcomes of Dr. Grad's research was that he received an award of recognition from the CIBA Foundation, an organization supported by funding from the CIBA pharmaceutical company. This was one of the few times in history a drug company actually applauded research into the physiological effects of laying-on-of-hands healing!

Dr. Grad's research spurred a number of other investigators to further explore the physiological reasons behind the effects of laying-on-of-hands healing. Dr. Justa Smith, a biochemist who had spent a great deal of time studying how strong magnetic fields could accelerate enzyme activity, was one of these inspired individuals. She wondered if laying-on-of-hands healing might actually result from the healer's ability to influence the body's enzyme systems. Dr. Smith set to work investigating her idea by enlisting a healer who would hold test tubes of the digestive enzyme trypsin to see if he could alter its biological activity (the speed at which the enzyme did its job). The experiments proved rewarding, especially when Dr. Smith compared this work with her earlier research on high-intensity magnetic fields. Like Grad, Dr. Smith discovered that there were certain qualitative similarities between healers' bioenergy fields and high-intensity magnetic fields. Both the healer's field and a strong magnetic field could accelerate enzyme activity.

Oddly enough, Smith observed that certain enzyme solutions held by the healer either speeded up or slowed down in their activity or in some cases remained unchanged altogether. Since the healer never knew exactly what was in the test tube he was holding, the intentionality of the healer could not be responsible for the differential changes in enzyme activity. Smith decided to investigate the matter further. She studied a biochemistry flowchart on her laboratory wall to see where each of the "treated" enzymes fit into the normal pathways of cellular energy metabolism. In a flash, she suddenly understood the meaning of the differential effect of healing energy upon different enzymes. When the "bigger picture" of cellular biochemistry was taken into account, it appeared that the change of enzyme activity produced by the healer was always in the direction of greater energy reserve and increased health of the body's cells! This suggested that healing energy seemed to possess some innate "biological wisdom" in knowing exactly which enyzmes to accelerate or decelerate in order to improve cellular energy metabolism. Earlier, we discussed Smith's finding that healers could even repair "sick" enzymes that had been structurally damaged by exposure to certain wavelengths of ultraviolet light. When held (in a test tube) in the healer's hands, solutions of damaged enzymes underwent spontaneous reconstruction as they refolded into their normal, active configuration. The longer the exposure of the damaged enzymes to the healer's field, the more their biological activity continued to accelerate. Interestingly, high-intensity magnetic fields were also able to repair the damaged enzymes.

Noting the striking similarity between the effects of healers and magnetic fields, Smith used a sensitive magnetometer in an attempt to verify her idea that magnetic fields might be the source of a healer's "power" to accelerate and repair enzymes. Unfortunately, the magnetometer registered no unusual magnetic-field activity around the healer's hands during the healing treatment. This negative result suggested that healers' fields were qualitatively similar to magnetic fields, but that they could not be measured with standard magnetic detectors. Ultimately, Dr. John Zimmerman, while working at the University of Colorado Medical Center, was able to measure weak pulsing magnetic fields emanating from the hands of healers by using ultrasensitive SQUID devices (extremely sensitive magnetic-field detectors). However, the magnetic fields coming from the healers' hands (as measured by the SQUID device) were too weak to explain their ability to accelerate enzyme activity, suggesting that another form of energy was really responsible for the observed effects. While Zimmerman failed to verify the theory that laying-on-of-hands healing was due to strong magnetic fields, the magnetic-field fluctuations detected in the hands of healers seemed to indicate that something unusual was going on at a very subtle level. Some researchers wisely asked, "Could there be some other form of magnetism responsible for the biological effects attributed to healers?" Further support for this "subtle" magnetism involved in healing was later noted by Dr. Glen Rein, an innovative subtle-energy researcher. In 1992, Dr. Rein worked with healer/physician Dr. Leonard Laskow to study the bioenergetic effects produced by hands-on healing. During their experiments, Laskow would enter into different states of consciousness associated with his healing work. As in the earlier work of Zimmerman, Rein noted unusual fluctuations in Laskow's magnetic-field emissions when Laskow consciously opened his crown chakra as a part of activating the flow of healing energy. One of the more unusual and distinct magnetic-field patterns was recorded when Laskow asked inwardly for the energies of "spirit" to flow through him. The research findings of Rein and Zimmerman suggest that something important is happening to healers at a very subtle vibrational and spiritual level when they begin to activate their special healing process, and that "subtle" magnetic energy is a key factor in this type of healing work.

While most of the healing research at the time of Smith's and Grad's studies was focused upon healing energy's effects upon plants, animals, enyzmes, bacteria, or physical systems, few had actually studied their effect on human beings. Dr. Dolores Krieger, a professor of nursing at NYU School of Nursing, began to carry out what would become seminal research into the nature of healing regarding its effects upon people, and also with regard to its potential as a learned skill. In order to study a healer's influence on actual sick human beings, Krieger brought together Grad's healer (Mr. Estebany) with two groups of patients on a farm outside New York. The two groups of patients had been diagnosed with a variety of illnesses. The only difference between the experiences of the two groups of patients was that one group received daily laying-on-of-hands treatment sessions with the healer while the other group did not.

In her attempts to prove that healers produced measurable physiological changes in sick individuals, Krieger searched for some parameter of body function whose change might indicate a healing response. In reviewing Grad's healing research with Mr. E, Krieger noted that plants exposed to healing energy were found to be richer in chlorophyll, a key molecule that is structurally similar to hemoglobin in human beings. Was it possible that humans might have a similar rise in their blood-hemoglobin levels following healing treatment sessions? Krieger measured hemoglobin levels in both the healer-treated and control groups. When compared with the control group, patients receiving healing from Mr. Estebany showed rises in their hemoglobin levels, and at statistically significant levels. What most impressed Krieger, however, were the follow-up reports from patients who had received healing treatments from Mr. E. The healer-treated group of patients had come with a wide range of medical diagnoses, including pancreatitis, brain tumor, emphysema, multiple endocrine problems, rheumatoid arthritis, and congestive heart failure. Krieger was astonished by both the medical reports and first-person accounts of patients who had experienced either an improvement in or a complete disappearance of their disease-related symptoms after being "treated" by Mr. E. Two subsequent replication studies produced similar rises in hemoglobin levels along with symptom amelioration in the healer-treated groups of patients. The tendency of healing energy to raise hemoglobin levels is sometimes so strong, it will occur in the face of medical factors that tend to *decrease* blood hemoglobin. To illustrate, some cancer patients who are receiving treatments from healers are also taking bone-marrow-suppressive anticancer drugs that tend to produce anemias (abnormally low hemoglobin levels). Surprisingly, some cancer patients receiving chemotherapy have been found to have dramatic increases in their hemoglobin levels, to the puzzlement of their oncologists (who are often unaware that these patients are also being treated by a healer).

Until this time, the healers that had been studied by scientists were "born healers" who had discovered their healing abilities during their youth. Many people, including Mr. Estebany himself, argued that the ability to heal was a skill you had to be born with and could not be learned. Krieger suspected that healing was probably an innate human skill, but one that was simply more naturally active in born healers. But could this type of healing skill actually be learned? Krieger sought out a method or system whereby she could learn and then teach hands-on healing skills to others. With the help of well-known clairvoyant Dora Kunz, who claimed that she could teach such healing skills, Krieger developed a healing curriculum for nursing students at NYU School of Nursing. Rather than call it laying-on-of-hands healing, a term fraught with controversy, Krieger referred to the practice as "Therapeutic Touch." Her first nursing students happily proclaimed their allegiance to this "different" healing approach by wearing T-shirts emblazoned with the logo KRIEGER'S KRAZIES. In an attempt to experimentally verify the ability of Therapeutic Touch to affect patients, Krieger carried out a multi-hospital study similar to her previous work with Mr. Estebany. She compared the hemoglobin levels of two groups of matched patients, a control group and a group that received Therapeutic Touch (TT) treatments from "nurse-healers" who had studied in Krieger's course at NYU. As in the earlier study, the results were very promising. Sure enough, hemoglobin levels in the patients of the TT-treated group rose to higher levels than those of the control patients, at statistically significant levels. This scientific proof was exactly what Krieger needed to verify her suspicion that healing was indeed an innate but learnable skill that could be taught to health-care professionals and interested laypersons. Her study also showed that nurse-healers possessed the same energetic qualities as "born healers," as confirmed by the hemoglobin rises in the TT-treated group of patients.

Since the early days of Krieger's first NYU class on healing, many nurses, doctors, health-care workers, and laypeople have since learned how to perform Therapeutic Touch on patients, family, and friends. A number of TT alumni nurse-researchers have even continued to look for further proof of the physiological benefits of Therapeutic Touch. Professor Janet Quinn, a nurse and TT practitioner, is among this group of researchers. She discovered evidence suggesting that Therapeutic Touch also enhances immune functioning in both healer and patient. In a small pilot study, Quinn looked at the effects of TT on patients suffering from severe grief following the death of a loved one. Even mainstream research has established that stress and depression can suppress normal immune functioning, making a depressed individual more susceptible to illness from bacteria, viruses, and cancer cells. In Quinn's study, blood samples were taken from deeply grieving patients before and after several forty-minute TT sessions. The goal of the sampling was to look at blood levels of circulating T suppressor cells, a white blood cell that actually decreases the body's immune function by suppressing antibody production. Quinn found a striking 18-percent drop in T suppressor cells after Therapeutic Touch treatments in all grieving patients, an effect that would result in an enhancement of normal immune functioning.

Therapeutic Touch has been documented to have additional healing effects besides raising hemoglobin and enhancing immune response. Just as Grad had discovered that healers could accelerate wound healing in mice, Therapeutic Touch was found to accelerate wound healing in human beings. One study by Daniel Wirth looked at the rate of healing of minor surgical wounds created on the shoulders of forty-four college students. Twenty-three students received TT treatments, while the other twenty-one did not. After eight days, the wounds in the TT-treated group had shrunk an average of 93.5 percent compared with 67.3 percent in the untreated group. After sixteen days, the TT group's wounds had been reduced by 99.3 percent compared with 90.9 percent in the control group. The differences in the rate of wound healing between the TT-treated and the control groups were found to be statistically significant.

Therapeutic Touch is also helpful for relieving pain and stress. Elizabeth Keller and Virginia Bzdek conducted a study to determine whether Therapeutic Touch could decrease the pain of muscle-tension headaches. The two researchers used sixty volunteers, including students, hospital staff, and community members who were suffering from tension headaches at the time of their participation in the study. The participants were randomly assigned to receive either TT or a form of placebo touch therapy ("mock" therapists mimicked the motions of TT practitioners while silently performing mental arithmetic). The headache sufferers were never actually touched by the TT practitioners or the placebo-touch therapists. The therapists brought their hands only to within six to twelve inches of the volunteers' bodies during either form of therapy. Each participant's assessment of his or her headache pain was evaluated using standard pain questionnaires. After comparing pre- and post-treatment levels of headache pain in both groups, the researchers noted that volunteers in the TT-treated group had a significant reduction in headache pain compared with the placebo-treatment group. The results were found to be statistically significant. There are those who suggest that one of the reasons for the headache relief might have been TT's producing greater levels of relaxation in the volunteers (since the headaches were caused by muscle tension). Even if that were the case, the ability of TT to relieve stress and promote relaxation was a significant achievement in itself. The relaxation-inducing effects of TT could not be attributed to simple touch, since "noncontact" TT was employed in the headache study. Later research by Wirth and Cram eventually went on to address this association of TT therapy with muscle-relaxation effects. Wirth and Cram looked at Therapeutic Touch's ability to influence the nervous system and levels of muscle tension. For experimental subjects, they used two groups of meditators who were not informed of the true nature of the study in order to control for the effects of expectation and belief. Each participant was wired with electromyography (EMG) electrodes on several points of the body in order to measure the amount of muscle tension experienced during the course of the study. Unbeknownst to the participants, as they meditated, TT therapists gave them noncontact-TT treatments, while the control group received only mock-TT treatments (without any conscious intention to produce a therapeutic effect upon the experimental subjects). Meditators given Therapeutic Touch treatments had a significant reduction in muscle-tension levels (as measured by electromyography), while meditators in the control group actually showed increases in muscle tension along the spine (suggesting that they unconsciously bristled at the nontherapeutic intrusion into their space during meditation).

In addition to promoting muscle relaxation, Therapeutic Touch has been shown to reduce anxiety levels as well. This finding has implications for the many patients who experience high anxiety levels when they find themselves hospitalized. Patricia Heidt, Ph.D., R.N., showed that TT was able to significantly reduce the anxiety levels of patients hospitalized in the coronary-care unit of a large New York medical center. The subjects were compared with a control group that received either casual touching or no touching by the therapists involved with the study. Janet Quinn also confirmed these results in her own study of TT's ability to reduce anxiety levels in hospitalized cardiac patients. She again used noncontact TT versus mock TT (mental arithmetic) performed by nurses to eliminate the effects of touch itself in relieving anxiety. Anxiety levels in cardiac patients were significantly lower in the TT-treated group. She later repeated this work with patients who had undergone open-heart surgery, again finding not only lower anxiety levels in the TT-treated group but also lower diastolic blood pressures.

The idea of using Therapeutic Touch and other forms of bioenergy healing as an adjunct to medical and surgical therapies is slowly catching on. Dr. Mehmet Oz, a cardiovascular surgeon at Columbia Presbyterian Medical Center in New York, has been using healers in the operating room for several years now with good results. Healers work not only to reduce anxiety but also to stabilize the patient's vital signs and cardiac function before, during, and after difficult surgical procedures. In other hospitals, TT-trained nurses have had success in treating premature infants, with their attendant medical problems. Parents are also being taught Therapeutic Touch as a healing treatment for their fragile babies. Preliminary studies suggest that premature infants who are given TT by parents or nurses appear to do better and make faster strides than those babies who do not receive this form of healing therapy. In Canada, there are now several hospitals that actually have their own separate Department of Therapeutic Touch.

But over the course of the last year or so, Therapeutic Touch has come under attack by various skeptical members of the health-care field. The controversy stems from erroneous conclusions about the validity of Therapeutic Touch assessment techniques. A recent study published in *The Journal of the American Medical Association* (*JAMA*) attempted to verify the claim that Therapeutic Touch practitioners can actually detect the human energy field by using hand-scanning. In order to test this claim, researchers set up a fairly simple experiment. TT practitioners were asked to sit behind an opaque screen with a hand outstretched while another individual would intermittently place a hand near the hand of the TT practitioner. The TT practitioners were then asked whether or not they could detect someone placing a hand close to their outstretched palm. Based upon the authors' simplistic understanding of hand-scanning and bioenergy-field assessment, it was assumed that an individual who could detect the human energy field would be able to tell if another person's energy field was close to their hand. The results of the study showed that TT practitioners could not reliably distinguish whether another person's hand was close to theirs in the absence of visual feedback. From this result, the *JAMA* study researchers concluded that there was no scientific basis for Therapeutic Touch and thus the claims of TT practitioners were highly suspect. It seems the critics did not quite understand the entire TT process. Healing touch is only one part of the TT approach. A session of TT starts with a Therapeutic Touch practitioner using a technique known as hand-scanning in order to assess a patient's energy field. TT practitioners are trained to use their hands to detect subtle differences in the "feel" of another person's energy field. The theory behind this practice is that disturbances in a person's bioenergy field are both a reflection of as well as a precursor to illness in the physical body. Often, TT practitioners direct their healing energies to areas of the body that have localized bioenergy-field disturbances. Such energy disturbances in a person's field may reflect blockages in subtle-energy flow to the body's tissues and localized areas of "energy congestion." To the Therapeutic Touch practitioner, the disturbed areas may feel like regions of unusual heat, tingling, prickly sensations, coolness, or other subjective sensations that register as "energy differences" from the rest of the patient's bioenergy field. TT practitioners scan with their hands (with palm sides facing the body), back and forth, from side to side, moving from the head gradually down toward the feet, looking for areas associated with unusual hand sensations. This information is used to guide practitioners in their healing work when they begin to actually do Therapeutic Touch. When TT practitioners scan a patient's field, they usually detect unusual sensations only when their hands go over an area of the body that has an associated field disturbance. A normal region of the body may not produce any unusual hand sensations at all.

Unfortunately, the authors of the *JAMA* study had no real understanding of the process they were attempting to evaluate. The authors of the *JAMA* article assumed that if TT practitioners could detect the human energy field, they should be able to detect the field of another person's hand. But this is not what happens in the hand-scanning phase of bioenergy-field assessment used by TT practitioners. What TT practitioners feel with their hands is not the presence or absence of an energy field but dynamic variations in the field related to localized disease processes within the physical body (as well as primary energy disturbances within the auric field itself). Hand-scanning of a person's bioenergy field over a normal region would not ordinarily provide any unusual sensations to the TT practitioner. Therefore, trying to assess the ability to detect pathological energy-field disturbances with the hands by bringing a normal hand, unaffected by illness, close to the practitioner's hand would be unlikely to evoke any unusual sensations. In other words, a negative result would have already been expected. Thus, the experiment, as devised by the researchers, would not give an accurate reflection of what TT practitioners do when they assess the human energy field with hand-scanning. The basic premise of their experiment, based on false assumptions, was totally flawed.

More important, the authors of the *JAMA* study did not actually attempt to measure the effects of TT upon human subjects before making sweeping claims about its lack of validity as a healing technique. The study's erroneous conclusions about TT were based upon information that had very little to do with Therapeutic Touch as it is taught and practiced. The claims of the *JAMA* study that TT was an ineffective or useless therapy were based upon experiments that had little to do with the technique supposedly being evaluated. Perhaps if the authors of the *JAMA* study had reviewed the extensive body of scientific research on the physiological effects of healing touch on bacteria, plants, animals, hemoglobin levels, enzyme activity, and the like, they might have designed a more thoughtful study. However, the *JAMA* article did generate radio and television news coverage which inspired many questions about Therapeutic Touch and the very nature of healing. Certainly, anything which helps to raise public awareness of vibrational medical approaches and the under-appreciated potential for healing that we each possess can only be of benefit in the long run.

Healing by the Laying On of Hands: How Does It Work?

We have examined some of the research suggesting that laying-on-of-hands healing (or "bioenergy healing") produces measurable physiological changes in human, animal, and even plant subjects at statistically significant levels. Since it seems that something other than the placebo effect is responsible for the healing changes brought about by bioenergy healing, it becomes crucial that we begin to ask, "What are the actual mechanisms behind the laying-on-of-hands healing process?" A definitive answer to this question is still hotly debated within the alternative-medicine community. While Bernard Grad's research convinced him that some kind of life-energy transfer was responsible for the observed healing effects, a more accurate explanation may actually be much more complex than a simple energy exchange. For example, the question has been posed as to whether a healer's stimulating effects are related to a simple "life-energy transfusion" or possibly some type of "energetic restructuring" of the patient's bioenergy field. We'll next examine variations of both proposed explanations.

A number of different models look at this form of bioenergy healing from the competing perspectives of energy exchange, information exchange, and various combinations of the two processes. To begin with, the simplest model of healing is sometimes referred to as the "jumper-cable analogy." In this model, the patient is seen as a bioenergy system with a severely "depleted" bioenergy battery, while the healer is someone who possess a "fully charged" bioenergy battery. By placing his or her hands on the patient, the healer encourages life energy to flow from the healer's stronger battery to the patient's weaker battery, essentially "recharging" the depleted battery. This might be considered a kind of "life-energy transfusion" that helps to raise life-energy to levels high enough to support the patient's own inner healing mechanisms. This transfer of life energy may also supply something besides more "voltage" to the batteries. For one thing, life energy is "negatively entropic" in nature. Remember, the tendency of negative entropy is such that life energy pushes biological systems toward states of greater internal order and organization. Dr. Smith's studies of healers' effects on damaged enzymes showed that healing energy was able to repair and reassemble damaged enzyme molecules back into their normal active configurations (a negentropic phenomenon). Smith also showed that life energy coming from healers' hands seemed to possess some innate "biological wisdom." During her experiments in which healers treated different types of enzymes, the life energy seemed to know innately which enzymes should be accelerated in their activity and which needed to be decreased in activity in order to promote greater health and energy balance of the body's cells.

As we have already discussed, healers demonstrate an ability to influence water, endowing "healer-charged" water with properties that promote healing in plants (and animals). More than 90 percent of the molecules that make up our bodies are water molecules. One theory which attempts to explain the healer's effect suggests that the bioenergy coming from a healer's hands is directly transferred to the water molecules of the body, and it is this "biologically activated" water that brings about inner healing changes. A number of Russian scientists studying the healing effects of conventional magnetism on human beings concluded that the magnet's healing power was likely due to such an energetic alteration in the water content of the body. Since life energy (or "animal magnetism") is a distant cousin of conventional magnetism, these same principles of influencing "biological water" might also apply to laying-on-of-hands healing.

There are also a number of alternative-medicine proponents who seek to understand healing in terms of something other than just an energy transfer. For example, some theorists, like Dr. Larry Dossey, author of *Prayer Is Good Medicine,* believe that healing is due to the effects of consciousness, or more specifically, our "nonlocal mind." This term "nonlocal" refers to various phenomena (remote viewing, distant clairvoyant observations, distant healing by prayer) that demonstrate that our minds can observe and even influence individuals, objects, and events widely separated from the observer by distance and time. We need only refer back to the studies that have enlisted prayer, a form of focused consciousness, to produce significant healing effects in plants, animals, and even people at great distances. Since our nonlocal minds may be able to affect healing changes in another human being at a distance through an act of prayer, they could also be the true mechanism behind the effects of laying-on-of-hands healing as well. That is to say, healing might be triggered by an act of consciousness distinct from the inner healing effects of the placebo effect or faith in a particular treatment.

After all, when most people pray for another person to be healed, their prayers usually involve invoking the energies of the divine. It is possible that both prayer and laying-on-of-hands healing might actually produce healing by the very same mechanism—the divine energies of the Creator. According the this "divine-energy hypothesis," the only difference between the two healing approaches are the methods chosen for invoking and directing these divine energies. To illustrate, if you ask most healers how they heal, they will usually tell you they are not doing the actual healing but instead are only acting as channels or vehicles for a higher divine healing energy. So in this view, healing by the laying on of hands might be considered a consequence of divine energies or, quite literally, an "act of God." As scientific as we would like to remain, it is sometimes hard to separate ourselves from the fact that we are all creations of God, a higher power, the divine light (or however else we wish to define our relationship to the Creator). This fact holds true for spiritual believers and atheists alike. Even great scientific thinkers such as Albert Einstein acknowledged the existence of God as an influence in the creation of the universe and the affairs of humankind. Therefore, we must consider the energies of the divine as a possible, credible mechanism behind these and other forms of healing.

The bioinformational model is another theory of healing that does not necessarily involve the direct transfer of energy from healer to patient. According to this theory, healing may be brought about by supplying appropriate bioinformational instructions to the human body in an effort to stimulate an individual's own inner mechanisms of healing. Specifically, remember how homeopathy uses a solution containing less than a single molecule of original plant substance to trigger a healing response. One explanation for how this takes place may have to do with the extraction and transfer of bioinformation from a healing plant into water (via the process of homeopathic potentization). This extracted bioinformation (in the form of a homeopathic remedy) may supply a therapeutic bioinformational pattern to the sick individual, resulting in the triggering of the body's inner healing systems, which then overcome the illness. The transfer of bioinformation from homeopathic dilution to patient may occur because of a vibratory-resonance process going on between the patient and the homeopathic remedy. This same effect could also occur when a healer's bioenergy field comes into close contact with the patient's field while the healer is focusing on healing. Through some resonance effect between the bioenergy fields of the healer and patient, the healer may resonantly transfer some type of healing bioinformational pattern directly into the patient's field. Once the patient's bioenergy field has been "restructured" into a healthier pattern, changes in the cellular patterns of the physical body would likely follow, resulting in a state of improved health. If this process actually does happen, it is likely that such resonance effects take place at the level of the etheric body. As you recall, the etheric body is believed to guide the growth and development (or distortion) of the physical body. Many healers and clairvoyants who can see auras and changes in a patient's bioenergy field will often note changes heralding impending illness in the auric field weeks to months before the development of any symptoms, supporting the notion that many illnesses probably do begin at the etheric level. Therefore, it's possible that healing might occur through a "resonant repatterning" of the patient's etheric field without the need for any actual "energy transfer" to take place.

Another potential mechanism behind the process of healing could be referred to as the "earthfield" connection and is related to the balancing energies of the earth's magnetic field. I suggest that healers may somehow connect their patients with the planetary geomagnetic field in a way that may trigger an inner healing response. How might this connection between the healer, the patient, and the earth's magnetic field come about? It may have to do with the process of resonance that is hypothesized to occur between healer and patient. For example, researchers who have studied healer/patient interactions at a physiological level have observed how the physiological rhythms of both healer and patient seem to go into synchrony during the act of healing. This synchrony between bodily rhythms is especially evident in the brain-wave patterns of the healer and the healee.

During the late 1970s and early '80s, English biofeedback researcher Maxwell Cade used a special brain-wave-biofeedback device he had developed, known as the Mind Mirror, to study the synchronization of brain rhythms between healers and patients. Cade's Mind Mirror is a unique biofeedback device capable of translating the squiggly lines of the EEG (the brain-wave recording of the electroencephalograph) into a dynamic visual graph (kind of like multiple bell-shaped curves) which shows the real-time frequency patterns of human brain-wave activity in both the right and left hemispheres of the brain. Experienced healers studied by Cade who were monitored during the act of healing produced a unique brain-wave pattern (known as the "state 5 pattern") consisting of three peaks of activity in three different frequency ranges, all occurring simultaneously. This subtle effect could easily be missed during a conventional EEG analysis. The three peaks of brain-wave activity seen in the state 5 pattern consisted of a peak in the alpha range (normally associated with relaxation), a peak in the alpha-theta range (associated with creativity), and a peak in the delta range (normally seen during deep sleep). The middle peak of brain-wave activity in the alpha-theta range occurred at about 7.8 cycles per second, the same frequency as the Schumann resonance, the predominant frequency of the earth's magnetic field. When a healer began the healing process, the patient's brain-wave pattern (as monitored by the Mind Mirror) would go from a usual alpha-beta range activity (associated with normal waking consciousness) into the three-peaked state 5 pattern that Cade had seen almost exclusively in healers. In other words, there seemed to be a kind of resonance between the healer's brain and the patient's brain activity (almost as if the healer was "driving" or "entraining" the patient's brain-waves to assume the state 5 pattern). The patient's three-peaked brain-wave pattern would be observed only during the healing session. After the session had ended, the patient's brain-wave patterns reverted either to their original range or to one of greater alpha activity and relaxation. In other words, patients would demonstrate a state 5 brain-wave pattern only when they were under the direct influence of the healer. Cade determined that healers who tended to have the best healing outcomes with their patients were also the ones who most consistently produced the "brain-wave synchronization effect." Interestingly, Cade also found that healers who seemed to tap in to their own personal energy reserves for healing frequently became depleted after doing multiple healings.

Cade's unique discovery of a healer-healee brain-wave synchronization with the state 5 pattern (the central peak of which is the Schumann resonance frequency of the earth's magnetic field) provides us with possible clues to the link between healers and the "earthfield." In Cade's research, volunteer healers were able to energetically entrain a patient's brain-wave rhythms into the 7.8-cycle-per-second pattern of the geomagnetic field. This means that once the healing process had begun, the healer, the patient, and the magnetic field of the earth were all oscillating at exactly the same frequency, which thus might allow a resonant-energy exchange to take place from earth to healer to patient. What may be occurring is that the healer, by shifting the patient's energy frequency to the same as the earth's field, opens up a kind of "resonant frequency window" through which the magnetic energies of the planet can flow from the healer directly into the patient's body. In other words, the healer might be "channeling" energy to the patient from the healing magnetic energies of the earth itself. Support for this idea of a magnetic connection between healer and patient might also be drawn from Dr. Grad's observations that healers' fields and magnets' fields both can accelerate plant growth.

An alternate interpretation of the "earthfield hypothesis" is that healers create a kind of resonant-frequency window through which higher vibrational (etheric and higher) energies may become "piggybacked" onto the geomagnetic currents flowing through the healer and into the patient's field. Thus, there may be additional (spiritual) healing frequencies of energy, coupled to the earth's magnetic energies, that contribute to healing the patient at a variety of multidimensional levels.

One other theory of healing that is rarely talked about in "scientific circles" involves the idea that "healing spirit guides" might also play a role in laying-on-of-hands healing. Such ideas need to be taken seriously, since many of the world's greatest spiritual healers frequently proclaim that they are only acting as a vehicle for a "higher" spiritual power. But whether the channeled power flowing through their hands is, as was mentioned earlier, the force of God, the divine energies of spirit, or some other higher-dimensional power is uncertain. While practitioners of Therapeutic Touch often talk about tapping in to some vast pool of universal life energy and intelligence that they channel into patients, other healers, such as those trained at the Barbara Brennan School of Healing in New York, sometimes attribute their healing work to spirit guides and discarnate beings who channel healing energy through them when they go into their "healing state of consciousness." "Spirit-aided" healing is based upon the premise that after death, healers and doctors who have passed on to "the other side" can still work through living healers who, through their inner prayers, literally "ask the universe" for healing help and assistance. Even Brennan admits to sometimes calling in her spirit guides as she prepares to do healing. She then observes the various colored healing energies her "guides" channel through her hands, never knowing in advance exactly what will happen. Simply put, Brennan mentally stands back and allows the process to occur. However, most other times, she consciously directs specific types and colors of healing energy to where she intuitively feels they are needed in the patient's physical and spiritual bodies, and in the auric field.

The concept that spirit guides work through healers is by no means a new one. Some famous psychic healers have actually claimed to have specific discarnate doctors or other "deceased" health professionals working through them from "the other side." One such healer was Arigo, the famous Brazilian psychic surgeon who was known as the "surgeon of the rusty knife." Arigo claimed that a German physician by the name of Dr. Fritz was working through him to accomplish his healing work. While Arigo did not do laying on of hands per se, he appeared to channel both medical information and some type of healing energy to patients. Arigo would operate on hundreds of patients a week, removing tumors and growths and draining abscesses under nonsterile conditions using only a Swiss Army knife. Most of his surgeries were nearly bloodless, and the vast majority of "patients" quickly healed without any sign of postoperative wound infections. Many doctors who observed Arigo work believed that the healing energy flowing from his "spirit-guided" hands decreased bleeding, prevented infection, and accelerated wound healing as well. Arigo also wrote prescriptions for various medicines that were often effective treatments for many illnesses. For a highly trained surgeon, such an effective healing rate might have been considered somewhat impressive, but for Arigo, a man with only a sixth-grade education, it was unbelievable!

Although Arigo's type of psychic surgery was not true laying-on-of-hands healing, it might be considered another form of "bioenergy healing" (which is "overshadowed" by higher spiritual forces and spirit guides). In support of the spirit-guide hypothesis, a number of Reiki healers who practice this ancient form of bioenergy healing also believe they are being assisted by healing spirit guides during Reiki healing sessions.

Perhaps the only way we will ever find out the true mechanism behind the various forms of hands-on healing will be through the use of both scientific instrumentation and clairvoyant observation to closely scrutinize what really happens during the healing process. In the future, when science and spirit move closer together, as they once were in times past, we will certainly have better answers to such tantalizing questions. For the present, we can only build upon the research that has already been done in stating that the laying on of hands is truly a powerful phenomenon with healing effects not easily explained solely on the basis of the placebo effect.

The Many Varieties of Healing-Touch Therapies

There has been a proliferation of bioenergetic healing therapies since the early work of Krieger and other Therapeutic Touch researchers. For example, in England, more than 8,500 legally registered healers are permitted to "give healing" if patients request it. At least 1,500 government hospitals in Great Britain have been given approval to use bioenergetic healing as an adjunct to conventional medical and surgical therapies. In fact, many bioenergetic healers are paid for their services under the UK's National Health Service. Healers in England are even able to purchase liability-insurance policies similar to those available to more traditional physicians. Of great interest is that fact that English physicians who attend courses in bioenergy healing receive postgraduate continuing-education credits. In Russia and Poland, where bioenergy healing is actually a part of conventional medical practice, some medical schools have courses in alternative healing as part of their curriculum. In addition, bioenergy healing is currently under study by the Academy of Science in Russia. In nearby Bulgaria, healers undergo rigorous examinations to assess their healing abilities before being "licensed" by a government-appointed medical board. But China seems to lead the rest of the world in research on the therapeutic effects of bioenergy healing, or more specifically, medical chi gong (also called qigong) therapy. In fact, departments of medical chi gong research exist in every college of traditional Chinese medicine throughout China.

While other countries may be well on the path toward incorporating bioenergy healing into mainstream medical treatment, the United States is making slow progress toward acceptance of this unique form of vibrational therapy. According to some estimates, there are some 50,000 healing practitioners in the United States providing about 120 million treatment sessions annually. Of the 50,000 healing practitioners, perhaps 30,000 have been trained in Therapeutic Touch. For some, bioenergy healing is their major vocation, while for others, it is a process occasionally used to help heal family and friends. Many healing practitioners have had no formal training in healing, having independently discovered their natural healing abilities and methods for producing therapeutic bioenergy effects. Still others have learned basic healing techniques from the handful of schools in the United States that teach various forms of bioenergetic healing. Interestingly, women far outnumber men among those who have enrolled in these healing schools. Throughout our country, there are no state licensing requirements for bioenergetic practitioners. Because of legal constraints in many states, most healing practitioners tend to avoid the terms "patient" and "treatment," substituting the terms "receiver" and "session" in describing their healing work. To avoid being charged with practicing medicine without a license, many healers work under the auspices of a healing church, sometimes describing what they do as "healing the spirit," which, in turn, produces healing changes in the physical body.

In addition, there are probably several thousand conventional health-care practitioners (such as doctors, nurses, psychotherapists, chiropractors, and massage therapists) who at least occasionally use bioenergetic therapy as an adjunct to their regular treatments. Also, a small number of hospitals are currently using Therapeutic Touch, healing touch, and SHEN therapy in programs for alcohol abuse, drug abuse, and codependency recovery. At least four different forms of bioenergy healing have been taught in conventional medical training programs, including Therapeutic Touch, healing science, SHEN therapy, and healing touch (a blend of Therapeutic Touch, healing science, and other healing approaches developed by the American Holistic Nurses Association).

Therapeutic Touch is probably one of the best-known forms of bioenergy healing. But the touch-based bioenergetic therapy known as SHEN is also very popular. SHEN (an acronym for Specific Human Energy Nexus) was developed by Richard Pavek of Sausalito, California, as an offshoot of his work with the bioenergetic fields of the body and how they respond to various types of emotional stress. Pavek's SHEN therapy is a healing-touch treatment with elements of both healing touch and Polarity Therapy. SHEN, or physio-emotional release therapy, is said to be useful for relieving pain due to the repressed emotions associated with migraine and cluster headaches, chronic low-back pain, anxiety attacks, chest pains related to heart disease, and various other forms of chronic pain. It is based partly upon the concept that there are various directional energy currents that flow throughout the right and left sides of the body (as described by therapists who use Polarity Therapy). Through his research, Pavek discovered how to release blocked energy currents caused by repressed emotions in the body. In his technique, the SHEN practitioner places his or her hands in a specific right-to-left orientation on certain areas of the patient's body, including major chakra points.

Early in his research, Pavek noted that placing his right hand on a patient's right hip while simultaneously placing his left hand on the patient's right shoulder caused the individual to become relaxed, sometimes to the point of falling asleep, after which the patient would awaken refreshed. When Pavek reversed his hand orientation (left hand on right hip with right hand on right shoulder), the patient's heart would beat faster and in time the patient would become disoriented and nauseous. Pavek believed that his initial hand placement was enhancing the normal energy flows through the body (with energy moving down the torso along the right side of the body). Pavek used hand placements in specific right-to-left orientations at particular locations on the body in order to enhance ch'i or subtle-energy flow through the bodies of his patients. He noted that subtle energy tended to flow from the right to the left hand in a manner not unlike magnetic-field lines flowing from the north to the south pole of magnets. When the natural energy flow from the hands was applied to specific locations on the body, there seemed to be an enhanced bioenergy flow through the patient's body. Over time, Pavek developed an organized system of rebalancing the body's bioenergy flows, which eventually became known as SHEN therapy. According to the theory behind SHEN, its practitioners attempt to move bioenergy across regions of increased energetic resistance in an effort to complete disrupted energy circuits that have become blocked by chronically held-in emotions. These blocked and suppressed emotions are believed to contribute to the pain and localized muscle spasms associated with various stress-related disorders. SHEN treatments appear to produce real physiological changes in patients, including decreases in PMS symptoms, reduced pain associated with the birthing process, and relief from chronic migraine headaches. In a preliminary study of SHEN therapy's effect on cancer patients, four volunteer cancer patients had a significant increase in their white-blood-cell counts during chemotherapy (which tends to decrease WBC counts) when SHEN therapy was directed to the region of the patients' thymus glands.

Besides SHEN and Therapeutic Touch, there are several other laying-on-of-hands therapies that have come to the United States from the Far East. One of these older healing techniques is known as Reiki, a Japanese word that means "universal life energy." In Reiki therapy, a trained Reiki healer acts as a channel for "universal life energy" that is taken into the healer's body from the surrounding environment and then directed to a patient in need of healing. The discovery of Reiki as a healing technique has been attributed to Dr. Mikao Usui.

During the early 1900s, Usui worked as both a principal of Doshisha University in Kyoto, Japan, and as a Christian minister. He developed Reiki after a lengthy search for the ancient methods by which Jesus and the Buddha did healing. Apparently, the healing knowledge of what today is called Reiki was known in India at the time of the Buddha (Gautama Siddhartha) and also in the mystical teachings of ancient Tibetan Buddhism.

While similar in some ways to Therapeutic Touch, which channels healing energy into the body through the hands, the practice of Reiki healing is somewhat more complex. Reiki healers often employ the use of special Japanese symbols (drawn by hand motions through the air). The symbols are said to give greater power to both local, hands-on healing and distant healing. For many years, trained Reiki masters have carefully guarded the Reiki healing symbols from the eyes of the uninitatied (in accordance with long-standing traditions). Only at the higher levels of initiation is a Reiki student instructed in their use. Reiki symbols may function by activating and tapping in to certain "healing thoughtforms" and healing spiritual energies that have been built up on the "higher planes" through many years of Reiki healing. The stages of Reiki training progress gradually through three primary degrees, called levels one, two, and three. Each successive degree represents an increasingly higher level of complexity and inner attunement. There also are a number of variations on the three levels, such as with the "Radiance technique," an offshoot of Reiki, which is said to have eleven separate degrees of training. Each of the attunements involves transmissions of subtle energy to the Reiki students (from the Reiki master) that are supposed to open up their inner healing channels and to allow more universal life energy to flow through them. At each attunement, a Reiki student experiences a cleansing process that works through the physical, emotional, mental, and spiritual levels, often resulting in self-healing, detoxification, and the releasing of old energy blocks. Reiki work can be an extremely powerful form of healing when performed by a highly trained Reiki practitioner. It is said to rapidly relieve aches and pains and to relieve many acute symptoms of illness. Reiki therapy can also recharge and rebalance the body's subtle energies while relieving the effects of stress. One study by Wendy Wetzel, published in *The Journal of Holistic Nursing,* found that Reiki therapy was able to increase hemoglobin levels in patients in a manner similar to Therapeutic Touch. In addition, Reiki therapy is said to have a positive influence on spiritual growth in both healer and healee.

Reiki is by no means the only healing therapy to come West by way of the Asian continent. Another form of laying-on-of-hands healing originating in the Far East is Johrei, a hands-on-healing approach developed by Mokichi Okada in Japan in the 1920s. The word "johrei" literally means "purification of the spirit." It is considered to be a form of spiritual healing that, while widely known in Japan, is only now gaining greater recognition as a valuable healing technique in the West. Still another healing method to come out of the Far East is medical qigong, also known as chi gong or chi kung. Chi gong, as mentioned earlier in this book, is part of a discipline that recognizes the importance of a normal and balanced flow of ch'i energy through the acupuncture meridians of the body. Chi gong exercises are practiced by millions of Chinese daily. The exercises are part of a system that uses various internal and external exercises to build up the body's natural ch'i energy, which in turn allows people to achieve optimal health by healing illness from within. Some advanced practitioners of chi gong, known as chi gong masters, are able to consciously move ch'i energy through the meridians of their body and project it at a distance to patients four or five feet away. By directing ch'i energy into the weakened systems of sick patients, a number of chi gong masters have been successful in healing illnesses that did not respond to more traditional medical approaches. In certain cases, chi gong masters have been able to get patients' paralyzed limbs to move for the first time while projecting bioenergy from a distance of several feet away. Some chi gong masters are reportedly able to anesthetize patients for surgery using only chi gong techniques. Research on chi gong masters has shown that when they project bioenergy, various types of weak ultrasonic and electromagnetic emissions can be detected near their hands. This recorded energy emission, while tantalizing, is probably only a secondary effect of a primary subtle energy that remains elusive to most scientific investigators.

Another form of healing somewhat similar to chi gong healing is known as pranic healing (mentioned earlier, in the chapter on color and light therapy). Instead of consciously directing ch'i through their acupuncture meridians and out through their hands as a chi gong master does, a pranic healer takes in and circulates prana through the chakra system and out through the hands and chakras. Some pranic healers, such as Master Choa Kok Sui, talk about moving ch'i as well as prana through their hands to energize patients' bodies during their healing work. Pranic healers seek to heal the physical body by transferring vital energy (in the form of prana) to a patient's physical and etheric or subtle bodies. According to Master Choa, there are three major sources of prana available to the healer and to all living things on the planet. These are "solar prana" (from sunlight), "air prana" (absorbed through the lungs as well as through the chakras), and "ground prana" (absorbed from the earth through the soles of the feet). Prana is taken into the body of the healer via special breathing techniques and by visualization methods that help to draw it up from the earth through the feet. From there, the absorbed prana is directed out through the pranic healer's hands via the palm and finger chakras (and through the major chakras as well).

Pranic healers use vital pranic energy in two basic ways: for cleansing and for energizing the patient's etheric body. First a trained pranic healer will cleanse or remove "diseased etheric matter" from any affected etheric chakra or diseased etheric organ. This cleansing is then followed by an energizing of the problem areas with pranic energy so that healing of the physical body can soon follow. In cases of serious illness, the entire etheric body may need to be cleansed of diseased etheric matter before and after pranic energizing is done. Pranic healers claim that the cleansing process is necessary because fresh prana needs to flow into the affected body region that is already filled with "diseased" etheric matter. Cleansing is said to remove the blockages that affect the flow of energy through the normal subtle-energy channels of the chakras (the thread-like nadis) and the meridians.

The pranic healer's method for removing diseased etheric matter is similar to a technique employed by healers trained at Barbara Brennan's school for healers. Brennan's technique, also known as "chelation," was originally developed by Reverend Rosalyn Bruyere, a healer and founder of the Healing Light Center in Glendale, California. When healers perform chelation, they focus energy through their hands to clear the auric field and lower subtle bodies of distorted energy patterns and auric debris, followed by an energizing of the auric field with healing energy. Pranic healers also focus pranic energy through their hands to heal specific dysfunctional chakras that may be contributing to the underlying physical disease process. Specific colors of healing prana are also consciously visualized and projected by the healer for treating different chakra-related problems and various illnesses.

As in Brennan's healing science, Therapeutic Touch, bioenergy healing, and other similar healing approaches, a pranic healer first assesses a patient's auric field for energy disturbances before attempting to project pranic energy. According to Master Choa Kok Sui, this healing approach can be useful for treating a variety of different health problems. Parents trained in simple pranic healing have been able to quickly bring down the temperature of children experiencing high fevers (often in a matter of a few hours). In many cases, pranic healing can rapidly relieve headaches, gas pains, toothaches, and even muscle pains. Healers using these techniques can supposedly clear up coughs and colds in a day or two, and diarrhea is said to respond in a few hours in most cases. More serious illnesses, such as heart, kidney, liver, and eye diseases, are said to be symptomatically improved in a few pranic healing sessions, with long-lasting cures achieved only after several months of pranic-energy treatments. Healers using this technique claim that recovery from strokes can also be accelerated by pranic healing, although the rate of recovery will depend upon the experience and expertise of the particular pranic healer. Asthma patients have been known to improve with pranic healing, although asthma caused by emotional factors appears to take longer to heal than does that caused by purely physical factors. Pranic healing is thought to accelerate the overall process of healing from most common illnesses by a factor of three times the normal rate of recovery. For instance, pranic healers can accelerate the healing of broken bones so less time will be required for a plaster cast to be worn. If a patient undergoes surgery, pranic healing (before, during, and after surgery) may help reduce bleeding, minimize postoperative infections, accelerate the rate of wound healing, and decrease the overall recovery time from surgery. Master Choa recommends a pranic-healing cleansing process for any "donated" organ (such as a kidney) prior to and after transplantation surgery in order to decrease the incidence of organ rejection by the body of the transplant recipient. Although pranic healing does not often completely heal cancer, it can relieve associated pain, improve a cancer patient's energy level, increase appetite, and sometimes halt or slow the progress of cancer-cell growth. In cases of "terminal" cancer, pranic healing can assist patients to die with peace and dignity if that is the patient's choice.

A most fascinating aspect of pranic healing relates to its use in healing psychological problems. Master Choa refers to this technique as "pranic psychotherapy." During pranic psychotherapy, a healer assesses the condition of the patient's chakras and etheric body. After an assessment of the energetic problems of the patient, the pranic healer uses sweeping motions of the hands to "clear" the patient's auric field. During this process, the healer attempts to clear away bioenergetic debris and thoughtforms, as well as other bits of psychic flotsam and jetsam that can be clairvoyantly appreciated in the auric fields of psychologically disturbed patients. Some psychiatrists in the Philippines have been trained in this technique and find it quite useful as an adjunct to conventional psychotherapy for treating depression and other psychiatric disorders. Of course, the concept that depression may not be strictly a biochemical health challenge is one that will be slow to catch on with conventionally trained psychiatrists. Nevertheless, response to pranic psychotherapy can sometimes be quite rapid, with changes in a person's mood and outlook appearing in a matter of days instead of the weeks and months usually required with the use of psychotherapeutic drugs.

Another aspect of pranic healing involves its use in healing both "on site" as well as at a distance. For example, many of the bioenergy healing techniques already mentioned usually require the presence of the healer in the same room as the patient. But with the more advanced levels of pranic and Reiki healing, a practitioner appears to be capable of producing significant distant-healing effects as well. This specific capability places advanced Reiki and pranic healing into a "subcategory" of hands-on healing that has elements of bioenergy healing. This subcategory is often referred to as "spiritual healing" and includes many of the distant-healing practices used by "natural" healers as well. The ability to do distant healing is the one aspect that truly distinguishes advanced Reiki and pranic healing from techniques like Therapeutic Touch. While Therapeutic Touch practitioners usually require either direct contact or at least close proximity to their patients, advanced Reiki healers can be effective in healing patients who are many miles distant from the healer. In addition, advanced pranic healers use a technique known as "divine healing" (considered a higher form of pranic healing) to heal distant patients. They do this by projecting healing energy through photographs of distant patients in a manner not unlike certain aspects of radionics. Many spiritual healers are also capable of performing successful distant healings in this fashion, with or without the photographs. Frequently, just the name of the distant patient is needed to intitate healing contact.

Successful distant healing by "natural" spiritual healers has been documented in a number of scientific research studies. Most notably, the late Ambrose and Olga Worrall, two extremely powerful spiritual healers, were able to heal sick individuals who directly experienced their healing touch, as well as patients who were hundreds of miles away during their healing sessions. The Worralls participated in many research studies with a number of different scientists who verified their ability to influence various living and nonliving systems both locally as well as at a distance. Olga Worrall was shown to produce measurable energy effects as a result of her potent healing energy. These effects included accelerated plant growth, increased enzyme activity, and even unusual cloud-chamber wave patterns. Worrall could produce the unusual and distinct wave patterns both when her hands were near the cloud chamber as well as when she projected healing energy to the chamber from her home, hundreds of miles away! The spiritual-healing force that Worrall was able to project was clearly unaffected by distance.

While most hands-on healers are able to rebalance the physical body, the meridians, and the etheric body, a number of spiritual healers claim that they can heal and restructure the higher spiritual-energy bodies as well. Probably the distinction between pure healing touch and spiritual healing is somewhat artificial in a sense, because some practitioners who claim to be working at one body-spirit level may also be working on healing higher levels of spiritual anatomy. It is entirely possible that some of the most successful body workers, massage therapists, Therapeutic touch practitioners, and gifted physicians may also be powerful spiritual healers who are unaware of the full scope of their healing gifts. After all, quite a few spiritual healers, including Olga Worrall, never entirely understood the mechanisms by which their healing was actually accomplished. They knew only that it did work in many cases where conventional therapies had failed. Spiritual healing seems to operate on many different bioenergetic and spiritual levels simultaneously. While spiritual healing and healing-touch therapies have many outward similarities, with only subtle differences to the eyes of a nonpsychic observer, both approaches could seem indistinguishable from one another unless one truly were clairvoyant.

Perhaps because of the inherent complexity associated with spiritual healing, a number of schools have been established to teach students a variety of spiritual-healing techniques. Some schools, such as Dr. Jaffe's School of Energy Mastery in Sedona, Arizona, instruct healers to work with the higher spiritual bodies and the auric field of their patients in order to diagnose illness and heal from a higher spiritual level. Similarly, at the Barbara Brennan School of Healing in New York (a center for learning Brennan's "healing science" approach), healers-in-training are taught to clairvoyantly observe and energetically manipulate the human spiritual bodies, the auric field, and the chakras of clients seeking healing. Students learn how to clairvoyantly perceive the different spiritual and energetic levels of multidimensional human anatomy so that they can make precise energetic adjustments in the physical bodies, the spiritual bodies and the chakras of the patients they work on. Students at Brennan's school are also instructed in the fine art of "etheric surgery," which is said to repair damaged chakras and diseased etheric organs that may be the real underlying cause of a patient's physical health problems.

Like Brennan, other healers have also developed their own school for teaching unique healing approaches. One of these healing schools is the nonprofit Wirkus Bioenergy Foundation in Bethesda, Maryland, established by healer Mietek Wirkus and his wife, Margaret. Wirkus discovered his natural healing abilities before the age of five, while he still lived in his native Poland. By merely placing his hands upon the chest of his asthmatic sister during an asthma attack, Wirkus could consistently provide her with immediate relief from wheezing and labored breathing. Later, when Wirkus began his formal healing work, he would often feel patients' pains or illnesses in his own body as he worked on them. As he grew older, he discovered techniques that would protect him from any energy bleed-through into his own body, to keep him from experiencing a patient's pain and distress. Wirkus learned to feel patients' energy fields with his hands and to clairvoyantly see the energy patterns of their auric fields and subtle bodies in order to diagnose a health problem without having to experience disturbing physical symptoms. He also received instruction in specialized breathing techniques and meditation from a monk steeped in the Tibetan tradition of healing. With meditative centering and a specific form of breathing, Wirkus was better able to direct healing energy from his hands into the bodies and energy fields of his patients. Before beginning bioenergy healing with a patient, he would assess each patient's physical and spiritual bodies to determine the true cause of the health problem. After determining where a patient had difficulties because of bioenergies that were of balance, congested, or merely weakened, Wirkus would then direct healing bioenergy from his hands to the afflicted areas of the patient's body. He became quite successful as a bioenergy healer, with many profound healings of both young and old patients alike. Wirkus eventually became licensed as a bioenergy therapist so that he could work cooperatively with doctors at a Polish medical clinic.

After immigrating to the United States, Wirkus became the subject of research studies by a number of different scientists, including Dr. Steven Farion and Dr. Elmer Green of the Menninger Foundation in Topeka, Kansas. During one series of experiments, Wirkus was hooked up to various electrodes and equipment to monitor electrical changes in his body during healing. The researchers were startled to find that Wirkus's body generated voltage surges of up to eighty volts or more, something most scientists had previously believed impossible!

Wirkus' clinical results with patients in the United States and Poland have been impressive, especially his work with sick and disabled children. Among his greatest successes with bioenergy healing have been a number of children with cataracts, recurrent heart problems (poorly responsive to multiple cardiac surgeries), and uncontrolled seizures. During the time Wirkus did healing at the medical clinic in Poland, he and colleague Dr. Stefania Szantyr-Powolny performed bioenergy healing on deaf children with hearing loss from a variety of causes. One five-year-old girl who had been born deaf acquired near-normal hearing after five months of bioenergy treatments. Another child whose hearing had been destroyed by streptomycin toxicity when he was two miraculously recovered his hearing after only four bioenergy treatments. Audiograms of the healed deaf children confirmed that their hearing had indeed returned.

Wirkus and his wife, Margaret (who also does bioenergy healing), have since brought their version of healing to the United States, where they continue to teach both laypeople and doctors the bioenergy healing techniques he has developed. At the Wirkus Bioenergy Foundation, bioenergy healing is taught along with the techniques of bioenergy field perception. One doctor who trained in bioenergy healing at Wirkus's foundation is Dr. Ursula Thunberg, a child psychiatrist with the Jewish Board of Family and Children's Services in New York. She found her bioenergy healing studies to be as technical and as systematic as anything she had experienced in medical school. After working with Mietek Wirkus, she became convinced that healers are using a naturally occurring energetic aspect of life that follows specific laws. Unfortunately, such specific natural laws remain unknown to the majority of current Western medical professionals. Dr. Thunberg noted that by following the laws of healing as taught by Wirkus, she was able to produce positive healing results in many of her patients.

As you might imagine, each unique form of healing touch or spiritual healing therapy has its own school of thought and philosophy of training that uses slightly different methods. But what is common to all of these methods is the use of conscious intention and visualization, combined with influencing the body, the chakras, and the auric field in highly specific ways, to bring about healing and rebalancing of the various energetic components of body, mind, and spirit. All the different hands-on-healing approaches share the commonality that the healing practitioner actively facilitates some kind of energy flow between the universe, the healer, and the patient in order to rebalance and recharge each patient's physical- and spiritual-energy systems.

We Are All Healers:
Our Innate Capacity for Healing Through Touch

What is most striking about the research validating the physiological effects of healing is the extraordinary capacity of human beings to heal one another. Further, this capacity for healing may be an inherent aspect of merely touching another person with loving compassion. The effects of touch upon human beings can sometimes be profound. It seems that just the simple act of innocent touching can produce unusual effects on the process of perception itself. A number of years ago, a study on the psychological effects of touch was conducted by observing its effect upon people checking out books from a library. The librarian checking out the books maintained the same demeanor and facial expression to all people but would randomly touch certain people as they left with their books. As the people who had checked out books were leaving the library, they were asked their opinion about how they had been treated by the librarian. Those individuals who had been innocently touched for less than a second were much more likely to comment that the librarian was extremely friendly and smiled at them, when in fact the librarian had not smiled at anyone. Still another study on the effects of touch found that people eating at a popular restaurant who had been innocently touched by their waiter were more likely to leave larger tips after finishing their meals.

As observed in the aforementioned two studies, the subtle effects of human touch are only now being appreciated by people in the business community. Therapeutic massage is finding increased use in hospitals and health centers, as well as in many businesses, due to a greater recognition of its stress-reducing and health-promoting benefits for both employees and patients. Besides basic massage, many forms of touch therapy using highly specific forms of physical manipulation are practiced also. Such therapies include Swedish and shiatsu massage, Rolfing, osteopathic and chiropractic manipulation, craniosacral work (a specialized form of osteopathic manipulation of the bones of the skull), and a variety of other specialized body-work disciplines. Each form of physical manipulation applies its own method of touching and manipulating specific parts of the body in order to create powerful restructuring, relaxing, and healing effects.

Until very recently, only the physical aspect of touching was believed to be responsible for any observed therapeutic benefits. However, as more medical research validates the physiological effects and therapeutic benefits of Therapeutic Touch, Reiki, and other forms of hands-on healing, the concept that the hands function as transmitters for healing energy has become increasingly accepted. Scientific research on the effects of hands-on healing has established that the healing benefits of different forms of physical touch can be extended even further by consciously directing healing energy out through the hands and into the body of another person. The discovery of the human capacity for channeling healing energy through the hands is almost like finding a missing chapter from the owner's manual for the multidimensional human being. Such a missing chapter could contain methods and techniques for cultivating the healing energy in all of us.

Learning to channel healing energy is actually fairly simple. But the degree to which we may each become successful healers will vary from person to person. For instance, most people can play a game of checkers or chess, but not everyone will graduate to the proficiency level of a chess master. The same holds true for healing. Healing is an innate ability possessed by all human beings. Some people seem to be born with a natural healing ability, while others have to take classes to learn how to do it. After learning a healing technique, some novice healers have actually become quite skilled in their healing abilities. Many healers, born or trained, will find that they tend to have certain strengths and weaknesses in their ability to heal particular medical conditions. Some healers are very good at relieving pain or inflammation but have little success with shrinking tumors. On the other hand, some healers have developed remarkable reputations in their ability to reverse even late stages of cancer, although these healers may be in the minority.

One point should be made here: There is no single "best" healing technique, since each approach has potential limitations. In fact, ignorance of the limitations of bioenergetic healing in reversing a particular difficult illness has sometimes produced unbelievably positive healing results. For instance, Reverend Rosalyn Bruyere worked with doctors at the UCLA Medical Center to help burn victims heal more quickly following skin-graft surgery. In at least one case, where a patient had severe burns all the way through skin, fat, muscles, and nerves, Bruyere attempted to promote healing of a skin graft. When the bandages were removed, the skin graft fell off. Where the skin graft had been in place, entirely new skin, nerve, and muscles cells had completely grown back. Such a miraculous healing from such a severe burn had never been seen by the UCLA doctors before. Had Bruyere known in advance that such a healing was "impossible," she might never have attempted it. There are also some case reports of healers able to reverse or markedly improve genetically inherited disorders, although there is as yet no scientific documentation of these types of cases.

With regard to the question of which healing techniques may be better than others, probably those healing approaches that address the spiritual causes of people's illnesses in addition to their physical and etheric problems are more likely to produce longer-lasting healings and "cures." In this sense, spiritual healing may have certain advantages over simple healing-touch therapies, although both approaches will provide symptom relief in the short term. Another important point to make about healing is that we are capable of healing through more than just our touch. For instance, when we listen and talk while attempting to make someone feel a little less anxious and more relaxed, we are actually healing with the energy of our voice. By the soft and caring tones we use in speaking to others, by the content of our speech, by our intention to help and heal another person, we have the power to bring comfort and healing. Through our healing intention and loving concern, we have the ability to tap in to and channel though ourselves higher spiritual energies that have the power to heal. When we project our thoughts and feelings toward others in the hope of comforting them, and accept them with unconditional love, we connect at the higher spiritual level where all life is interconnected and interrelated. This level exists outside of time and space as a place that psychologist and healing researcher Lawrence LeShan has referred to as the "clairvoyant reality." When we allow ourselves to connect to other human beings at the level of clairvoyant reality, we activate a profound spiritual energy and information exchange capable of miraculous healings.

For those individuals who would like to explore their own capacity for healing others, the first step involves learning to center yourself. When your mind is filled with anxiety, worries, and distractions, very little healing can take place. Therefore, before even attempting healing, it is important to clear the mind and center the emotions in order to adopt a more conducive state of consciousness. There are many ways to go about doing this, from simple progressive-relaxation techniques to the repetition of a mantra over and over again to quiet the mind. Probably the easiest approach is to use your breathing as a focus for relaxation. Listen to your breathing as you breathe in slowly for a count of four and then breathe out slowly to a similar count of four. After several minutes of listening only to the sound of your breathing, it is possible to become deeply relaxed and centered. You must also try to ground yourself. One easy method is to visualize tying a cord around your waist, perhaps a cord of gold or even rainbow colors. Then envision inserting the long end of the cord down through the spine and deep into the core of the earth. Another part of preparation for healing involves invoking protection from potentially negative energy influences that might be picked up during the healing interaction. Try visualizing yourself surrounded by a large sphere of bright white, blue, or gold light that extends at least two feet above your head and two feet below your feet. Most important for the process of centering and protection is to connect with your own higher spiritual source—your inner light, so to speak—because this is the part of you that will actually carry out the healing work. Your conscious personality merely sets the stage so that your higher spiritual self can perform the real healing. Intentionality is again important here. You might say an inner prayer such as "I pray that I may be a clear, pure channel of healing energy and light, and that I may bring through only those energies that come from the highest spiritual source."

After centering, grounding, and protecting oneself, the next step in simple healing is to make an energy connection between yourself and the person you wish to assist in healing. Stand behind the person while he or she sits in a straight-backed chair, back facing toward you. Place your hands on the person's shoulders. Create a heart to heart connection between the two of you by sending an imaginary beacon of golden-white light between your heart center and the other person's heart center. As you send this beacon of light from heart to heart, imagine that you are sending unconditional love, acceptance, and forgiveness of the other person for all his or her positive and negative aspects. Try to imagine the person you are sending energy to as a divine spiritual being who is locked into a physical body that is not serving the person as well as it could. If you have a hard time unconditionally accepting the other person, try envisioning him or her as a beautiful infant, spiritually pure and innocent, the same way the person first entered this physical life. As you attempt to connect with the other individual, remember that you are trying to attune to the highest spiritual aspect of the person's being, not his or her physical personality.

Once you have made a heart-to-heart connection, it is time to activate a flow of healing energy. Remember, though, as researcher Maxwell Cade discovered, you will not be of much use as a healer if you use up your own bioenergy reserves in the process. Instead, envision tapping in to an unlimited source of energy, which might include the sun, Buddha, Christ, Mohammed, Kwan Yin, gods and goddesses of different healing traditions, angels and archangels, saints and higher spiritual beings, devas and nature spirits, or even the earth itself. Some healers have actually imagined that they were plugging themselves in to an electrical socket of a large power grid, like the type that supplies electricity to our homes and businesses. Whichever specific form of healing energy you select is not really important. What is important is that there be a sense of an energy connection between that source and yourself. Whether or not you actually believe in the existence of such spiritual beings, or wish to consider them only as a metaphor for a higher spiritual power, invoking such symbolic imagery strongly activates the right-brain connection to the higher self, allowing us to tap in to these healing energies with greater ease.

After selecting a source of healing energy, you must "channel" or bring through this energy. Either channel the energy down through the head and crown chakra and into the heart or, conversely, draw up energy through the feet, the spine, and into the heart. Energy is best sent through the heart because it becomes "conditioned" by the loving energy of the heart chakra before it is sent out through the hands. In the case of channeling the healing energy of the earth (referred to as "ground prana"), many earth healers will work by drawing up earth energy through the soles of their feet, up their legs, up their spine, through their heart chakra, then out through their arms and the two minor palm chakras in the center of their hands. When channeling the energy of higher spiritual sources or the sun, visualize a continuous flow of healing light, like a flowing waterfall of white light, flooding in through the top of your head and down your spine. While visualizing the healing energy moving down your spine, see it pooling in a small bubbling fountain in your heart center, and from there, imagine it flowing out through your arms and hands.

As they attempt to move energy through their body, many healers will use their breathing pattern as an aid. Breathing slowly and regularly at the beginning of healing allows you to enter a centered, meditative state. However, when moving energy into and through the body and out through the hands, we must use our breathing in a slightly different fashion. Try to imagine drawing in healing energy (from above or below) while inhaling, and then, as you breathe out in a slow but forceful exhalation, see and feel the energy flowing out through the arms and the palms of your hands. Before even attempting to do any healing, first try this simple exercise. Imagine sending energy out through the palms of both your hands, with the palms oriented so that they face each other about two to three inches apart. Begin by rubbing your palms together rapidly a few times, as if you were trying to warm your cold hands on a winter morning. Rubbing the palms together in this fashion helps to stimulate the flow of subtle energy from the hands. As you attempt to breathe out energy through your hands during each long, slow out breath, imagine that you are creating a ball of healing energy or a sphere of white light in the space between your palms. After a few minutes of doing this, you may begin to feel your hands getting a little warmer. Slowly and gently try to move your hands toward each other, seeing if you sense anything unusual between your palms. In time, you will be able to feel a springy repulsion between your palms, as if there were some kind of magnetic energy subtly pushing your hands apart. As you continue to send energy through your palms with your breathing, you will find that the region where you can sense energetic repulsion begins to expand from, at first, a few millimeters between the palms and then, gradually, to an inch or two as the energy intensifies. This exercise, often performed in healing classes, is a good method for learning to feel subtle energy, as well as for moving energy out through your hands in tune with your breathing.

After having practiced this exercise for awhile, it will be time to begin your healing work. Go through the various stages of attunement. Center yourself using your breathing. Ground yourself with your grounding cord. Envision yourself surrounded by a protective sphere of white light. Visualize yourself becoming linked to the other person, heart to heart, higher self to higher self, so that you are spiritually connected at a higher level. Next, link with your chosen source of spiritual healing energy. See and feel it, either flowing in through the top of your head or drawn up through your feet and legs. Stand behind while the individual you're attempting to heal sits in a straight-backed chair facing away from you. Place your hands on the person's shoulders, then activate your intention to heal. Sometimes it is best to say a brief prayer that invokes the healing energy and helps to focus your intention.

Envision the two of you surrounded by a sphere of white light or some type of energy field that becomes connected to all parts of your body, mind, and spirit. As you do this, it might also be helpful to hold in your mind the thought that both of you are completely balanced and whole on all physical, emotional, and spiritual levels of your being. If you wish to focus on healing a specific problem area of the body, you can place your hands above or touching that body region as you perform the aforementioned sequence of steps. Hold in your mind the thought that the energy you are sending is correcting all imbalances that in any way influence the problem area. After sending energy to a problem area of the other person's body, return to visualizing the individual surrounded by a sphere of white, blue, or gold light that automatically heals and rebalances that person on all physical, emotional, and spiritual levels simultaneously. Allow the healing energy to flow through you for as long as it feels appropriate. This might take ten minutes or as much as a half hour or more. The absolute amount of time spent doing healing is not critical. After all, some healers accomplish miraculous results after spending only a few minutes with a client.

After allowing the healing energy to flow through you, it is important that you mentally and energetically disconnect yourself from your client. If this is not done, your client will continue to draw energy from you, even after the healing session has ended. This is accomplished by taking your hands away from the patient and mentally telling yourself that you are now energetically disconnected from the other individual. After disconnecting, spend a few moments imagining and feeling yourself becoming reenergized, rebalanced, and whole again as well. While some healers may feel depleted of energy after doing healing, others tend to feel energized. This is one of the reasons for not using your own personal bioenergy reserves in doing healing. Otherwise, you might become weakened and more susceptible to illness.

Developing the ability to do healing is sort of like exercising a muscle. Like a well-trained muscle, the more you exercise them, the stronger your healing skills will become. You may surprise yourself at how fast you are able to relieve family members' headaches or improve their back pain merely by doing a quick healing on them. The important point to be made here is that we are all potential healers. It's just that we were never told healing was one of our natural abilities. Perhaps in the future, training in the laying on of hands and even spiritual healing will be an integral part of the life-education process for all of us as we move closer to realizing the full potential of twenty-first-century vibrational medicine. Just imagine what wonders we will accomplish.

Eleven

VIBRATIONAL MEDICINE AND ITS IMPLICATIONS FOR PERSONAL AND GLOBAL SPIRITUAL TRANSFORMATION

THROUGHOUT THIS BOOK we have looked at the various ways in which human beings are really more than mere biomachines. We have explored the uniquely dynamic chemical, electrical, light-based, biomagnetic, spiritual, and subtle magnetic-energy systems that must work together in harmony in order to maintain the health of our minds and bodies. We also have seen that the integral relationship between our minds and bodies is influenced by the level of balance within various emotional-energy systems of the body, including our physical brain, our chakra system, and our emotional and astral bodies. When this emotional-energy balance becomes disturbed, we can develop physical changes and weaknesses that contribute to disease susceptibility in the different organs and tissues of our bodies. As we have learned, our emotions are related to a form of subtle magnetic energy that strongly affects the health of our physical bodies in a variety of ways. There must be an optimal balance on all the various energetic levels—physical, emotional, mental, and spiritual—for a person to truly experience that state of being we call "wellness."

Mainstream medicine attempts to promote wellness primarily by healing illness through the correction of faulty biomechanisms. The more all-encompassing viewpoint of vibrational medicine provides us with a framework for understanding the relationship between physical (biomechanistic) life processes and the higher spiritual forces and life-energy systems that nourish and animate the physical body. By viewing human beings from a wider perspective that includes the dimensions of life energy and spirit, we are being led to new avenues for exploring healing. By harnessing the energies of light, magnetism, pulsed electromagnetic fields, the life energies of plants (captured in flower essences, homeopathic remedies, herbs, and essential oils), the environmental energies of ch'i and prana, as well as the healing energy of the life force itself (via healing-touch therapies), we are expanding the physician's armamentarium for healing illness with simple, powerful, and cost-effective methods. Perhaps it is only when we start to examine healing from a higher vibrational perspective that we will eventually learn to cure some of the more difficult diseases that plague modern society.

In recent years, we have become increasingly aware of the hazards of pollution (chemical as well as electromagnetic) and the health hazards they bring. But pollution is only one of many factors that filter into the multifactorial human equation of who will become sick. Some doctors dream of a molecular-biology approach to illness, solving all of humanity's woes through mapping the human genome to figure out (and eliminate) the genetic basis of all disease susceptibility. While it is true that we inherit a tendency toward a particular disease, there must also be a host of other factors working together in unison for that disease to become fully manifest. Genetics, diet, chemical exposure, radiation exposure—these are all quantifiable physical factors that go into the human equation of who gets sick and who doesn't. But often two people who have these same factors operating in unision will have two entirely different experiences, with one becoming ill while the other stays healthy. Why should there be such a difference between the two outcomes in health?

We cannot ignore the growing body of research in psychosomatic medicine, psychoneuroimmunology, and cardioneuroimmunology that has taught us that our consciousness and our emotions play a much greater role in disease causation than once previously thought. However, the physiology of the mind/body connection is much more complicated than just a stress-induced neurochemical imbalance that leads to depression, impaired immunity, or elevated stress-hormone levels that weaken the organs of the body. The vibrational-medical model teaches us that our mind/body/spirit complex, the true integrated human system, is positively or negatively influenced not only by physical health factors but also by the health of our thoughts and our emotions and the amount of love we allow to flow through our hearts.

We humans are unique energy-processing systems at many different levels. We are constantly taking in, processing, and putting out various forms and frequencies of energy. Part of the energy we take in comes from the earth itself, the geomagnetic field, as well as from various cosmic energies that bathe the earth in sunlight, starlight, and subtle stellar forces whose actions we are only now beginning to comprehend. Through the ch'i that flows through our meridians as well as the solar prana we absorb through our chakras and our breathing, we are in an energetic equilibrium with the energies of the universe around us. Sometimes our exposure to fluctuations in solar, geomagnetic, and even cosmic background energies subtly determines when an illness in a susceptible person will begin to manifest. Vibrational medicine provides a scientific model to suggest possible ways that health or illness might relate to our energetic relationship with the world around us and our relationship to the energies of the spiritual universe within ourselves. While attention to such inner and outer energy imbalances is not yet a part of standard medical practice, the time is coming when such issues will be given greater attention by physicians and health-care workers from all disciplines.

The number of doctors, nurses, and allied health-care workers in the U.S. medical community who practice vibrational medicine is certainly on the increase, though. And the rapidly growing segment of health care frequently referred to as alternative medicine, is becoming an area of great interest to health-care consumers everywhere. Because of growing public interest, conventionally trained physicians are paying closer attention to their patients' use of herbs, vitamins, and even homeopathic remedies. But perhaps more important, there is a growing debate within conventional medicine as to the role of spirituality and spiritually based factors in the human equation of illness versus wellness. Exactly how mainstream doctors will deal with all of the alternative-health-care issues using only the biomechanistic approach of conventional medicine still remains to be successfully resolved. Certainly, the fact that mainstream physicians are prescribing relaxation techniques, biofeedback training, stress-reduction classes, imagery-healing exercises, nutritional supplements, and herbal medicines shows that change is gradually making its way into the conventional doctors' offices. But while there is great value in supplements, good nutrition, relaxation and stress-reduction techniques, and the like, the approach of vibrational medicine goes well beyond these basic, somewhat physical techniques.

Vibrational medicine also provides a rationale for using inexpensive healing approaches that tap in to the very energies of Mother Nature herself. As we have seen, simple, inexpensive acupuncture needles, tiny neodymium magnets, and pocket lasers emitting brilliant beams of light are all capable, when properly applied, of quickly relieving pain and promoting rapid healing. Vibrational medicine also uses the vibrational energies of the plant kingdom (as harnessed through homeopathy, flower essences, essential oils, and herbal preparations), along with other natural healing substances, to cheaply and effectively rebalance body, mind, emotions, and spirit in order to heal illness and promote wellness at many different levels simultaneously. Vibrational medicine provides healing tools, both ancient and modern in origin, that can not merely relieve pain, balance body energies, and alleviate physical symptoms, but also heal emotionally and spiritually related maladies that can be "cured" only by addressing the highest level of disease causation. It is, in fact, this question of where disease actually originates—at the spiritual or the physical level—that forms the "higher focus" of vibrationally oriented practitioners. This question will likely be a subject of great scientific and philosophical debate for many years to come.

These are indeed interesting times. We are living during a historical period that will someday be characterized as one of the greatest philosophical and spiritual shifts in human thinking to come along in nearly a thousand years. We are poised on the doorstep of the twenty-first century. Within this next millennium, the knowledge of vibrational medicine may allow us to extend modern medicine to a new golden age of healing. Perhaps more important, vibrational medicine may provide us with the tools to reestablish our harmony with the planet we live on. Vibrationally related techniques such as radionic pest control, paramagnetic building materials, agricultural homeopathy, and dowsing are inexpensive, nontoxic methods that, if properly used, have the potential to increase food-crop yields to a level at which it might be possible to end world hunger while simultaneously eliminating toxic pesticides from the food chain and from our environment. When we, as a global culture, truly begin to use the knowledge of vibrational medicine to appreciate our place in the greater scheme of things, and to understand and respect the spiritual evolution of all living beings on this fragile planet earth, we will start to heal on many different physical, social, emotional, and spiritual levels. But it is up to each of us as individuals to make that first step, that leap in consciousness to a vibrational and spiritual level of understanding. The power to create long-lasting and meaningful change is finally within our grasp. Perhaps we will have the compassion and understanding to use our knowledge of vibrational medicine wisely as we attempt to create a new approach to healing and a new spiritual philosophy for the twenty-first century.

Appendix 1

A HOMEOPATHIC RESOURCE GUIDE

Homeopathic first-aid kits, also known as homeopathic self-care kits, can be obtained from local health-food stores that stock homeopathic remedies or by ordering from the following companies and resources. Kits come in large sizes with many remedies, as well as in smaller sizes with fewer, more commonly used homeopathic remedies.

Arrowroot Standard Direct
85 E. Lancaster Avenue
Paoli, PA 19301
Phone: (800) 234-8879

Homeopathic Educational Services
2124 Kittredge Street
Berkeley, CA 94704
Phone: (800) 359-9051 or
(510) 649-0294
Fax: (510) 649-1955
Web site: www.homeopathic.com
(Also offers homeopathic books, tapes, and software programs)

Boiron USA
6 Campus Boulevard, Bldg. A
Newton Square, PA 19073
Phone: (800) 264-7661

Washington Homeopathic Services
4914 Del Ray Avenue
Bethesda, MD 20814
Phone: (800) 336-1695

National Homeopathic Products
518 Tasman Street
Madison, WI 53714
Phone: (800) 888-4066

National Homeopathic Products is one of the few homeopathic companies in the United States that sells ultrahigh-potency homeopathics (in the 50M, 1MM, and 10MM potencies) and will produce custom homeopathic remedies as well. They also sell prevaccination preparatory kits that provide homeopathic remedies to minimize potential side effects from vaccinations against DPT, polio, tetanus, measles, mumps, and hepatitis (a controversial issue among homeopaths.)

Dolisos America, Inc.
3014 Rigel Avenue, Suite C
Las Vegas, NV 89102
Phone: (800) 365-4767 or
(702) 871-7153
Fax: (702) 871-7153

Apex Energetics
1701 E. Edinger Avenue, Suite A-4
Santa Ana, CA 92705
Phone: (800) 736-4381 or (714) 973-7733
Fax: (714) 973-2238
Web site: www.apexenergetics.com
E-mail: comments @apexenergetics.com

Apex Energetics sells a wide range of combination homeopathic remedies, some of which also contain flower essences. They have formulas for detoxification from chemicals, pesticides, and other environmental stress factors in addition to formulas for allergies, metabolic problems, menstrual and PMS disorders, and many others. They also offer some teaching audiocassettes and videotapes on various homeopathic approaches.

Homeopathy Overnight
Phone: (800) 276-4223

Provides fast delivery of homeopathic remedies (Monday through Friday, 9:00 A.M. to 5:00 P.M. EST).

When you are seeking out a homeopathic physician, sometimes the better-stocked health-food stores that carry extensive lines of homeopathic remedies will have the names of local practitioners. If such resources are not easily accessible, the following organizations may be contacted to obtain a referral list of homeopathically oriented health professionals:

National Center for Homeopathy
801 North Fairfax, Suite 306
Alexandria, VA 22314
Phone: (703) 548-7790
Fax: (703) 548-7792

American Institute of Homeopathy
925 E. 17th Avenue
Denver, CO
Phone: (505) 989-1457
Fax: (505) 989-3236

Many books, tapes, videos, and software on homeopathy can be ordered through Minimum Price, phone (800) 663-8272, Web site: www.minimum.com

EAV/EDI-Based Homeopathy (Electrodermal Information Gathering Systems)

BodyScan 2010
Phazx Systems
710-A North Weber Street
Colorado Springs, CO 80903
Phone: (719) 632-0991
Fax: (719) 632-9811
Web site: www.phazx.com
E-mail: phazx@usa.net

Although there are quite a few EAV-type/EDI systems out there that can test the body for compatibility with different homeopathic remedies, one of the best EDI systems I have found is the BodyScan 2010, produced by Phazx Systems. One interesting advantage of this system is that in addition to allowing "traditional" acupoint testing, the device also has a "practitioner-independent" mode of testing in which patients are hooked up to the system via special wrist, ankle, finger, and headband electrodes (essentially eliminating the practitioner from the information-gathering loop). While the patient is hooked up to the device, the body's bioenergetic-stress responses to hundreds and thousands of remedies, environmental chemicals, nosodes, and other agents are taken in a completely automated fashion. Remedies and homeopathic medicines that test positive in this passive screening phase can then be individually checked for acupoint response during a second, more traditional acupoint-probe testing phase (which is also able to best determine the homeopathic potencies that will be most appropriate for the body). The device is FDA-registered as a Class II Medical Device (which means that the practitioner can use it mainly as a "biofeedback device" to determine the body's bioenergetic-stress responses to different agents). The prescription of specific homeopathic medicines by a practitioner using such a device is not solely based upon feedback from the instrument, but more from the clinical interpretation of the bioenergetic profile of the individual that the device presents along with other clinical information. The people at Phazx Systems have done their best to maintain good compliance with the FDA, so this is one of the few EDI devices on the market that I can highly recommend. Their Web site also provides access to listings of health-care practitioners who own BodyScan 2010 systems.

Appendix 2

AN ACUPUNCTURE RESOURCE GUIDE

Different states have varying requirements for the legal practice of acupuncture. In some states there is no licensing of acupuncturists, while in other states, such as Maryland, nonphysician acupuncturists can practice freely as long as the patient is referred by a physician. In California and several other states, acupuncturists are considered primary-care health providers and are able to treat patients without the need for referrals. In states where acupuncture is legal, acupuncturists must graduate from an approved school and pass a state licensing exam. In order to find a qualified acupuncturist, it is sometimes helpful to ask friends if they know of anyone who has had positive results. Health-food stores often have booklets listing alternative-health-care practitioners in the area and can be consulted for information. The following organizations can also provide information helpful in locating acupuncture therapy.

National Commission for the Certification of Acupuncturists
1424 16th Street NW, Suite 601
Washington, DC 20036
Phone: (202) 232-1404

American Association of Acupuncture and Oriental Medicine
4101 Lake Boone Trail, Suite 201
Raleigh, North Carolina 27607
Phone: (919) 787-5181

The American College of Addictionality and Compulsive Disorders
5990 Bird Road
Miami, Florida 33155
Phone: (305) 661-3474

National Acupuncture Detoxification Association
3115 Broadway, Suite 51
New York, New York 10027
Phone: (212) 993-3100

On the Internet, access to acupuncturists in different regions of the country can be obtained by going to www.Acupuncture.com and clicking on the appropriate state for a list of practitioners.

The HealthPoint Home Treatment System is a combination acupoint finder and electrostimulator, perhaps one of the best on the market. It is unique in that it helps in electronically locating the correct acupoints and also gives direct feedback when the true acupoint is being stimulated. The feedback is a mild discomfort, warmth or stinging. If there is no sensation during electrostimulation, you are not on the acupoint. The device also has high- and low-frequency settings and different intensity settings for stimulating ear or body acupoints. Since the device is a kind of acupuncture-based TENS or acu-TENS device, there is the possibility that certain insurance companies may even reimburse patients if they purchase the device for relief of a chronic pain condition. There is an optional treatment manual for use with the HealthPoint acupoint stimulator that should also be obtained when the device is purchased. The manual includes anatomical diagrams of the body that give clear locations of specific points to stimulate for different types of disorders, including various pain syndromes. The device also comes with a special ear-clip electrode that provides ear-acupoint stimulation to treat addiction disorders and insomnia. In the United States, the device and manual are available through several catalogs. It can be ordered through the Explorations catalog at (800) 456-1139 (items number 73-0043 and 81-0053 for the device and manual respectively), and through the Inner Balance catalog at (800) 482-3608 (items number 42-0005 and 42-6002) and sells for about $352, including shipping and handling.

Appendix 3

A FLOWER-ESSENCE RESOURCE GUIDE

The following is a list of flower-essence manufacturers. The Flower Essence Society, run by Richard Katz and Patricia Kaminski, also offers training in the use of their FES essences. In addition, Perelandra, Ltd., offers a two-year training program in the use of flower essences and gem elixirs. With regard to the Bach flower remedies, there are several distributors of these essences. Nelson Bach USA, Ltd., distributes the traditional Bach remedies, which are labeled exactly as such. The other main distributor of Bach remedies in the United States is Ellon USA (formerly Ellon Bach USA). Ellon manufactures and distributes all of the thirty-eight Bach remedies under the name Traditional Flower Remedies from Ellon. Dr. Bach's Rescue Remedy is available from health-food stores under that name, as well as under the names Calming Essence and Nature's Rescue (produced by Ellon).

Nelson Bach USA, Ltd.
100 Research Drive
Wilmington, MA 01887
Phone: (978) 988-3833

Korte Phi Essenzen Orkid
Alpenstrasse 25
D-78262 Gailingen
Germany
Phone: (011) 49-77-74-7004

Flower Essence Services (FES Essences)
P.O. Box 1769
Nevada City, CA 95959
Phone: (800) 548-0075 or
(530) 265-0258
Fax: (530) 265-6467
Web site: www.floweressence.com
E-mail address:
orders@floweressence.com

Bush Flower Essences
8a Oaks Avenue
Dee Why
NSW 2099
Australia
Phone: (011) 61-2-972-1033

Alaskan Flower Essence Project
P.O. Box 1329
Homer, AL 99603
Phone: (907) 235-2188

Flower Essence Pharmacy at Centergees
2007 NE 39th Avenue
Portland, OR 97212
Phone: (800) 343-8693
Fax: (503) 284-7090
Web site: www.FlowerEssences.com
E-mail: info@floweressences.com

The Flower Essence Pharmacy carries one of the largest variety of flower essences available in the United States, with essences from over seventy different manufacturers, including Andreas Korte's Orchid Essences as well as the Australian essences. They are also one of the few flower-essence distributors carrying multiple brands of essences that are easily accessible through the Internet.

The Vita Florum House
Box 876
Banff, AB
Canada
TOL OCO
Phone: (403)-762-2673

Ellon USA, Inc.
644 Merrick Road
Lynbrook, NY 11563
Phone: (516) 593-2206

Perelandra, Ltd
P.O. Box 3603
Warrenton, VA 22186
Phone: (703) 937-2153

Pegasus Products, Inc.
P.O. Box 228
Boulder, CO 80306
Phone: (800) 527-6104
(outside Colorado)
Phone: (303) 667-3019 (in Colorado)
Web site: www.pegasus.com

Living Essences of Australia
Box 355
Scarborough
Perth
W. Australia 6019
Phone: (011) 61-9-244-2073

Desert Alchemy Essences
P.O. Box 44189
Tucson, AZ 85733
Phone: (602) 325-1545

Appendix 4

~~~

A COLOR- AND LIGHT-HEALING RESOURCE GUIDE

College of Syntonic Optometry
1200 Robeson Street
Fall River, MA 02720-5508
Phone: (508) 673-1251

Organization of optometrists who use syntonic light therapy, a form of Ocular Light Therapy. They can also make referrals to practitioners in your area.

John Downing, OD., Ph.D.
1955 West College Drive, #107
Santa Rosa, CA 95401
Phone: (707) 526-1881

Developer of the Lumatron Ocular Light Stimulator, Dr. Downing has spent more than twenty years researching the various applications of colored-light therapy delivered through the eyes.

Applied Light Technology
(Ernest Baker, Jr., CEO)
79 Belvedere Street, Suite 11
San Rafael, CA 94901
Phone: (415) 456-5046
Fax: (415) 456-4708

Sells the Lumatron Ocular Light Stimulator, Dr. Downing's unique eye- and brain-stimulating color-therapy instrument, which uses eleven colors and adjustable strobe flicker rates varying from 0.1 to 60 hertz.

Universal Light Technology
Jacob Liberman, O.D., Ph.D.
P.O. Box 520
Carbondale, CO 81623
Phone: (303) 927-0100 or
(800) 81-LIGHT
Fax: (303) 927-0101
~~~

Developer and distributor of color-therapy device known as the Color Receptivity Trainer, which is capable of delivering light therapy to either the eyes or the body. It comes with a color wheel containing thirteen different color gels.

Tools For Exploration: An amazing catalog of light and sound machines, biofeedback instruments, full-spectrum lights, and even monochromatic light sources such as the Light Shaker and the Tri-Light (items number LS101 and LS102 respectively. They also sell the Shealy Relaxmate under the name RelaxEase (item number NS101). "TFE" bills itself as the worlds largest one-stop-shopping resource for alternative therapies and subtle-energy healing tools. Highly recommended. Everyone should get this catalog. It is an invaluable resource.

Tools For Exploration
9755 Independence Avenue
Chatsworth, CA 91311-4318
Phone (U.S.): (888) 748-6657
Phone (International): (818) 885-9090
Fax: (818) 407-0850
Web site: www.toolsforexploration.com
E-mail: toolsforexploration@yahoo.com

Pauline Willis
The Oracle School of Colour
9 Wyndale Avenue
Kingsbury
London, England
NW9 9PT
Phone/Fax: 011-4-181-204-7672

For information on the reflexology crystal torch and other color-therapy products.

Dinshah Health Society
100 Dinshah Drive
Malaga, NJ 08328
Phone: (609) 692-4686

Organization led by Darius, the son of Dinshah Ghadiali, promoting the value of color therapy. They also publish a book, *Let There Be Light,* about Dinshah's color-healing system. However, they do not give referrals to color practitioners.

Edmund Scientific Company
101 Gloucester Pike
Barrington, NJ 08007
Phone: (609) 573-6250

Edmund's catalog sells the entire line of Roscolene colored-gel filters recommended for Spectro-Chrome Tonation color therapy as originally devised by Dinshah Ghadiali. Order item number F7068 to receive a book of the forty-four Roscolene gel filters in an eight- by ten-inch size, suitable to adapt for light projectors. Alternatively, order the individual colored gel filters according to the following color formulas developed by Dinshah:

Red: #818 with 828
Orange: #809 with 828
Yellow: #809
Lemon: #809 with 871
Green: #871
Turquoise: #861 with 871
Blue: #866
Indigo: #818 with 859 and 861
Violet: #832 with 859 and 866
Purple: #832 and 866
Magenta:#818 with 828 and 866
Scarlet: #810 with 818 and 861

Six additional colors were mentioned for use by Dinshah. These color gradations were used if a particular single-color energy was too powerful for the patient, allowing a more flexible color treatment.

Red-Orange: #809 with 818
Orange-Yellow: #809 with 826
Yellow-Lemon: #809 with 878
Lemon-Green: #810 with 871
Green-Turquoise: #871 with 877
Turquoise-Blue: #871 with 866

Appendix 5

A MAGNETICALLY ORIENTED RESOURCE GUIDE

BEMI
Dr. John Zimmerman
2490 W. Moana Lane
Reno, NV 89509
Phone: (702) 827-9099

The Bioelectromagnetics Institute (BEMI) is an organization run by Dr. John Zimmerman. For $10, he provides an extensive information packet on over two dozen companies selling therapeutic magnetic products, ranging from the two largest, Nikken and Japan Life, all the way to smaller companies with less expensive magnetic systems. The money for the packet goes to fund research at BEMI. BEMI also publishes an excellent newsletter on related research.

North American Academy of Magnetic Therapy
28240 West Agoura, Suite 202
Agoura, CA 91301
Phone: (800) 457-1853

This organization sells magnetic products, teaches classes in magnetic therapy, and organizes yearly conferences on magnetic healing.

Bioelectromagnetic Society (BEMS)
120 Church Street
Frederick, MD 21701

This society publishes *The Journal of Electromagnetic Therapy* and holds annual conferences on clinical research in magnetic and electromagnetic therapies. For those who are more technically oriented.

Psychophysics Labs
1803 Mission Avenue, Suite 24
Santa Cruz, CA 95073
Web site: www.buryl.com

This lab, run by Dr. Buryl Payne, Ph.D., at one time published "Solar and Planetary Influences," a newsletter providing forecasts of solar activity, geomagnetic activity, special weather patterns, and their possible effects upon human health and behavior. They also manufacture and import pulsed-magnetic field devices such as the Pulsar and Power Pulsar. Custom magnetic-field devices can be specially built for research purposes.

Earth Changes Report
Matrix Institute, Inc.
P.O. Box 336
Chesterfield, NH 03443-0336
Phone: (800) 783-4903 or
(603) 627-2997
Web site: www.earthchanges.com

This is a fascinating bimonthly report on changes occurring in geological weather patterns, solar-flare activity, geomagnetic shifts, and other earth changes that have relevance to both human health and the stability and functionality of our electronic technologies (Y2K issues aside). Interesting and often accurate precognitive/intuitive information is provided by publisher Gordon Michael Scallion on future trends in global weather patterns, solar-flare activity, economic ups and downs, and new technologies, all of which make this an indispensable little guide to the (probable) future. Their Web site also provides updates on currrent solar-flare activity. (Recent solar-flare bursts caused a major communications satellite to cease functioning which resulted in loss of service to millions of beepers and other satellite-dependent technologies.)

Enviro-Tech Products
17171 Southeast 29th Street
Choctaw, OK 73020
Phone: (800) 445-1962 or
(405) 390-3499
Fax: (405) 390-2968

This service provides self-help information about magnet healing, information for physicians, as well as information and guidance for magnet-related research projects under the Institutional Review Board of the Bioelectromagnetics Institute in Reno, Nevada. Also sells various magnetic products.

Magnetism and Youthing Foundation
5336 Harwood Road
San Jose, CA 95124-5711
Executive director: Dewey Lipe, Ph.D.
Fax: (408) 264-9659
E-mail: DeweyL@ix.netcom.com

This organization publishes a newsletter and provides guidelines for individuals wishing to be a part of their ongoing research project to use magnetic therapy for healing and rejuvenation. Membership fee is $50 and newsletter is $25.

Sources of Magnetic-Healing Products

The Harmony catalog is a wonderful collection of ecologically safe products for health care and home use, including magnetic systems, water-filtration systems, full-spectrum lights, and a variety of other useful products. They carry small neodymium magnets for use over joints and acupoints (item 08-0160, Super Neo Dot Magnets) along with magnetic lumbar braces, joint supports, magnetic keyboard wrist supports, magnetic mattress pads (standard strength and high density), and even a magnetically charged Water pik–type dental-flossing system that has been shown effective in reducing dental plaque and mineral deposits (item #15-0054, Hydro Floss Dental Irrigator), all at fairly reasonable prices. For individual pricing, request a catalog from:

Harmony: Products in Harmony
with the Earth
360 Interlocken Boulevard, Suite 300
Broomfield, CO 80021
Phone: (800) 869-3446
Fax: (800) 456-1139
Web site: www.harmonycatalog.com

The Stress Less catalog offers a variety of stress-reducing products, including some interesting magnetic appliances such as a soft magnetic-collar appliance for neck pain, magnetic wristbands for injuries and carpal tunnel syndrome, magnetic bracelets, magnetic back pads for low-back pain, dot magnets, and even magnetic seat cushions. Request their catalog by writing or calling:

Stress Less
P.O. Box 699
Holmes, PA 19043-0699
Phone: (800) 555-3783
Fax: (610) 532-9001
Web site: www.stressless.com

For individuals wishing to use magnet stimulation of acupoints, some of the cheapest neodymium rare-earth magnets can be found at most Radio Shack stores. They are sold as Rare Earth Magnets, catalog #64-1895. A package of two powerful dot magnets sells for only $1.59 (plus sales tax). If these magnets are not available in your local Radio Shack, they can easily special-order them for you if you give them the appropriate catalog number. You can get a ten-pack of tiny potent magnets with fifty circular adhesive patches from the Self Care catalog for only $10 (plus shipping and handling). The magnetic product is sold under the name Magnetic Adhesive Spot and lists as product #A7353. The Self Care order line takes orders twenty-four hours a day at (800) 345-3371. Once purchased, the magnets should be labeled as to their north and south sides, determined by placing them near the north or south pole of a regular, pole-labeled magnet. The side that is attracted to the south pole of a labeled magnet will be its north pole, since opposites attract. It is usually helpful to place a dot of white nail polish on the north side (or a dot of red nail polish on the south side) in order to ensure that the correct pole (the north side) faces the body when the dot magnet is placed on the acupoint. One of the easiest ways to "stick" the magnets to the body is by using flexible-fabric type Band-Aids. Cut off the central gauze pad at the center of the Band-Aid and use the two remaining adhesive-covered strips to adhere two dot magnets to the appropriate acupoints. This adhesive will often keep the magnets stuck to the body for at least twenty-four to forty-eight hours (with careful bathing). These magnets are quite effective for relieving pain when placed over the correct acupoints and are among the most inexpensive neodymium magnets available for acu-magnet therapy (some catalogs may charge up to $40 or more for a set of four acupoint magnets). Books such as Dr. Julian Whitaker's *The Pain Relief Breakthrough: The Power of Magnets* (see Recommended Reading) or Leon Chaitow's *Acupuncture Treatment of Pain* can serve as pictorial guides for acu-magnet placement on the body to help specific painful conditions.

Body Magnetics
871 Thrall Avenue
Suffield, CT 06078
Phone: (203) 231-2377

Breakthrough Media, Inc.
5065 SW 153rd Avenue
Beaverton, OR 97007
Phone: (800) 321-5641

The Cutting Edge (catalog)
P.O. Box 5034
South Hampton, NJ 11969
Phone: (800) 497-9516

Japan Life (multilevel marketing)
One Executive Drive
Fort Lee, NJ 07024
Phone: (201) 944-7790
Fax: (201) 944-5507

LHASA Medical, Inc.
539 Accord Station
Accord, MA 02018-0539
Phone: (800) 722-8775 or
(617) 335-6484
Fax: (617) 335-6296

Young Living Essential Oils
250 S. Main
Payson, UT 84651
Phone: (801) 465-5400

For those wishing to experiment with magnets and essential oils, this company produces some of the highest-quality essential oils I have come across. The company's founder, Dr. Gary Young, also has an appreciation of the vibrational qualities of different essential oils, in addition to their medicinal healing properties. A number of books on essential-oil therapy, including one by Dr. Young, are available through this organization.

Magnetic Therapy Products
4926 Indian Springs Court
Plant City, FL 33565
Phone: (813) 757-0508

Magnetic Wellness Centers
9711 Montgomery Road
Cinti, OH 45242
Phone: (800) 484-7964 (code 1956)

American Health Service
694 S. Waukegan Road, Dept. F
Lake Forest, IL 60045
Phone: (800) 544-7521

Nikken, Inc. (multilevel marketing)
10866 Wilshire Boulevard, Suite 250
Los Angeles, CA 90024
Phone (800) 669-8859 or
(310) 446-4300

Oriental Medical Supplies, Inc.
1950 Washington Street
Braintree, MA 02184
Phone: (800) 323-1839 or
(617) 331-3370
Fax: (617) 335-5770

Appendix 6

A RADIONICS RESOURCE GUIDE

United States Psychotronics Association
409 Marquette Drive
Louisville, KY 40222
Phone: (502) 423-1188
Fax: (502) 425-6465

This organization is one of the best resources for radionic instruments, practitioners, and related subtle-energy therapies. They hold a conference every summer covering a wide spectrum of radionics-related topics by credible researchers and practitioners. The conference always has an interesting array of healing-device technologies, demonstrated by various manufacturers, as well as an excellent on-site bookstore featuring many hard-to-find books.

The American Dowser's Society
Danville, VT
Phone: (802) 684-3417
Fax: (802) 684-2565

One of the best organizations oriented toward various forms of dowsing and medical radiesthesia. Holds a yearly conference with a focus on subtle energies and dowsing-related topics. A good resource for finding dowsers (as might be used for pinpointing geopathic-stress sites around the home).

Radionic Resources on the Internet

THE SANJEEVINI REMEDY CARDS

The Sanjeevini remedy cards are a unique (and free) radionic-healing resource based on the energy of prayer adapted to modern radionic-symbol technologies developed by Malcolm Rae. All of the individual remedy patterns along with information on their use can be downloaded, or, alternatively, order a report on the Sanjeevini remedy cards from the G-Jo Institute in Hollywood, Florida (which contains all of the several hundred remedy cards). The cost of the cards and report are $23, postpaid.

Web site: www.saisanjeevini.org/
G-Jo Institute
P.O. Box 848060
Hollywood, FL 33084
Phone: (954) 791-1562

INFINITY: FORMS OF YELLOW REMEMBER

This very strange-sounding group of remedies and vibrational tools represents another form of radionics that comes from Australia. Like the Sanjeevini cards, they use an aspect of consciousness to energetically imprint their remedies. While their Web site uses certain radionic-type symbols with healing and transformational effects similar to those of the Sanjeevini system, most of the Forms of Yellow Remember products involve a different form of radionic technology, using imprinted liquids in sealed plexiglass tubes. The tubes can be used to repeatedly imprint water. Alternatively, they can be used with special devices known as "Stargate boxes" that can radionically broadcast healing-energy patterns from the sealed liquid-remedy vials to individuals at a distance. The individual remedies have unusual, metaphorical names such as Full Head of Steam, Iron Heart, Mighty Atom, and Deep Peace. Other interesting devices produced by this organization are said to radionically broadcast certain beneficial and protective energies around homes and businesses with these objects strategically placed within them. Their Web site also offers a free radionic-healing experiment (under the section "free gift"), which is worth checking out.

Web site:
www.infinity-formsofyellow.com
American contact:
Infinity: Forms of Yellow Remember
2701 University Avenue, Suite 489
Madison, WI 53705
Phone: (608) 827-7153
Fax: (608) 827-7867

A p p e n d i x 7

~~~

A HEALING RESOURCE GUIDE

Nurse Healers Professional Associates
1211 Locust Street
Philadelphia, PA 19107
Phone: (215) 545-8079

NHPA is an international organization for nurses that will provide referrals to both teachers and practitioners of Therapeutic Touch.

Haelan Works
3080 Third Street
Boulder, CO 80304
Phone/fax: (303) 449-5790
E-mail: janetquinn@aol.com

Provides training in Therapeutic Touch and other healing approaches (for both laypersons and health-care professionals). A video home-study course on Therapeutic Touch is also available.

Institute for Integrative Healthcare Studies
P.O. Box 1139
Pine Bush, NY 12566-1139
Phone: (800) 364-5722

Provides home-study courses in Therapeutic Touch and other healing methods for continuing education credits.

The Barbara Brennan School of Healing
P.O. Box 2005
East Hampton, NY 11937
Phone: (700) 432-5377
Phone: (for non-AT&T network carriers) 10-288 (700) 432-5377
Fax: (700) 465-4448

Excellent school for healers. Offers a four-year college-level training course in healing science that includes on-site
~~~

classes and home-study segments. Continuing-education credits are available for nurses, massage therapists, and acupuncturists.

Wirkus Bioenergy Foundation
9907 Fleming Avenue
Bethesda, MD 20814
Phone/fax: (301) 652-3480

Provides training in Mietek Wirkus's approach to bioenergy healing.

The Biofield Research Institute
20 YFH Gate Six Road
Sausalito, CA 94965
Phone: (415) 332-2593
Fax: (415) 331-2455

Provides training and certification in SHEN Therapy, a unique form of bioenergy therapeutics.

National Federation of Spiritual Healers
1137 Silent Harbor
P.O. Box 2022
Mount Pleasant, SC 29465
Phone: (803) 849-1529

The Jewish Association of Spiritual Healers
106 Cabrini Boulevard
New York, NY 10033
Phone: (212) 928-4275

A nondenominational organization affiliated with the British Alliance of Healing Associations. Provides training for healers.

Institute for Inner Studies, Inc.
Vice President
Evekal Building
855 Passay Road
1200 Makati, Metro Manila
Philippines
Phone: (632) 819-1874 or (632) 810-2808 or (632) 818-2562
Fax: (632) 731-3828

For those who wish to take Master Choa's Pranic Healing Course, it is suggested that you first read Choa's books on pranic healing listed in the Recommended Reading section. For those who do not wish to travel to the Philippines for training, pranic-healing workshops in the United States are also available through:

The American Institute of Asian Studies
Mr. Stephen Co
U.S. Certified Pranic Healing Instructor
P.O. Box 1605
Chino, CA 91708-1605
Phone: (909) 465-0967

The Reiki Alliance
P.O. Box 41
Cataldo, ID 83810-1041
Phone: (208) 682-3535

Reiki Outreach International
P.O. Box 609
Fair Oaks, CA 95628
Phone: (916) 863-1500

RECOMMENDED READING

CHAPTER 1

The Chakras and the Human Energy Field. Karagulla, Shafica, M.D., and van Gelder Kunz, Dora. Wheaton, Ill.: Theosophical Publishing House, 1989. An amazing book describing the correlations between clairvoyant observation of patients' chakras and auric fields and their medical diagnoses in a variety of different cases. Highly recommended.

The Miracle of Birth: A Clairvoyant Study of a Human Embryo. Hodson, Geoffrey. Wheaton, Ill.: Theosophical Publishing House, 1981. This is a truly remarkable book, written by noted English clairvoyant Geoffrey Hodson. Hodson was one of the few clairvoyants to have his psychic observations validated by extensive scientific testing. The book describes the creation of a new human being from a clairvoyant perspective, noting how the baby's etheric and spiritual bodies are actually formed before the development of the physical body. Provides a whole new perspective of the true nature of the multidimensional human being. Highly recommended.

The Inner Life. Leadbeater, Charles, W. Wheaton, Ill.: Theosophical Publishing House, 1978. Another remarkable book by a famous clairvoyant, Reverend Charles Leadbeater, describing the higher-dimensional aspects of everyday human life as seen from a spiritual perspective. Includes discussion of the role of the astral, mental, and higher spiritual bodies, as well as the nature of karma.

Ultimate Journey. Monroe, Robert. New York: Doubleday, 1994. An amazing book about the higher-dimensional nature of human beings as seen through the eyes of pioneering astral traveler Robert Monroe. A very insightful book. Highly recommended.

CHAPTER 2

Miracles Do Happen: A Physician's Experience with Alternative Medicine. Shealy, Norman C., M.D., Ph.D. Rockport, Mass.: Element Books, Inc., 1995. An excellent guide to practical aspects of nutrition, self-regulation, intuitive medical diagnosis, spiritual healing, acupuncture, GigaTENS, and many other aspects of alternative medicine by a pioneer in the field of healing and energy medicine. Highly recommended.

The Elements of Feng Shui. O'Brien, Joanne, with Man Ho, Kwok. Rockport, Mass.: Element, Inc., 1991. A guide to the ancient geomantic science of feng shui, the Chinese art of energetically based design and placement in order to promote health and prosperity. Co-written by a prominent European feng shui master.

Earth Harmony: Siting and Protecting Your Home—A Practical and Spiritual Guide. Pennick, Nigel. Covent Garden, London: Century Hutchinson, Ltd., 1987. A guide to ancient Northern European geomantic principles of design and architecture used to enhance the flow of beneficial earth energies through one's home. Fascinating reading.

The Complete Book of Chinese Health and Healing. Reid, Daniel. Boston: Shambhala Publications, Inc., 1995. A marvelous guide to the various types of ch'i energy, traditional Chinese medicine philosophy, and the Taoist approach to medical self-care. Highly recommended.

Anatomy of the Spirit: The Seven Stages of Power and Healing. Myss, Caroline, Ph.D. New York: Harmony Books, 1996. Medical intuitive Carolyn Myss's seminal work on chakra dynamics and their relevance to health and illness. Includes many case histories. Highly recommended.

The Creation of Health. Shealy, Norman C., M.D., Ph.D., and Myss, Carolyn, M.A. Walpole, N.H.: Stillpoint Publishing, 1988. A fascinating overview of human health and illness from both a medical and a clairvoyant perspective. Highly recommended.

Love and Survival: The Scientific Basis for the Healing Power of Intimacy. Ornish, Dean, M.D. New York: HarperCollins Publishers, 1998. A scientific look at the healing power of love by a pioneering cardiologist. Highly recommended.

Soul Psychology: Keys to Ascension. Stone, Joshua David, Ph.D. Sedona, Ariz.: Light Technology Publishing, 1994. A fascinating guide to multidimensional human functioning, with discussions of the chakra system, the spiritual bodies, tools for healing the emotions, and the nature of karma. Highly recommended, especially for those on a spiritual path.

CHAPTER 3

QiGong: Traditional Chinese Exercises for Healing Body, Mind, and Spirit. An excellent instructional video on using various qigong (chi gong) exercises for health by Ken Cohen, a health educator, China scholar, and master of qigong healing. Available through Sounds True, P.O. Box 8010, Boulder, CO 80306-8010. Phone: (800) 333-9185.

T'ai Chi for Health. A good video guide to the basics of t'ai chi as a workout for mind and body. Uses slow movements of the body combined with special breathing exercises. Also *T'ai Chi for Seniors.* A perfect low-impact exercise regimen for seniors. Has been shown to decrease incidence of falling and hip fractures when practiced regularly. Both videos are available from the Explorations catalog. *T'ai Chi for Health* is item # 80-0443, *T'ai Chi for Seniors* is item #80-0107, Phone: (800) 720-2114.

A.M. and P.M. Yoga for Beginners. An excellent video guide that provides two short and effective beginning-yoga workouts for morning and evening taught by two leading yoga instructors, Rodney Yee and Patricia Walden. Includes gentle yoga poses, stretches, and guided meditation. Available through Explorations catalog, item # 80-0683, Phone: (800) 720-2114.

Energy Medicine: Balance Your Body's Energies for Optimum Health, Joy, and Vitality. Eden, Donna, and Feinstein, David. New York: Jeremy P. Tarcher/Putnam, 1998. An amazing book on balancing the body's energies by a medical intuitive and healer. Includes techniques to balance meridians, chakras, the auric field, as well as muscle kinesiology testing for health assessment. Highly recommended. Video instruction on Donna Eden's techniques are also available in a ninety-minute video entitled *An Introduction to Energy Healing* or a full six-hour training program on three videocassettes entitled *Energy Healing With Donna Eden.* Both instructional videos are highly recommended. The tapes are available from Innersource, P.O. Box 213, Ashland, OR 97520. Phone: (800) 835-8332, Web site: www.innersource.net.

Freeze-Frame: Fast Action Stress Relief, A Scientifically Proven Technique. Childre, Doc Lew. Boulder Creek, Calif: Planetary Publications, 1994. (Excellent guide to one of the best techniques available for stress reduction, empowerment, and tuning in to the inner wisdom of the heart. Highly recommended. Available through Planetary Publications, P.O. Box 66, Boulder Creek, CA 95006, Phone: (800) 372-3100 or (408) 338-2161, Fax: (408) 338-9861.

Why People Don't Heal and How They Can. Myss, Caroline, Ph.D. New York: Harmony Books, 1997. Discusses the dynamics of the chakras and the symbolic perspective of the higher self in seeing the deeper meaning behind illness and life events. A must read. Highly recommended.

Three Levels of Power and How to Use Them, Myss, Carolyn, Ph.D. An excellent tape series on using the three levels of power (external power, internal power, and symbolic power) to view one's life and to maintain good health. Highly recommended. Available through Explorations catalog. Videotape version is item # 80-0554, audiotape version is item # 83-0044, Phone: (800) 720-2114.

New Chakra Healing. Dale, Cyndi. St. Paul, Minn.: Llewellyn Publications, 1996. A fascinating look at the emotional dynamics and developmental patterns of the chakra system and the auric field. One of the few books to discuss the different emotional energetics of the fronts versus the backs of the chakras in relating to conscious versus unconscious life issues. Highly recommended.

A Handbook for Light Workers. Cousens, David. Bath, England: Barton House Publishing, 1993. An amazing book by an English healer and clairvoyant detailing insights on soul evolution, karma, thoughtforms, and highly specific visualization exercises for healing and rebalancing our physical and spiritual bodies. Contains many valuable imagery techniques for inner healing work, chakra opening and closing, energy protection, cleansing (including karmic cleansing), and releasing toxic, negative energies from the body. A helpful guide for anyone following a spiritual path. Highly recommended. The book is available through Barton House, 51 Audley Park Road, Bath, England, BA1 2XJ, or through Atrium Distribution, 11270 Clayton Creek, Road, P.O. Box 108, Lower Lake, CA 95457.

The Life You Were Born to Live: A Guide to Finding Your Life's Purpose. Millman, Dan. Tiburon, Calif.: H.J. Kramer, Inc., 1993. A wonderful guide to finding your true soul purpose, based on Millman's insightful Life-Purpose-System. Highly recommended.

Healing Words: The Power of Prayer and the Practice of Medicine. Dossey, Larry. New York: HarperCollins Publishers, 1993. A thoughtful look at the healing power of prayer and the spiritual dimensions of consciousness. Fascinating reading. Highly recommended.

Synchrodestiny: Discovering the Power of Meaningful Coincidence to Manifest Abundance in Your Life. An eight-audiotape (or eight-CD) program from renowned author/physician Dr. Deepak Chopra on using the power of consciousness, intention, and synchronicity to shape your life in positive and meaningful ways. One of the best discussions of the theory and dynamics of synchronicity that I have found anywhere. Highly recommended. The tape or CD series is available through Nightingale Conant, 7300 North Lehigh Avenue, Niles, IL 60714. Phone: (800) 525-9000; fax: (800) 647-9198. (Tape series is item #18390AF, CD series is item #18390CDF.)

The Secret Language of Signs: How to Interpret the Coincidences and Symbols in Your Life. Linn, Denise. New York: Ballantine Books, 1996. A wonderful guide to interpreting the higher meaning of signs, symbols, and synchronistic events in your life. It also has an extensive symbol dictionary. Highly recommended.

CHAPTER 4

Everybody's Guide to Homeopathic Medicines. Cummings, Stephen, M.D., and Ullman, Dana, M.P.H. Los Angeles: Jeremy P. Tarcher, Inc., 1991.

The Family Guide to Homeopathy: Symptoms and Natural Solutions. Lockie, Andrew. New York: Prentice Hall Press, 1993.

Healing with Homeopathy: The Complete Guide. Jonas, Wayne B., M.D., and Jacobs, Jennifer, M.D., M.P.H. New York: Warner Books, Inc., 1996.

Homeopathic Medicines at Home: Natural Remedies for Everyday Ailments and Minor Injuries. Panos, Maesimund B., M.D., and Heimlich, Jane. Los Angeles: Jeremy P. Tarcher, Inc., 1981.

CHAPTER 5

The Web That Has No Weaver: Understanding Chinese Medicine. Kaptchuk, Ted. New York: Congdon and Weed, 1992. A classic text on the nature of Chinese medicine and acupuncture therapy. Highly recommended.

Between Heaven and Earth: A Guide to Chinese Medicine. Beinfield, Harriet, L.Ac.; and Korngold, Effrem, L.Ac. O.M.D. New York: Ballantine Books, 1991.

Acupuncture: Its Place in Western Medical Science. Lewith, George T. Wellingborough. Northamptonshire, England: Thorsons Publishers Ltd., 1982.

Acupuncture for Americans. Wensel, Louise Oftedal, M.D. Reston, VA: Reston Publishing Company, Inc., 1980. An excellent guide for physicians and patients to specific acupoint formulas for treating a wide variety of medical problems. Points listed could be used in conjunction with electroacupuncture systems like the HealthPoint System for needleless self-therapy to treat minor medical problems.

The Acupuncture Treatment of Pain: Safe and Effective Methods for Using Acupuncture in Pain Relief. Chaitow, Leon. Rochester, Vt.: Healing Arts Press, 1990. An excellent resource to have on hand to guide the use of needleless electro-acupuncture, magnetopuncture, or similar acupoint-based therapies.

CHAPTER 6

The Encyclopedia of Flower Remedies. Harvery, Claire, and Cochrane, Amanda. Great Britain: Thorsons: An Imprint of HarperCollins Publishers, 1995. An excellent guide to the therapeutic uses of different varieties of flower essences from around the world, including Bach, Pegasus, Korte Orchid Essences, and many others. Highly recommended.

Flower Power: Flower Remedies for Healing Body and Soul Through Herbalism, Homeopathy, Aromatherapy, and Flower Essences. McIntyre, Anne. New York: Henry Holt and Company, 1996.

Water Magic: Healing Bath Recipes for the Body, Spirit, and Soul. Muryn, Mary. New York: Fireside Books, 1995. An excellent guide to many unique bath therapies, with accompanying visualizations to enhance their effectiveness.

Bach Flower Therapy: Theory and Practice. Scheffer, Mechthild. Rochester, Vt.: Thorsons Publishing Group Ltd., 1986.

Pocket Guide to Bach Flower Essences. Hasnas, Rachelle. Freedom, Calif.: The Crossing Press, 1997. A wonderful little pocket guide to the use of the Bach remedies with examples of personalities typifying each flower remedy taken from television, movies, and history.

New Bach Flower Therapies: Healing the Emotional and Spiritual Causes of Illness. Kramer, Dietmar. Rochester, Vt.: Healing Arts Press, 1995.

New Bach Flower Body Maps: Treatment By Topical Application. Kramer, Dietmar, and Wild, Helmut. Rochester, Vt.: Healing Arts Press, 1996. These two books by Kramer provide the keystone to using Bach Flower Skin Zone Therapy and are highly recommended.

Flower Remedies Handbook: Emotional Healing and Growth with Bach and Other Flower Essences. Cunningham, Donna. New York: Sterling Publishing Co., Inc., 1992.

Flower Essences and Vibrational Healing. Gurudas. San Rafael, Calif.: Cassandra Press, 1989. This book is an excellent guide to many of the Pegasus Flower Essences, with extensive descriptions of how the essences affect different cellular and subtle-energetic systems of the multidimensional human energy system. Highly recommended.

Flower Essence Repertory: A Comprehensive Guide to North American and English Flower Essences for Emotional and Spiritual Well-Being. Kaminski, Patricia, and Katz, Richard. Nevada City, Calif.: The Flower Essence Society, 1994. This is an excellent guide to the use of the FES essences and is available through the Flower Essence Society, listed earlier in Appendix 3.

Flower Essences: Reordering Our Understanding and Approach to Illness and Health. Small Wright, Machaelle. Jeffersonton, Va.: Center for Nature Research, 1988. This book is a guide to the use of the Perelandra Rose and Garden Essences and offers a fascinating discussion of open communication with the nature spirits or plant devas that oversee plant and flower growth.

Findhorn Flower Essences. Leigh, Marion. Scotland: Findhorn Press, 1997. A guide to the use of the Findhorn Essences, as well as discussion of communication with overseeing flower devas regarding specific essence applications. Findhorn Press also offers several other titles, including guides to the use of Andreas Korte's Orchid Essences and the Australian Bush Essences. Findhorn Press: The Park, Findhorn, Forres IV36 0TZ, Scotland. Phone: (011) 441-309-690-582. Fax: (011) 441-309-690-036. Web site: www.gaia.org/findhornpress/.

Healing with Flower and Gemstone Essences. Stein, Diane. Freedom, Calif.: The Crossing Press, 1996. This is a guide to the use of specially prepared combinations of flower essences and gem elixirs developed by Diane Stein, as well as an information source on how to order these remedies. Her fascinating group of flower and gem remedies are to be used mainly to heal dysfunctional emotional states and to rebalance chakras and subtle bodies, provide psychic protection from outside influences, and to assist in spiritual transformation.

Cause, Effect, and Treatment. Ken Gillemo. Bridgwater, Somerset, England: Bigwood and Staple Printers at The Mill, 1988. This book is a self-published guide to the use of the Vita Florum flower products. Both the book and the various Vita Florum products can be obtained through the Vita Florum House, listed in Appendix 3.

CHAPTER 7

Light Years Ahead: The Illustrated Guide to Full Spectrum and Colored Light in Mind-Body Healing. Light Years Ahead Productions. Berkeley, California: Celestial Arts, 1996. This is perhaps the most informative book available on the wide spectrum of clinical applications of color and light therapy. The book is a collection of articles based upon material presented at the Light Years Ahead conference. Highly recommended.

Colour Therapy: The Use of Colour for Health and Healing. Willis, Pauline. Rockport, MA: Element Books, 1993. Discusses color therapy, reflexology, and the use of the color reflexology torch.

Reflexology and Color Therapy: A Practical Introduction. Willis, Pauline. Rockport, MA: Element Books, 1998. Excellent guide to color therapy, reflexology, and the use of color meditations for inner healing.

Color Me Healing/Colorpuncture: A New Medicine of Light. Allanach, Jack. Rockport, MA: Element Books, 1997. A fascinating guide to Peter Mandel's approach to healing via directing color energy into the acupuncture points of the body, including case histories.

Healing with Color Zone Therapy. Corvo, Joseph, and Lilian Verner-Bonds. Freedom, California: The Crossing Press, 1998. A slightly different approach to healing using color combined with zone therapy.

The Colour Therapy Workbook: A Guide to the Use of Color for Health and Healing. Gimbel, Theo. Rockport, MA: Element Books, 1993. A more esoteric guide to healing with color and light by one of Europe's leading color therapy practitioners.

CHAPTER 8

Magnetic Healing: Advanced Techniques for the Application of Magnetic Forces. Payne, Buryl, Ph.D. Twin Lakes, Wis.: Lotus Books, 1997. One of the most intelligent resources for practical applications of magnetic healing around, with extensive discussion of various types of magnets, polarity uses, and magnetic-healing research. Also includes information on ordering the Pulsar, a pulsed-magnetic-field therapy system.

The Pain Relief Breakthrough: The Power of Magnets. Whitaker, Julian, M.D., and Adderly, Brenda, M.H.A. Boston: Little, Brown and Company, 1998. An excellent guide to using magnets for pain relief, including many wonderful diagrams of select acupoints to stimulate for magnetically treating various types of arthritic and other pain syndromes. Also includes an extensive discussion of the magnetic-healing research literature.

Magnet Therapy. Holzapfel, E.; Crepon, E.; and Philippe, C. Wellingborough, England: Thorsons Publishing Group, 1986. A nice little book, translated from the French, that discusses acu-magnet stimulation for relief of pain, stress, and in treating various illnesses. Includes diagrams of the body showing key acupoints to stimulate for different health problems.

Healing with Magnets. Null, Gary, and Koestler, Vickie, R. New York: Carroll and Graf Publishers, Inc., 1998. A good guide to magnetic-healing research. Also includes a section describing interviews with various alternative-medicine practitioners and their specific uses of magnets for healing.

Magnet Therapy: The Pain Cure Alternative. Lawrence, Ron, M.D., Ph.D.; Rosch, Paul, M.D., F.A.C.P.; and Plowden, Judith. Rocklin, Calif.: Prima Health, 1998. Details many magnetic-healing studies as well as case histories of individual patients treated with magnetic therapies.

Magnetotherapy Self-help Book. Bansal, H. L., and Bansal, R. S. New Delhi, India: Jain Publishers, Ltd., 1976. A book on magnetotherapy from India that also details the uses of magnetized water as well as the placement of magnets beneath the feet and hands for healing in a fashion similar to magnet researcher Don Lorimer's use of magnets for polarity balancing.

The Body Electric: Electromagnetism and the Foundation of Life. Becker, Robert, M.D., and Selden, Gary. New York: William Morrow and Company, 1985. One of the best books on the biological effects of magnetic and electromagnetic fields on health and healing, by a pioneering researcher.

The Cycles of Heaven: Cosmic Forces and What They Are Doing to You. Playfair, Guy, and Hill, Scott. New York: Avon Books, 1978. A fascinating book detailing some of the harder-to-find European research on solar and geomagnetic influences upon human health and behavior.

Paramagnetism: Rediscovering Nature's Secret Force of Growth. Callahan, Philip, S. Metairie, La.: Acres U.S.A., 1995. Describes some of Phil Callahan's theories about paramagnetism, the Irish Round Towers, and a variety of other magnetic influences on plant growth.

CHAPTER 9

Future Science: Life Energies and the Physics of Paranormal Pehnomena. White, John, and Krippner, Stanley, eds. Garden City, N.Y.: Doubleday and Company, Inc., 1977. Contains a number of fascinating chapters on the physics of healing, radionics, and related energy phenomena. Although it is sometimes difficult to find copies, this book is well worth the search.

Radionics: Science or Magic? An Holistic Paradigm of Radionic Theory and Practice. Tansley, David, D.C. Essex, England: C. W. Daniel Company Ltd., 1982. One of the most intelligent books written on the subject of radionics.

Radionics—Interface with the Ether-Fields. Tansley, David, D.C. North Devon, England: Health Science Press, 1975. Includes some excellent radionic photographs taken from blood spots, including the famous image of a fetus photographed from the blood spot of a pregnant woman many miles away.

Healing with Radionics—The Science of Healing Energy: The Theory and Practice of Radionic Therapy. Baerlin E., and Dower, A. L. G. Wellingborough, England: Thorsons Publishers, Ltd., 1980. An excellent little book on the history and practice of radionics by two therapists who were able to repeatably produce radionic photographs using the Mark VI DeLaWarr Camera.

CHAPTER 10

Healing Research: Holistic Energy Medicine and Spirituality—Volumes 1–4. Benor, Daniel J. United Kingdom: Helix Editions, Ltd., 1993. Perhaps the best compilation and analysis of international healing research available in a single resource. Highly recommended.

The Uncommon Touch: An Investigation of Spiritual Healing. Harpur, Tom. Toronto, Canada: McClelland & Stewart, Inc., 1994. A fascinating account of the history of spiritual healing and various aspects of scientific research in healing.

Hands of Light: A Guide to Healing Through the Human Energy Field. Brennan, Barbara, New York: Bantam Books, 1988. Remains one of the best guides to human multidimensional anatomy and the process of healing available. Highly recommended.

Therapeutic Touch: A Practical Guide. Macrae, Janet. New York: Alfred A. Knopf, 1987. A short, nicely illustrated guide to the basics of Therapeutic Touch.

Essential Reiki: A Complete Guide to an Ancient Healing Art. Stein, Diane. Freedom, Calif.: The Crossing Press, Inc., 1995. One of the best introductions to Reiki healing available. Also includes pictures of many of the Reiki healing symbols that have never before appeared in print. Highly recommended.

An entire Reiki healing workshop, conducted by Diane Stein, author of *Essential Reiki,* is available on a four-videotape set. The program includes information on many different Reiki levels of life-force healing. Highly recommended. The video workshop is available through the Explorations catalog (item #80-0701). Address: Explorations, 360 Interlocken Boulevard, Suite 300, Broomfield, CO 80021. Phone: (800) 720-2114. Fax: (800) 456-1139.

The Power of Reiki: An Ancient Hands-On Healing Technique. Honervogt, Tanmaya. New York: Henry Holt and Co., 1998. A good guide to Reiki healing, including self-healing techniques. Beautifully illustrated with photographs and color diagrams.

The Complete Healer: How to Awaken and Develop Your Healing Potential. Furlong, David. London: Judy Piatkus (Publishers) Ltd., 1995. A wonderful book on how to develop one's healing abilities, complete with exercises and illustrations. Highly recommended.

Pranic Healing. Sui, Choa Kok. York Beach, Maine: Samuel Weiser, Inc., 1990. An excellent introduction to pranic healing as taught by Master Choa. Also includes fascinating discussions of eleven major and minor chakras, complete with color chakra illustrations and diagrams of healing techniques. Highly recommended.

Pranic Psychotherapy. Sui, Choa Kok. York Beach, Maine: Samuel Weiser, Inc., 1993. One of the few books to discuss how specific healing techniques can be used for healing mental and emotional problems. A unique approach to subtle-energetic psychotherapy.

The Health Professional's Handbook of SHEN Physioemotional Release Therapy. Pavek, Richard R. Sausalito, Calif.: The SHEN Therapy Institute, 1987. A good introduction to SHEN therapy, a specialized bioenergy-healing approach, available through the Biofield Research Institute, address in Appendix 7.

Miracles of Mind: Exploring Nonlocal Consciousness and Spiritual Healing. Targ, Russell, and Katra, Jane, Ph.D. Novato, Calif.: New World Library, 1998. An exploration of healing from the perspective of nonlocal consciousness and information exchange as opposed to directed energy transfer. A fascinating book that explores an alternative viewpoint of healing.

RESOURCE GUIDE TO BOOKS

Although many of the books listed under recommended reading can be obtained at most bookstores, some of the more esoteric books (such as those on radionics and healing) may be difficult to find except at New Age–oriented bookstores. Three of the best resources for locating books are:

Mayflower Bookstore
2645 W. 12 Mile Road
Berkley, MI 48072
Phone: (248) 547-8227
E-mail: mflower@aol.com

Mayflower specializes in rare, unusual and esoteric books on healing, vibrational medicine, radionics, metaphysical philosophies, and spiritually oriented topics. They can also special order hard-to-find books and are used to shipping around the United States. One of the best bookstores of its kind.

East West Books
Phone: (212) 243-5994

Another excellent resource for finding books on healing, New Age philosophy, yoga, spirituality, and alternative medical practices.

Amazon.com

This Internet-based bookstore offers many books on healing, including capsule reviews of books submitted by readers. They will also do special searches to locate out of print books such as some of the more difficult to find radionic texts.

INDEX

Page numbers in *italics* refer to charts.